Principles
of
ENGINEERING

Principles
of
ENGINEERING

James J. Duderstadt
Glenn F. Knoll
George S. Springer

College of Engineering
The University of Michigan
Ann Arbor, Michigan

1807 1982

John Wiley & Sons, Inc.

New York   Chichester   Brisbane   Toronto
Singapore

Cover photo by Susan Berkowitz

Library of Congress Cataloging in Publication Data:

Duderstadt, James J., 1942–
    Principles of engineering.

    Includes Index.
    1. Engineering.   I. Knoll, Glenn F.
II. Springer, George S.   III. Title.
TA147.D8   620   81-10450
ISBN 0-471-08445-X   AACR2

Printed in the United States of America

10 9 8 7 6 5 4 3 2 1

To Anne, Gladys, and Susan

PREFACE

The challenges faced by modern society are almost overwhelming. They have become a familiar part of our everyday vocabulary: the population explosion, the energy crisis, pollution, the arms race, the loss of identity and human dignity in a world increasingly dominated by automation, and the cultural shock caused by technological change. These are serious problems that threaten to disrupt civilization as we know it.

But these are also the challenges for the profession of engineering. The role of engineers is to apply their knowledge of science and technology to meet the needs of society, to solve its problems, and to pave the way for its future progress. As our society has become increasingly dependent upon science and technology, it has also become more dependent on engineers, not only for its prosperity but for its very survival. Many of the major problems faced by the world today cannot be solved by social or political actions alone; they require the knowledge, skills, and discipline of engineers.

Engineering students must prepare for this world and these challenges. In a world of change and diverse opportunity, the key to success in engineering is a broad education based on fundamental scientific concepts, coupled with the development of a disciplined approach to problem solving that is characteristic of engineering. But this is not enough. The technological and social aspects of the challenges faced by modern society have become closely intertwined. It is essential that an engineer's education also include studies in the arts, humanities, and social sciences.

Herein lies a dilemma. A broad-based education tends to concentrate technical subjects in engineering into the final two years of the undergraduate program. The first two years are given over almost entirely to basic courses in mathematics, physics, chemistry, the humanities, and social sciences. Many students find this program quite frustrating since they seek exposure to engineering and engineering faculty at the earliest possible moment. They wish to learn what engineering is all about, what its opportunities and challenges are, and how one prepares to become an engineer.

For this reason many engineering programs have introduced engineering courses in the freshman year. Varied approaches have been taken, including survey courses sampling various engineering fields, motivational courses designed to stimulate interest in the engineering profession, and even courses on modern computer methods (primarily programming) taught by engineering faculty. Of most interest have been those more ambitious attempts to introduce material on engineering problem solving in the freshman year, to begin development of the intellect and discipline required by the engineering profession.

This text is directed toward this latter type of course. Our particular goals are both diverse and ambitious:

1. To expose first-year students to engineering methods, to motivate them and provide orientation for further course work.

2. To demonstrate the engineering approach to problem solving using concepts from conventional science courses, but applying them to the complex, open-ended, and ill-defined problems so characteristic of the engineering profession.

3. To provide a satisfying experience in actually solving relevant problems at an early stage, thereby building confidence in one's ability.

4. To introduce the principal tools of engineering, including a survey of fundamental scientific concepts, mathematics, computing, and statistical and testing methods.

This text is intended to help students develop their own skills in self-teaching, engineering synthesis, the solution of open-ended problems, and other intellectual activities of engineering. The difference between problem solving in formal courses and in engineering practice is illustrated by applying concepts from basic science courses to a variety of relevant and practical problems taken from real-life situations. Of particular concern are an introduction to the art of problem definition, problem solving under constraints, the development of solution algorithms, solution verification and evaluation, modeling, simulation, and optimization.

The text has been organized into four major parts: the engineering *profession*, the *approach* to engineering problem solving, the *tools* of engineering, and *constraints* on engineering practice. In Part I we introduce the student to the excitement of engineering, the variety of roles and activities of the engineer, and the profession of engineering. In Part II we examine the principal intellectual activity of the engineer, problem solving, and develop a general procedure for attacking the array of complex engineering problems that arise in practice. This approach is illustrated by considering in detail the most important application of engineering problem solving: engineering design.

With this background, we proceed to Part III where we examine the tools of modern engineering. First, the essential tools of mathematical analysis and scientific principles are outlined. These chapters provide a foundation for further studies of the engineering student. They are also intended to provide the student with sufficient knowledge to confront many challenging engineering problems and to sample the satisfaction (or frustration) that accompanies attempts to solve such problems. Students are introduced to the modern tools of engineering such as digital methods and computing, computer-aided design (CAD/CAM), and experimental and testing methods. We have also included a chapter on a most essential skill of engineering, communication through written, oral, and graphical means.

Part IV of the text is concerned with the constraints that complicate and restrict engineering practice. We introduce students to the important subject of engineering economics. We then discuss relations with people, with particular attention directed to the subjects of management activities and the legal aspects of engineering. Finally we examine interactions with the institutions of society such as government and the new constraints these interactions impose through regulation and technology assessment.

The text is organized into compact, self-contained units. Each subsection ends with a short summary that is followed by an example illustrating the material and a number of exercises to allow the student to master the material by actual practice.

The exercises vary greatly in both purpose and degree of difficulty. Most sets of exercises begin with straightforward problems designed to stress text material and develop student confidence. But we have also included several more complex problems of an open-ended, ill-defined nature typical of engineering practice. This latter approach is particularly evident in Chapters 3 and 4 where we have provided the student with an opportunity to apply mathematical and scientific concepts to engineering problems. We have also included problems designed for solution by various types of calculators and computers.

We introduce the SI system of units at an early stage and use this system throughout the text. Although other unit systems are not treated in detail, discussion and practice in converting from one system to another are given.

Our survey of introductory engineering courses used in various programs has revealed more diversity than commonality. Different programs have chosen to emphasize different goals, approaches, and topics. We have addressed this diversity by presenting a wide range of material in as flexible and independent a fashion as possible. In a sense we have solved the problem of what to cover by attempting to cover everything. It is hoped that the text will be appropriate for most introductory engineering courses with a suitable selection and sequencing of topics. The instructor's manual accompanying the text suggests several possible approaches and provides more detailed lesson plans.

Our efforts in developing this textbook have benefited greatly from the advice, encouragement, and assistance of many faculty and students at the University of Michigan. In particular we acknowledge the technical assistance of Joe G. Eisley and Maurita Holland, the assistance in problem preparation and solution provided by Keith Hampton and Tom Sutton, and the artistic contributions of Mike Manley. We acknowledge the considerable efforts of the editorial and production staff at Wiley, with a particular note of thanks to Carol Beasley and Irene Zucker. Of most importance are the contributions of the many students who have toiled through earlier versions of the text and provided us with valuable insight and assistance.

James J. Duderstadt
Glenn F. Knoll
George S. Springer

Contents

APPENDICES

Principles
of
ENGINEERING

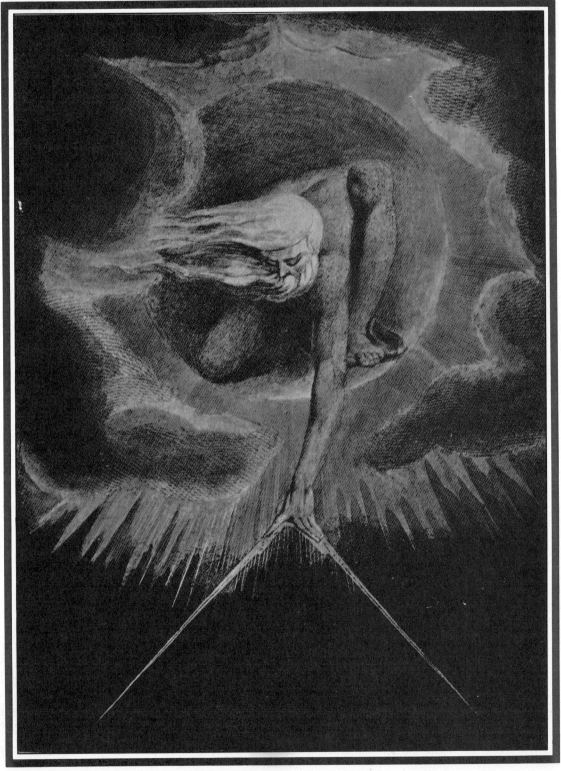

FIGURE I.1. The first engineer (William Blake's frontispiece for *Europe, A Prophecy*, 1794).

PART I
The Profession

In his hand
He took the golden Compasses, prepar'd
In Gods Eternal store, to circumscribe
This Universe, and all created things:
One foot he center'd, and the other turn'd
Round through the vast profunditie obscure,
And said, thus farr extend, thus farr by bounds,
Thus be thy just Circumference, O World.

Milton, *Paradise Lost, Book VII*

What are engineers? What do they do? What does it take to become one? These are the first questions that usually occur to prospective engineering students, and these are also the first questions that should be addressed by any introductory text in engineering.

In a general sense engineers are *creators* of ideas and concepts, *builders* of devices and structures, and above all, *problem solvers*. They apply their knowledge of science and technology to meet the needs of society. These needs are often so complex and intertwined with social, economic, and political issues that engineers must be far more than problem solvers. They must also develop a broad perspective that will allow them to assess the impact of their activities on society and their natural environment. They must be prepared to face a world of change, of new and ever varying challenges.

In Chapter 1 we survey the various roles and activities of engineers and the role of the engineering profession in the modern world. We give the educational requirements for careers in engineering and engineering technology. In addition, we introduce the important concept of engineering as a learned profession, governed by laws of professional registration and guided by a code of ethics. Finally, we preview some of the challenges that will confront future engineers.

CHAPTER 1

The Profession of Engineering

Most living creatures are locked into a Darwinian pattern. They must either adapt to their changing environment for survival or perish. Only the human species is an exception. Only we possess the intellect to modify our environment, to adapt it to meet our needs. On the most primitive level, this role is most characteristic of the engineering profession. The roots of engineering lie with our ancestors' first use of tools in their struggle for survival. Throughout history engineers have sought to apply knowledge in the form of science and technology to meet the needs of society. Their accomplishments have had as much impact on history as any social, political, or artistic accomplishment. In a very real sense, the engineer has paved the way for the progress of civilization.

Today we are in a new era, as our population strains against the limits of its natural environment, threatening to disrupt nature, perhaps even to make this planet uninhabitable. Future historians may refer to the twentieth century as that moment in time when we first recognized the finite nature of the earth's resources, even as we rapidly approached their exhaustion. Today our expanding numbers and activities have the potential to destroy in an instant an environment produced after millions of years of evolution. It is against this ominous backdrop that we must examine the role of the modern engineer.

We are surrounded by and dependent on the work of engineers. Engineering supplies the food we eat and the clothing we wear; it provides us with shelter and energy, the means for transportation and communication. It has replaced human toil with machines, and today it is replacing tedious human mental tasks with computers, thereby providing us with the opportunity for other pursuits.

FIGURE 1.1 Engineers must face the challenges posed by an increasing population straining against the limits of its natural environment. (Courtesy National Aeronautics and Space Administration)

Our society has become increasingly dependent on science and technology, and therefore upon engineering, not only for its prosperity but, indeed, for its very survival. Problems of serious proportions loom before us today. How can we provide food and energy for an ever-growing population in the face of dwindling resources? How do we protect our natural environment from the impact of human activities? How do we preserve our dignity and freedom, even as new technologies such as genetic engineering and automatic data processing threaten our integrity and individualism, the very nature of our humanity? These problems cannot be solved by social or political actions alone. They will require the efforts and talents of engineers. Yet many social and technological problems have become so closely intertwined that modern engineers must be acutely sensitive to social and political factors. They must be concerned not only with the efficient use of our natural resources, the preservation of our natural environment, but also with the importance of human dignity in modern society.

The basic role of engineers has not really changed throughout history. Engineers continue to be innovators and problem solvers, applying science and technology for the benefit of society. However, as these needs have become more complex, they have required the development of more sophisticated and powerful scientific tools. The needs and problems of modern society also require an extraordinary combination of skill, experience, and knowledge, all of which must be acquired and applied by today's engineer.

Modern engineering remains both a science and an art. Before the basic principles of mathematics and science can be applied to engineering problems, engineers must identify, isolate, and analyze the essential features of these problems. They then must generate practical and acceptable solutions within the constraints posed by modern society. In this latter sense engineering problem solving

FIGURE 1.2. The activities of the modern engineer are varied and numerous: Columbia Space Shuttle (NASA-UPI); BART's Rockridge station in Oakland; Chesapeake Bay Bridge, Maryland (courtesy Bethlehem Steel); a nuclear power plant in Arkansas (Babcock & Wilcox); Air France's SST Concorde; Chevrolet's 1982 Cavalier Hatchback.

5

is an art, an expression of intellectual creativity every bit as intense and imaginative as that of the artist. Except that engineers, instead of working with media such as paint on canvas, apply technology to mold the fabric of our society. The impact of their efforts and decisions may extend far beyond the moment to influence the future of civilization.

Engineering is a challenging profession. It combines the precision and discipline of science with the creativity and imagination required by the complex problems faced by modern society. The options available to prospective engineers are almost overwhelming. They can choose a field of specialization from among traditional disciplines such as civil, mechanical, chemical, or electrical engineering or from newer areas such as environmental, aerospace, or nuclear engineering. The range of activities within each field is also immense. These include research, development, design, production, operations, sales, and management.

This array of possibilities may seem bewildering to the student contemplating an engineering career. Therefore it is natural in discussing engineering as a profession to begin by considering various engineering roles from the perspective of both discipline and function. This will help us establish the educational requirements for each of these engineering careers.

SUMMARY

*Throughout history engineers have sought to apply the knowledge of science and technology to meet the needs of society. They have paved the way for the progress of civilization. Engineers are innovators and problem solvers, applying their skill, experience, and knowledge to serve society. Engineering is a challenging profession. It combines the precision and discipline of science with the creativity and imagination required by the complex problems faced by the **modern** world.*

Exercises

Historical Background

1. Trace the evolution of engineering from the earliest records of stone age man through the great civilizations of early history down to modern times. In particular, prepare a short list of what you regard as the most significant achievements of engineers throughout history. Then attempt to analyze the impact of these achievements on the course of civilization (perhaps drawing on your related studies in history).

2. Prepare a short list of those you consider to be the most prominent

engineers of the twentieth century (e.g., Edison, Steinmetz, Von Karman, Von Braun) and briefly describe their accomplishments.

3. We frequently think of the evolution of technology as the product of the genius of individuals, such as Edison's invention of the electric light or Gutenberg's invention of the printing press. James Burke has advanced a somewhat different perspective of technological history in his book *Connections* (Little, Brown, and Company, 1978). He proposes that progress consists of a series of events triggered by recurring factors that manifest themselves as a constant product of human behavior. Using this "cause-effect" approach, one can find antecedents far back in history for any modern invention. By referring to Burke's book, trace the "connections" leading to one of the following modern innovations: the computer, the production line, telecommunications, the airplane, the atomic bomb, the guided missile, and television.

4. Identify the various stages in the production and use of energy through the course of civilization. In particular, attempt to extrapolate this development through the twenty-first century.

5. Briefly describe what changes you would expect to find in the modern world if engineers had not developed (a) the internal combustion engine, (b) the digital computer, and (c) nuclear energy.

6. What would you project as the most significant engineering breakthroughs of the next twenty years?

7. We noted that engineering is both an art and a science. Give one example of an engineering accomplishment that rivals art in the degree of its originality and creativity.

1.1.   WHAT IS AN ENGINEER?

Engineering is the application of science and technology to meet human needs. Because of the vast diversity of engineering fields and functions, it is difficult to provide a simple, concise definition of the word "engineer."

Historically, the term probably traces back about 2000 years to a military device, the *ingenium*, a type of battering ram used by Roman legions to attack walled defenses. During the middle ages the soldiers operating such machines became known as *ingeniators*. This word gradually evolved into "engineer." It was applied to those individuals concerned with the technology of war, with "military" engineering. Within the past two centuries the use of this term broadened to include nonmilitary activities, civilian or "civil" engineering, as well.

A formal definition has been adopted by the engineering profession itself. The Accreditation Board for Engineering and Technology provides the definitions

Engineering is the profession in which a knowledge of the mathematical and natural sciences gained by study, experience, and practice is applied with judgment to develop ways to utilize, economically, the materials and forces of nature for the benefit of mankind.

Engineers are persons who, by reason of their special knowledge and use of mathematical, physical, and engineering sciences and the principles and methods of engineering analysis and design, acquired by education and experience, are qualified to practice engineering.

However, the most useful approach is to simply regard the engineer as a doer, a problem solver, who applies science and technology to solve the problems and to meet the needs of society. The engineer is an innovator and creator of new products and processes aimed at solving problems in a practical and economic fashion.

The problems of society have become so complex and the tools of technology so advanced and sophisticated that one person can no longer master them alone. Engineering has become a team activity. It combines the efforts of many individuals with knowledge and talents in different and diverse areas. This team includes *scientists* who search for the fundamental laws of nature, *engineers* who translate science into usable forms through innovative design, *engineering technologists* and *technicians* who apply and execute engineering designs, and *craftsmen* who apply their skills to produce the materials and devices required by the design.

The Scientist

Scientists stand at one end of the spectrum of technical activities. Their efforts are aimed at basic research, that is, the understanding of nature for its own sake. Many scientists also become involved in activities with a more practical bent such

FIGURE 1.3. The engineering team: scientist, engineer, engineering technologist, engineering technician, and craftsman. (Courtesy Accreditation Board for Engineering and Technology)

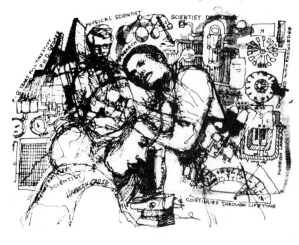

Scientist   Engineer

The engineering team

Engineering technologist
or technician

Craftsman

as device design and development. However their primary interest is scientific knowledge, without regard to its ultimate use. Such scientific research requires a thorough grounding in basic concepts and mathematics. It also requires advanced training and education, often to the level of a doctorate degree.

The Engineer

Engineers are also concerned with basic scientific principles. However, unlike scientists, the efforts of engineers are directed more at the application than the discovery of these principles. Perhaps the most distinguishing activity of engineers is design, the use of their knowledge of science to synthesize a device or process to answer a specific need. As the famous engineer Theodore Von Karman once observed, "The scientist explores what is; the engineer creates what had not been." Like scientists, engineers must possess a thorough foundation in basic science and mathematics. However they also require training in engineering sciences such as structural and fluid mechanics, thermodynamics, and electrical science. Their education includes topics from the humanities, social sciences, and communication arts that assist them in addressing the complex problems they will encounter in modern society. Although the accepted entry point into the engineering profession today stands at the four-year baccalaureate level (e.g., a Bachelor of Science (B.S.) degree in engineering or a Bachelor of Engineering (B.Eng.) degree), certain types of engineering practice may well require studies beyond this point to the level of Master of Science (M.S.) or Doctor of Philosophy (Ph.D.)

The Engineering Technologist

The development of the ideas of scientists and engineers into tangible results frequently involves other members of the engineering team known as engineering technologists and technicians. An engineering technologist deals more directly with applications of established scientific and engineering principles than with the development of these principles. More precisely, engineering technologists are concerned with the part of technology that requires the application of scientific and engineering knowledge and methods together with technical skills in support of engineering activities. Their position in the spectrum of activities characterizing the engineering team lies between the engineer and the craftsman. Engineering technologists will participate in the organization and management of engineering projects and the planning, construction, and operation of engineering facilities. They may even be called upon to perform limited design activities. Therefore their education is similar in many ways to the engineer's. It consists of a four-year baccalaureate program emphasizing science and mathematics as well as more specialized technical training in engineering applications. The corresponding degree is sometimes called Bachelor of Science in Engineering Technology or B.S.E.T. degree.

The Engineering Technician

Many activities in engineering require fewer analytical or conceptual skills, but more detailed knowledge and skills in particular applications. The training for these activities can sometimes be provided by a two-year (associate degree) educational program emphasizing practical techniques, coupled with on the job training. Graduates of these programs are known as engineering technicians. Although the general activities of the engineering technician are more routine than those required of the engineer or the engineering technologist, they are nevertheless of a highly skilled nature. Engineering technicians might be involved in the actual fabrication of prototype devices designed by engineers. Other possible activities include preparing drawings, assembly of electronic circuits, estimating costs, as well as service and maintenance activities. Frequently the engineering technician is the vital link between an idea on paper and its realization in practice.

The Craftsman

Craftsmen possess the manual skills necessary to produce the components specified by engineers. Therefore they need not be concerned with the basic scientific principles involved in design. Their training is usually acquired on the job (e.g., by serving an apprenticeship). Welders, machinists, electricians, carpenters, and model builders are examples of the specialized jobs of craftsmen.

In general, individuals primarily interested in pursuing knowledge for its own sake, that is, in following their own sense of curiosity without regard to application, would be most satisfied in a scientific profession. People with strong scientific and mathematical interests and attitudes, but with primary interests in problem solving and practical application, are well suited to engineering. Those who prefer a greater involvement with actual technical devices and processes than with theoretical concepts would probably be more satisfied as engineering technologists. Individuals less theoretically minded and preferring a practical involvement in a supportive technical role would be suited for a career as an engineering technician. Those interested in developing a high degree of skill in the use of tools or machinery are most likely to prefer work as craftsmen.

The occupational definitions listed above should not be interpreted as rigid. There is considerable overlap among the activities of members of the engineering team. For example, many engineers are involved in basic research, and engineering technologists and technicians may participate in design activities. Nevertheless this classification is useful in identifying the various educational requirements for each career in engineering. Table 1.1, which was prepared by the Accreditation Board for Engineering and Technology, compares these educational requirements and activities in greater detail.

SUMMARY

Although formal definitions can be given for the profession of engineering, it is simplest to think of the engineer as a problem solver who applies science and technology to meet the needs of society. Engineering is a team activity involving scientists who search for the fundamental laws of nature, engineers who translate scientific knowledge into usable forms through design, engineering technologists and technicians who apply and execute engineering designs, and craftsmen who apply their special skills to produce the materials and devices required by the designs.

Example   A mechanical engineer is faced with the task of improving the efficiency of a photovoltaic solar power collector by developing an inexpensive lens to focus sunlight on the photocells. The engineer notes that a scientific research group has recently developed the concept that a lens can be fabricated from thin plastic sheets. This scheme is based on an old idea developed for lighthouses known as the Fresnel lens in which a convex lens is sliced into thin bands so that it can be constructed on a flat surface. The engineer determines that this lens can be applied to improve the efficiency of the solar collector and designs a process to produce such lenses inexpensively by molding thin plastic sheets into the shape of Fresnel lenses. The engineering technologist takes this design and provides detailed specifications in the form of design drawings. The components that can be obtained directly from stock are ordered. Craftsmen fabricate the more unique components of the lens production, such as dies, and then assemble the actual devices.

Exercises

The Engineering Team

1. Compare the definitions of "scientist," "engineer," "engineering technologist," and "engineering technician" given in three different books discussing engineering careers. (See the reference list at the end of this chapter for possible suggestions.)

2. Trace the possible involvement of various members of the engineering team in the development of the following products: (a) the Xerox machine, (b) a wind turbine generator, (c) the pop-top beer can, and (d) an air-convection popcorn popper.

3. Engineers and engineering technologists frequently find themselves involved in group or team activities. What would you suppose are essential attributes for participating as an effective member of such teams?

4. Identify which member of the engineering team would most appropriately handle the following activities:

TABLE 1.1 The Engineering Team

ENGINEER	ENGINEERING TECHNOLOGIST	ENGINEERING TECHNICIAN
Preparation		
The first professional degree from an accredited engineering program: • One-half year of mathematics beginning with differential and integral calculus • Basic physical sciences • Engineering sciences • Interrelate engineering principles with economic, social, political, aesthetic, ethical, legal, and environmental issues	Four-year degree from an accredited engineering technology program: • Applied science and mathematics (through concepts and applications of calculus) • Technical sciences and specialty areas • Field orientation • Apply technological methods and knowledge, with technical skills, to support engineering activities	An associate degree in an accredited engineering technology program: • Mastery of technical skills • Training in specific instruments and equipment • Perform operational tasks, following well-defined procedures, to support engineering activities
Career Goals		
Research Conceptual design System synthesis and development Product innovation Operations management	Hardware design and development Product analysis and development System operation Process management Technical sales and services	Drafter Laboratory operations System maintenance Machine operations Data collection
Descriptors		
Conceptualizer Innovator Planner/predictor Designer Developer Systematizer Judge Decision maker Producer of standards Synthesizer	Operator of systems Translator of concepts into hardware and systems Director of engineering technicians and craftsmen Implementer Applier of established techniques and methods Producer Analyzer	Performer of operational tasks User of proven techniques and methods Builder of components Operator Tester Collector of data Maintainer of components Preparer of technical drawings

(a) Devise an experiment to study the interaction of molten metal and water.
(b) Design the test facility for conducting such an experiment.
(c) Prepare the detailed blueprints for each of the components of the experimental apparatus.
(d) Fabricate and assemble the apparatus.
(e) Record the data taken during the course of the experiment.
(f) Develop a computer program to analyze the results of the experiment.
(g) Run the computer program to analyze the data.
(h) Interpret the results of the data analysis.

5. Interview an engineer, an engineering technologist, and an engineering technician to determine what they regard as the principal rewards and disappointments of their work.

1.2. FIELDS OF ENGINEERING

The engineering profession is characterized by the number and variety of its fields of specialization. Engineering applications range from the microscopic world of integrated circuit design for digital computers to mammoth civil constructions projects such as dams, bridges, or entire cities. The traditional engineering disciplines include civil engineering, mechanical engineering, chemical engineering, and electrical engineering. Additional fields of engineering such as aerospace engineering and nuclear engineering have developed as a consequence of new scientific discoveries. Moreover, a variety of new engineering disciplines, such as environmental engineering, biomedical engineering, and systems engineering, have been stimulated by the complex nature of the problems that now face modern society (Figure 1.4).

Civil Engineering

Civil engineers are builders who are involved in the planning, design, and construction of projects such as bridges, dams, harbors and waterways, highways and railroads, buildings and mass transit systems. Civil engineering is the oldest of the engineering professions. The term ``civil'' was introduced during the eighteenth century to distinguish a group of engineers more concerned with public or civilian projects from ``military'' engineers who designed and constructed structures for military purposes. In fact the first ``civil engineers'' were machine builders. However, this term became gradually associated with the construction of structures such as buildings, roads, and bridges.

The civil engineer can choose today from many areas of concentration. These include construction engineering and management; environmental, sanitary, and water resources engineering; geodetic engineering (surveying); highway engineering; hydraulic engineering; municipal engineering; geotechnic engineering (soil

| Civil |
| Mechanical |
| Chemical |
| Electrical |

Aerospace	Computer	Mining
Agricultural	Environmental	Nuclear
Architectural	Industrial	Petroleum
Atmospheric	Manufacturing	Sanitary
Automotive	Materials	Systems
Bioengineering	Metallurgical	Transportation
Biomedical	Marine	Water Resources

FIGURE 1.4. The fields of engineering include the four traditional disciplines of civil, mechanical, chemical, and electrical engineering in addition to other more specialized fields.

mechanics and foundations); structural engineering; and transportation and traffic engineering. Opportunities for civil engineers are quite extensive and include construction, manufacturing, transportation, power generation, and environmental control activities in both industry and government. Civil engineering is also a primary activity of many consulting engineering firms.

Mechanical Engineering

Mechanical engineers are concerned with machines and mechanical processes such as energy generation and conversion. This aspect of engineering can be traced back to the industrial revolution in the late eighteenth century when steam-powered engines first began to revolutionize manufacturing and locomotion. Today, mechanical engineering has become one of the broadest and most popular areas of engineering activity. Mechanical engineers play a major role in energy production, conversion, and utilization, in both traditional forms such as

FIGURE 1.5. Mechanical engineers testing the heat resistance of a nuclear fuel element. (Courtesy E G & G, Idaho)

fossil-fuel combustion and newer forms such as nuclear or solar energy; in transportation such as the automobile and aircraft industries; in the design and fabrication of machinery with applications ranging from industrial production to home appliances; and in heating, air conditioning, refrigeration, and cryogenics. Mechanical engineers participate in almost every type of industrial activity. Their background includes extensive training in thermodynamics, fluid mechanics, heat transfer, solid mechanics, dynamics, materials, and electronics, as well as more specialized topics such as machine design and power generation.

Chemical Engineering

Chemical engineers apply principles of chemistry, physics, and engineering to the design and operation of plants and processes for the production of materials that undergo chemical changes during their manufacture. In its broadest sense, chemical engineering involves the creative application of chemistry to meet the needs of society. Chemical engineers develop processes for producing plastics, synthetic fibers, pharmaceuticals, paper, paints, and many other products of importance. The work of the chemical engineer encompasses many industries, from the synthesis of chemicals and the refining of petroleum to nuclear energy and space technology. General areas of specialization include: chemical process design, electrochemistry, corrosion, environmental engineering, biochemistry, molecular chemistry, polymer chemistry, nuclear chemistry, petroleum, and general chemical systems engineering. Chemical engineers are in demand not only in chemical industries, but also in nearly all types of manufacturing. Chemical engineers build on a solid foundation in chemistry and physics with engineering sciences of particular importance to chemical processes, including chemical reactions, thermodynamics, fluid mechanics, and materials science. Their strong training in the design and analysis of processes provides them with a general perspective of major importance in many fields.

Electrical Engineering

Electrical engineering is concerned with electrical devices, circuits, and systems. Traditionally electrical engineers have specialized in areas such as power generation and transmission, electrical machinery, electronics, communication, control, and circuit design. In the past several decades this field has expanded considerably with the development of the transistor, integrated circuits and microelectronics, the laser, and the digital computer. Many electrical engineers now work in new areas such as quantum electronics (microelectronics), quantum optics (lasers), and computer engineering. Electrical engineering graduates are sought by all major industries. This field requires perhaps the strongest mathematical background of the traditional engineering disciplines, particularly in the more theoretical areas such as communication and control. Newer areas such as microelectronics and laser development require a thorough foundation in modern physics.

FIGURE 1.6. The efforts of electrical engineers have made possible the dramatic advances in communication as illustrated by this comparison of early and advanced communication satellites. (Courtesy Hughes Aircraft Company)

SUMMARY

The traditional engineering disciplines are civil, mechanical, chemical, and electrical engineering.

Civil engineers are builders. They plan, design, and construct projects such as roads, buildings, and cities.

Mechanical engineers are concerned with machines and mechanical processes including energy production and utilization.

Chemical engineers design chemical processes and plants for the production of materials.

Electrical engineers are concerned with electrical phenomena, with electrical devices, circuits, and systems as well as with more abstract areas such as communication and control.

In addition to the four traditional engineering disciplines, the engineering student can choose from a wide variety of more specialized fields:

Aerospace (Aeronautical and Astronautical) Engineering

Since the early developments of manned flight, aeronautical engineers have played the key role in the design and development of aircraft. This field has broadened to encompass astronautical engineering as the space program in the mid-twentieth century rapidly expanded (hence the new title "aerospace" engineer). More recently it has been extended to include other areas such as high-speed ("ground effect") transportation, hydrofoil ships, deep ocean vehicles, and wind-powered electrical generators. Aerospace engineers acquire a solid foundation in fluid dynamics (particularly as applied to aerodynamics and propulsion), structural mechanics, flight and celestial mechanics, and guidance and control. They find many employment opportunities in aircraft or spacecraft development, manufacturing, and assembly, dealing with almost every aspect of both manned and unmanned flight, from light planes and helicopters to satellites and deep space probes. The advances developed by aerospace engineers have given rise to many spinoff developments that benefit society.

FIGURE 1.7. Perhaps the most dramatic achievement of aerospace engineering was the lunar landing of Apollo 11. (Courtesy National Aeronautics and Space Administration)

Agricultural Engineering

Agricultural engineers apply engineering principles to farm and food production industries. They are involved in every phase of agriculture from the production of plants and animals to the final processing of food, feed, and fiber products. This field makes use of both the physical and biological sciences. It has become particularly critical in the light of the rapidly growing world population's need for food and fiber. The agricultural engineer might become involved in the development of mechanized farm equipment and machinery, the application of modern methods of irrigation, erosion control, and land and water management, feed and crop processing methods, the design of specialized structures for farm use, and the processing and handling of food products.

A closely related area of engineering activity is management of forest and mineral resources. The new field of *resource engineering* deals with the efficient use of natural resources, including forest, productive crop lands, and mineral deposits. These engineers apply modern engineering techniques to improve the traditional methods of harvesting and conserving natural resources.

Architectural Engineering

Architectural engineers, like civil engineers, are concerned with the design and assembly of structures. However, their background allows them to interact closely with the architect, to apply engineering methods of structural analysis and materials selection to realize the particular artistic goals of the architect. Such engineers often work in established architectural or construction firms. Students in this field usually acquire a strong background in the traditional engineering subjects such as structural mechanics. But they also need an appreciation for the aesthetics of architecture. Of particular importance is familiarity with recent developments in computer-aided design.

Atmospheric and Oceanographic Engineering

In recent years engineers have directed their attention toward the earth's atmosphere and oceans. Atmospheric science and engineering is essentially applied meteorology, including problems associated with climate and weather, air pollution, industrial plant location and processes, and wind-loading considerations in the design of structures. Many important decisions in the design of transportation systems, whether by land, water, or air, depend critically on meteorological factors.

Oceanographic engineers attempt to understand the physical processes that give rise to the observable behavior of the oceans and then apply this knowledge to the wise use of this important natural resource. They develop methods of aquaculture and techniques to harvest the mineral resources of the oceans. The applied oceanographer is also concerned with water supply and control, water pollution, wave action on structures and beaches, and biological and geological processes in the ocean. More recently, oceanographic engineers have

FIGURE 1.8. Atmospheric engineers use high altitude balloons to study the upper atmosphere. (Courtesy University of Michigan College of Engineering)

helped to develop floating power stations that exploit temperature variations in ocean currents.

Automotive Engineering

Automotive engineers are responsible for the design, development, testing, manufacturing, and application of vehicles and their components for use as transportation. Representative fields include passenger cars, trucks, off-road vehicles, fuels and lubricants, construction and industrial machinery, engine design, occupant safety, and emissions control. Usually automotive engineering is offered as a speciality option within a mechanical engineering program, although it may also be regarded as a separate discipline in its own right.

Bioengineering or Biomedical Engineering

Biomedical engineers combine the methods of engineering with their knowledge of the biological and medical sciences. They participate in the development of medical instrumentation, artificial limbs or organs, and instrumentation for the diagnosis or treatment of disease. A closely related area is biochemical engineering, which is involved with the commercial development of products and processes required for human existence. This particular area of chemical engineering includes work with antibiotics, vaccines, biodegradation of insecticides, wastewater utilization, prosthetic materials, synthetic foods, biomedical devices, and artificial organs.

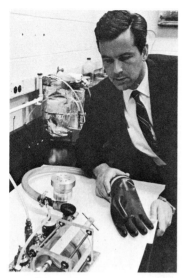

FIGURE 1.9. The design of an artificial respirator provides an example of bioengineering. (Courtesy University of Michigan College of Engineering)

Bioengineering is yet a third option; it is generally concerned with the protection of human, animal, and plant life from toxicants and pollutants or with designing systems that are more suited to the anatomy and function of the human body.

Computer Engineering

One of the most rapidly growing areas of engineering is concerned with computer development and application. Although computer engineering was originally a facet of electrical engineering, it has become a discipline of its own. Computer engineers may design computer "hardware", develop new computer components and computer architecture or peripheral equipment. They may also be concerned with computer applications, with the development of computer "software," the complex instructions that tell the computer the task it is to perform. At the core of this discipline are topics such as microelectronics, logic theory, digital systems, programming, computer linguistics, and numerical methods. Computer engineers are frequently responsible for computer system design and operation. They play an increasingly important role in many areas of society.

Environmental Engineering

Our concerns about the impact of human activities on our natural environment have stimulated a new area known as environmental engineering. Environmental engineers combine their knowledge of the natural, social, and physical sciences to analyze and improve the interaction between society, technology, and the environment. These engineers require a strong background in the natural sciences such as chemistry, botany, and zoology, in addition to traditional engineering subjects such as chemical engineering and systems analysis. Since most efforts to reduce environmental impact are tied to government regulation, some knowledge of law (environmental legislation) is also useful.

FIGURE 1.10.   Computer engineers are responsible for the development, care, and feeding of large computer systems. (Courtesy University of Michigan College of Engineering)

Industrial (Operations) Engineering

Industrial engineering is concerned with the design and analysis of industrial processes such as production, management, and marketing. It deals with the design, improvement, and installation of integrated systems of people, materials, and equipment. Industrial engineers draw upon a background in science, mathematics, and the social sciences, together with the principles and methods of engineering analysis, to predict and evaluate the results to be obtained from such systems. The industrial engineer is primarily interested in problems that involve economy in the use of money, materials, time, human effort, and energy. Management engineering, computers and information systems, human performance and safety engineering, operations research, and manufacturing engineering are particular areas of interest. Industrial engineers not only require a strong background in traditional engineering subjects, but also a background in systems analysis, economics, business, and finance. They should be trained in personnel administration and in the relations of people and machines to production.

Manufacturing Engineering

In many industries manufacturing engineers are responsible for taking a newly designed product and then determining how to produce it at an economical cost. These engineers organize men, materials, and machines so that reliable products can be produced efficiently. They transform the ideas and plans of the designer into a quality product that can be produced economically. Generally, manufacturing engineers are responsible for the development, design, analysis, planning, supervision, and construction of the methods and equipment needed for the production of industrial and consumer goods. Many universities offer undergraduate or graduate programs in manufacturing engineering; these programs may also be included as options within mechanical or industrial engineering. Of particular importance are courses dealing with machine design and operation, personnel management, and economics.

Materials and Metallurgical Engineering

Materials and metallurgical engineers specialize in the development, production, and utilization of the metallic, ceramic, and polymeric (plastic) materials that are used in all fields of technology. Engineers have developed many new materials such as ultra high-purity semiconductor materials for electronic devices; high-strength alloys for use in jet and rocket engines; strong, light metals for aerospace applications; specialized glasses and ceramics having high thermal, mechanical, and chemical stability for use in the chemical industry; and a host of polymeric materials that are replacing metal, glass, wood, and natural fibers in numerous applications. These engineers develop methods to recycle materials. In addition, new and better materials will be required to meet the needs of our advancing technology. Metallurgical engineers (metallurgists) concentrate their attention on locating and developing deposits of metal ores and then refining and fabricating the metal or metal alloy into various machines or metal parts. Ceramic engineers focus on ceramic materials suitable for high temperature, high stress conditions, such as jet aircraft engines.

FIGURE 1.11. Materials engineers are responsible for the development of exotic new materials such as this fiberglass mesh. (Courtesy Hughes Aircraft Company)

Marine Engineering

Marine engineering is concerned with the use of the world's oceans, lakes, and rivers as avenues of transportation. Of primary concern is the design of marine vehicles, ranging from massive ocean freighters to off-shore oil platforms to tiny deep ocean submarines. Marine engineers design ship hulls and power plants. They focus on the form, strength, stability, and seakeeping qualities of ship hulls and the various types of machinery available for ship propulsion. The marine engineer requires a strong background in mechanics, structural design, hydrodynamics, and energy conversion in addition to exposure and practical experience with marine systems.

Mining Engineering

Mining engineers concern themselves with the discovery, exploration, and development of mineral deposits. These engineers are involved in all phases of the recovery and processing of minerals (e.g., iron, coal, or uranium). Their activities include not only the design, construction, and operation of mining facilities using shaft, strip, and hydrological mining methods, but also the financing of these

FIGURE 1.12.   Marine engineers testing ship hull designs in a towing tank. (Courtesy University of Michigan College of Engineering)

ventures and the marketing of the crude minerals and mineral products. Mining engineers possess a strong background in many aspects of civil engineering, including structural analysis, soil mechanics, and hydrology. A background in geology and chemistry is also necessary.

Nuclear Engineering

Nuclear engineers apply basic scientific and engineering methods to the design, operation, and use of systems based on nuclear energy. Most of their attention has been directed toward the development of power systems based on nuclear fission reactors. These nuclear power systems are used to produce heat and electricity in central station power plants and propel submarines, ships, and spacecraft. Nuclear engineers are also concerned with the use of radiation in both industrial and medical applications. At the current forefront of nuclear engineering is the development of new forms of nuclear energy such as the breeder reactor and controlled thermonuclear fusion. Nuclear engineers have a thorough foundation of physics and mathematics coupled with knowledge of more traditional engineering subjects in mechanical and electrical engineering.

FIGURE 1.13.   The efforts of nuclear engineers to develop new sources of energy are illustrated by this experiment in controlled thermonuclear fusion. (Courtesy Los Alamos Scientific Laboratory)

Petroleum Engineering

Petroleum engineers combine chemistry, physics, and geology with engineering methods in the development, recovery, and field processing of petroleum. They are concerned with finding deposits of oil and gas in quantities suitable for commercial use and in the economic extraction of these materials from the ground. The petroleum engineer will design methods (e.g., pipelines) for transporting oil and gas to suitable processing plants or to places where they will be used. The growing scarcity of petroleum reserves has placed great importance on this branch of engineering.

Sanitary and Water Resources Engineering

Water resources engineers develop and maintain sources of water for various purposes including irrigation, residential, and industrial use. A primary concern is water transport. Sanitary engineering is concerned with waste water transport and treatment. This latter field has become particularly critical in water quality programs. Both water resources and sanitary engineers should acquire strong backgrounds in chemical engineering and civil engineering, with particular emphasis on process design.

Systems Engineering

The complexity of modern technology has grown so rapidly that special engineers are now called upon to design and analyze systems. For example, the system of interest might be a miniaturized component of a computer, or a massive hydroelectric dam, or perhaps a model of the interaction of the economic market with the production capabilities of the automobile industry. The system is treated as a unit and analyzed in terms of its input, output, and control parameters. Systems engineers are adept at developing abstract mathematical models of systems that can be analyzed to predict and improve performance. They rely heavily upon both mathematics and computer applications in their activities.

Transportation Engineering

Transportation engineers focus on the movement of people and products. For instance they are responsible for highway or railroad design and construction. Recently mass transit systems have attracted a great deal of interest. Transportation engineering is commonly included in many civil engineering programs.

Interdisciplinary Areas

Many other types of engineering fields that overlap with fields in the physical, biological, or social sciences are available to the student. For example, the engineering physicist works at the interface between basic physics research and the needs of technology. Engineers also work in the area of technology assessment and environmental impact analysis in close coordination with social scientists.

Engineering students are sometimes overwhelmed by the confusing array of fields available as possible engineering majors. Should they choose a traditional area such as civil or mechanical engineering, or strike out in one of the newer directions such as aerospace or biomedical engineering? Can they make a commitment to one of these fields and later change their minds? As we shall find later, the strong similarity between all engineering programs in the early years usually allows the student to postpone making such a decision until the sophomore year (or perhaps later). More significant, the engineer can assume a variety of roles in any of these fields, ranging from research and development to marketing and management. In fact, these roles are probably more important in an engineer's choice of career than the actual field of specialization.

SUMMARY

A variety of more specialized engineering fields have developed in response to the needs of society:

Aerospace engineers are concerned with the design and development of aircraft and spacecraft.

Agricultural engineers apply engineering principles to farm and food production industries.

Architectural engineers interact closely with architects in the design and construction of structures.

Atmospheric and oceanographic engineers are concerned with society's use of and impact upon the earth's atmosphere and oceans.

Automotive engineers are responsible for the design, development, manufacture, and application of land transportation vehicles.

Biomedical engineers combine the methods of engineering with the biological and medical sciences.

Computer engineers are concerned with the design and utilization of computer systems.

Environmental engineers combine knowledge of the natural, social, and physical sciences to analyze and improve the interaction between society, technology, and the environment.

Industrial engineers are concerned with the design and installation of integrated systems of people, materials, and equipment.

Manufacturing engineers are responsible for developing economical methods to manufacture engineering designs.

Materials and metallurgical engineers specialize in the development, production, and utilization of metallic, ceramic, and plastic materials employed in technology.

Marine engineers design marine vehicles and structures.

Mining engineers are concerned with the recovery of minerals using various mining methods.

Nuclear engineers design and operate systems utilizing nuclear energy.

Petroleum engineers develop methods for the recovery and field processing of petroleum.

Sanitary and *water resources engineers* design water transport and treatment facilities.

Systems engineers design and analyze the behavior of complex systems.

Transportation engineers are concerned with the transport of people and products.

Exercises

Fields of Engineering

1. Visit your college placement center and obtain a list of 10 companies actively recruiting engineers in your field of interest.

2. Leaf through the employment opportunities section of a large newspaper and determine the number of job openings in your field of interest.

3. List the types of engineers you feel might have played the key role in the following technological developments: (a) the jet turbine engine, (b) the digital computer, (c) the laser, (d) laundry detergent, (e) the Bay Area Rapid Transit System (BART) in San Francisco.

4. What would you consider as a major engineering accomplishment for each of the engineering fields discussed in this section?

5. What types of engineers do you feel would have played a significant role in the development of each of the following: (a) wind turbine generators, (b) nonstick frying pans, (c) satellite-based telecommunications systems, (d) microcomputers? (Remember, usually several engineering fields are involved in a given development.)

6. Describe the possible role of (a) mechanical engineers in the computer industry, (b) electrical engineers in the aerospace industry, (c) civil engineers in the nuclear power industry, and (d) chemical engineers in the solar power industry.

1.3. WHAT DOES THE ENGINEER DO?

One frequently distinguishes among engineering activities by field. For example, a civil engineer is commonly thought of as a builder, a mechanical engineer as the designer of machines, and so on. However even within a given field, the engineer's activities may span a wide range. For example, civil engineers might be involved in basic research concerned with the behavior of high-strength concretes or with designing or supervising the construction of a highway. They might also sell earth moving equipment or manage a large construction firm employing hundreds of employees. The interests, experience, and professional duties of engineers are distinguished by their diversity and versatility.

To be more precise, let us classify some of the various roles or functions of engineers by considering the variety of activities which typically are required in applying science and technology to meet a particular social need.

Research

Both the *research engineer* and the scientist seek to discover and interpret new facts about natural phenomena. However, while the scientist is often content with the discovery itself, with knowledge for its own sake, the research engineer must carry the investigation further and identify the potential practical application of the discovery. For example, a scientist might analyze the behavior of certain molecules when subjected to high temperature and pressure. The research engineer could then utilize this knowledge by applying it to the production of plastics. Or the principles a scientist discovered while studying the motion of electrons inside metals might enable an engineer to construct an electronic device.

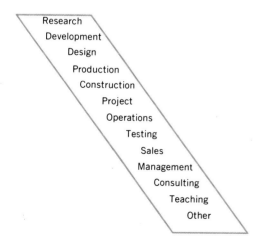

FIGURE 1.14. The roles of the engineer.

FIGURE 1.15. Research engineers make use of sophisticated tools such as electron microscopes. (Courtesy University of Michigan College of Engineering)

Research engineers work at the forefront of science. They often must pursue investigations beyond the frontiers of existing knowledge. Therefore a premium is placed on intelligence, perceptiveness, and ingenuity, traits so essential to the process of discovery.

Development

The research engineer's discovery of a new scientific principle sets into motion a complex process aimed at a practical application of this discovery. The first phase of this process is the responsibility of *development engineers*. These engineers begin by designing a device or process that represents a working model of the scientific discovery (and also of the intended application). At this stage more concern is directed toward the operation of the device rather than its commercial value or production, that is, with function rather than form. The device itself may be a machine or a structure or perhaps even a process.

Development engineers usually demonstrate only the feasibility of a design concept. They may do this by carrying out a design on paper or by building a prototype. This prototype frequently will bear little resemblence to the final product. For example, an electric circuit might be developed by first arranging individual circuit elements on a "breadboard". Only after the satisfactory operation of the circuit has been verified will the engineer proceed to design the actual circuit layout.

The design is then translated into a working prototype to test its suitability for eventual production and implementation. Usually, many practical problems become obvious only after such a working prototype is built. Parts do not fit, components do not function, and circuits fail. The development engineer is responsible for finding the causes of these malfunctions and correcting them, in other words, for removing the bugs from the design. Most prototypes will require many redesigns, each incorporating improvements and avoiding the pitfalls discovered in earlier versions.

The separation of the development from the research and design phases of an engineering project can be rather blurred. In many organizations research and development ("R & D") activities are conducted by the same group of engineers. Moreover, engineering design is important in translating a scientific discovery into a working prototype for further development work. In this sense the development engineer may participate in both research and design.

Design

Design has the important function of translating a scientific discovery into a working model for development. It is also essential to the successful transformation of the development model into a device or product suitable for implementation. *Design engineers* bridge the gap between the laboratory and practical application. In the manufacturing industry they assume responsibility for final details of making the device suitable for production. In construction industries they develop the detailed designs of structures such as buildings, bridges, or dams.

In general, design engineers work within the constraints posed by the state of the art in engineering materials, production facilities, and economic considerations in their task of designing a production model suitable for practical application whether as product, process, or structure.

Production

The responsibility for the mass production of the design falls to *production engineers*. They start with the design engineer's drawings and supervise the assembly of the device as it was conceived. In this capacity production engineers oversee the production line, establish the production schedule, and see that this schedule is met. They are responsible for ordering raw materials at the optimum times, setting up the assembly line, and handling and shipping the finished product. They see to it that all machinery functions properly, that operators perform their duties satisfactorily, and that product quality is maintained.

Production activities involve a variety of people including engineers, technicians, craftsmen, and assembly line workers. Production engineers supervise and lead this team. They give direction to the members of the team, assign their jobs, and answer their technical questions.

Construction

The counterpart of the production engineer in the building industry is the *construction engineer*. These engineers design and build large structures such as buildings, plants, bridges, and highways. In most construction work competitive bidding is used to award contracts. Hence the construction engineer begins by evaluating the project specifications provided by the customer. On the basis of

FIGURE 1.16. A project engineer is usually assigned the primary responsibility for a construction project. (Courtesy Hewlett-Packard Company)

these specifications and preliminary design plans, the engineer assists in preparing and submitting a bid for the project.

Once a bid has been accepted, it is customary to assign a *project engineer* to assume overall responsibility and supervision of the work. This individual will lead a team of construction and design engineers who specialize in more specific aspects of the project such as mechanical, civil, and electrical design. This team of engineers and draftsmen prepare a detailed set of technical specifications for the project (usually detailed design drawings or models). They also set up a construction schedule.

The project engineer then supervises construction of the project and coordinates the work of craftsmen and laborers at the site. He or she must be able to work effectively with the construction and trade labor unions involved in the project and has final responsibility for completing the project on schedule, according to the original specification.

Operations (Plant)

Industrial plants often cover many acres of land and consist of numerous buildings that house both production facilities and administrative offices. Some plants even provide their own utilities, power generation, and water supplies. The operation of this complex and vast facility is the responsibility of the *operations* or *plant engineer.*

Plant engineers may participate in the design and layout of the various buildings, utilities, and machinery of the plant. They also allocate space for equipment and offices and supervise the procurement of the necessary equipment and furnishings. During construction the plant engineer supervises and coordinates work of the various building contractors.

The plant engineer and staff have the primary responsibilities of operating and maintaining the facility after it has been built. They maintain the buildings, equipment, grounds, and utilities such as lighting, heating, ventilation, and air conditioning systems. In short, they are responsible for ensuring that the entire facility operates smoothly and that both production and office workers are provided with suitable working environments.

Testing

Testing is an important aspect of engineering. It is performed during all phases of research, development, and production. During research, testing takes the form of experiments designed to examine the validity of scientific conjectures or hypotheses. Prototype testing is a key aspect of the development of a new product or process. During production, a product must be tested frequently to ensure its quality and acceptability. The responsibility for designing and conducting appropriate test procedures lies with the *test engineer*. Test engineers must be capable of planning a test program, establishing the appropriate test conditions, the proper measurements, and the correct number of tests. They must then perform these measurements and collect data. Finally they must analyze the data and draw conclusions from them.

Engineers concerned with maintaining quality control during manufacturing or construction activities are sometimes called *quality assurance engineers*. Quality assurance (Q/A) has become a vital aspect of both the manufacturing and construction industries. Q/A engineers are responsible both for setting up appropriate testing and inspection programs as well as effective quality control procedures. Such programs are frequently mandated by government regulations.

Sales and Marketing

Although one seldom associates engineering with sales and marketing, engineers often perform these important functions. Of course all engineers are involved in selling to some degree since they must persuade management that resources should be allocated for the development of particular concepts or expansions of facilities. Indeed the increased sophistication of technology has added importance to an adequate technical background in sales, and more and more engineers are finding roles in sales and marketing, particularly in the area of industrial products.

Sales engineers seek out and establish contact with potential markets for the product. They evaluate the requirements of the user and offer a product that satisfies the customer's engineering and economic requirements. In this capacity sales engineers are an important liaison between the company and the customer. Sales engineers must clearly explain engineering functions, operations, and advantages of their products over competing items. They must be able to answer technical questions raised by customers. After the sale they frequently help in the installation of the product and train personnel in its use.

Management

Modern engineering requires a team approach. Since any team must have strong leadership to be successful, it is not surprising that many engineers find themselves involved in management activities. Indeed, surveys have shown that more than half of all engineering graduates are involved in some form of management within five years after graduation. Management plays a particularly important role in engineering research, development, and design. Since all managers at this level must have strong technical abilities, they are usually selected from engineering personnel involved in these activities. As engineers advance in management, they are inevitably less involved with details of technical operations. Their attention is directed instead toward the overall operations of the company—in activities ranging from research through production to marketing.

In the past most business managers and executives were selected with professional backgrounds in the social sciences, in law, accounting, or economics. Today there is an increasing trend in both industry and government to choose high level managers and executives with an engineering background. This is particularly true in industries relying heavily on advanced technology and large volume production. Today engineers often occupy executive positions in the automobile, aerospace, electronics, and manufacturing industries. The technical education and disciplined approach of the engineer provides an excellent background for managerial and executive functions.

Consulting

Many engineers act as independent consultants. Such *consulting engineers* usually do not work for any single company; instead they act as a source of information on specific projects for a number of companies or agencies. Consulting engineers usually possess specific skills in addition to several years of experience. They may advise and work on engineering projects either on a part-time or full-time basis. For example, industry may seek out a consultant to solve a particular problem. Government agencies also make frequent use of consultants. A growing consulting activity is supported by the legal profession, which uses consultants as expert witnesses in lawsuits involving technical matters such as product liability.

Some companies specialize in consulting services. These companies may consist of only one or two engineers, or they may employ a large staff of engineers and draftsmen. Some of them specialize in narrow fields and restrict their activities to designing and erecting buildings, water treatment plants, or petroleum refineries. Other consulting engineering firms may undertake a wide range of engineering projects.

Consulting engineers must be able to solve difficult technical problems. Their skills are developed through years of training and experience. They also should be adept at business since they must sell their services, estimate the cost of the project, and ensure the financial success of their work.

Academe

Many engineers choose teaching as their career and become college faculty members. Naturally engineering faculty must have a mastery of their particular areas of specialization. But knowledge of the subject does not in itself guarantee teaching effectiveness. Strong communication skills coupled with a genuine concern for students are also essential. In addition to teaching, engineering faculty at universities are extensively involved in research. Their research activities help them to keep abreast of the latest developments in their fields. Engineering research in universities also provides an ideal opportunity to introduce students to research methods and provide them with research experience. The student striving to become a teacher of engineering should pursue graduate level studies and obtain an advanced degree at the doctorate level.

Other Functions

Engineering is recognized as an excellent background for a wide variety of professions because of the technical abilities and discipline engineers acquire. For example many engineers continue their studies to become attorneys and physicians. The analytic approach to problems developed in an engineering education is a sound foundation for a wide spectrum of other careers.

FIGURE 1.17. Many engineers find rewarding careers in teaching.

Each of the various functions of engineers requires different skills. Research engineers have strong backgrounds in fundamental scientific principles. Managers and business executives are primarily concerned with financial problems and the supervision of personnel. Thus, depending on their function, engineers are involved to different extents in abstract technical concepts, practical engineering solutions, or in managerial and economic decisions. In each of these activities engineers rely on a thorough education in both scientific and engineering principles and also in the social sciences and communication arts. In addition they should learn how to identify problems and how to obtain practical solutions. The problems faced by engineers may or may not be of a primarily technical nature. For example, the engineer may be called upon to design a new device (e.g., an improved coolant level gauge for a nuclear power plant), to solve a complex management problem involving personnel (e.g., how to modify procedures to ensure that nuclear reactor operators will routinely monitor the new coolant level gauge), or deal with the complexities of government regulation (e.g., determine a way to introduce new instrumentation or operation procedures into the design of a nuclear power plant without invalidating the existing operating license).

Although the basic skills of problem solving are common to all engineering roles, engineers should also have the specific skills required in their particular field. These skills are generally acquired through advanced courses or on-the-job experience. Ultimately engineers must use their background, skills, and resources to arrive at a practical solution to the complex problems they will encounter in modern society.

SUMMARY

The spectrum of engineering activities is very broad.

Research engineers seek to discover, interpret, and identify possible applications for new scientific phenomena.

Development engineers take scientific discoveries and design and construct working prototypes to test their practical applications.

Design engineers translate the prototype into a device or process suitable for mass implementation.

Production engineers take the design engineer's specifications and supervise the mass production of the device as it was conceived.

Construction engineers are responsible for the design and construction of structures such as buildings, plants, and bridges.

Project engineers assume the overall responsibility for supervising the work on an engineering project.

Operations or plant engineers are responsible for the operation and maintenance of plant facilities.

> *Test engineers* design and conduct tests during the research, development and production phases.
>
> *Sales engineers* determine and develop markets for products while acting as liason between their company and customers.
>
> Many engineers apply their skills to **management,** supervising engineering activities and personnel.
>
> *Consulting engineers* provide assistance on specific technical matters to industry and government.
>
> Engineers may become involved in education as university **faculty** engaged in teaching and research.
>
> Engineers also become involved in a host of related activities that build on their engineering skills such as law, medicine, and government service.

Example   One of the major scientific discoveries of the twentieth century has been the laser, a device that produces extremely powerful light beams by using a physical process known as stimulated emission. Since the first scientific demonstration of lasing action in 1960, progress in laser development has been quite rapid. During the 1960s *research engineers* learned that by discharging an electrical current through a gas mixture containing carbon dioxide, intense beams of infrared light could be produced. *Development engineers* constructed and operated a series of progressively more powerful prototype carbon dioxide lasers during the 1960s. During this development period they encountered and solved many problems such as preventing electric arcing in the gas and designing suitable lenses and mirrors to focus the beams. Prototypes were built and tested with beams sufficiently intense to melt or even vaporize metals. Laser *design engineers* directed their efforts toward developing lasers for use in manufacturing and in metal cutting. When the laser moved out of the laboratory and into an industrial environment, reliability, cost, and ease of fabrication were at a premium. *Production engineers* were involved in designing and monitoring the assembly line manufacture of the laser units and in integrating the units into existing assembly processes. *Plant engineers* learned to master the new technology of high powered lasers and adapt to new safety standards. *Test engineers* developed new sampling and testing procedures to maintain the close tolerances and high reliability of the laser units. *Sales engineers* for industrial products had to become familiar with the new technology. Many of the engineers and scientists involved in the early development of lasers moved into *management* positions as these systems moved to the production phase. Other engineers played an important role as *consultants* to the growing industrial laser industry. New material was included in the engineering curricula on industrial uses of lasers, and new *faculty* were added to teach and conduct research in this area.

Exercises

Engineering Functions

1. Identify the various functions of engineers that will be involved in (a) the successful development and implementation of solar photovoltaic cells to power homes, (b) a manned mission to Mars, (c) the development, production, and marketing of an electric automobile.

2. Engineers will usually find themselves involved in a variety of functions during their professional careers. Project the various roles you see yourself playing in your own future career.

3. What functions might be performed by engineers in government service?

4. Rank the various engineering functions in the order of the degree to which they make use of abstract scientific principles.

5. Choose a typical industry (or company) and identify the various roles played by engineers.

1.4. HOW DOES ONE PREPARE TO BECOME AN ENGINEER?

The early ancestors of the engineer were artisans and craftsmen, who passed on their knowledge of building, of creating, from generation to generation by apprenticeship. Certainly this aspect of engineering education should not be overlooked, for the advice and experience of seasoned engineers can prove quite valuable. But the complexity of modern engineering problems requires more. Engineers must possess a thorough understanding of the scientific principles basic to their field. In particular, they must acquire a sound grounding in mathematics, physics, and chemistry. Some background in the social sciences and humanities is also vital if they are to interact effectively with society.

But mere knowledge alone will rarely suffice in engineering. This profession requires a highly disciplined and thorough approach to problem solving that is most uncharacteristic of other fields. It is this methodology of engineering problem solving, coupled with the basic scientific knowledge that supports this activity, that is the principal focus of the formal education of the engineer.

1.4.1. High School Preparation

Education for a career in engineering must start very early, long before the student enters a college level engineering program. Students interested in careers in engineering (or science) should take all of the high school mathematics and science courses available, as well as college preparatory courses in writing, literature, and history (Table 1.2).

TABLE 1.2   Basic Admissions Requirements for Most College Engineering Programs

SUBJECT	HIGH SCHOOL COURSE UNITS (ONE YEAR)
English	3 (4 recommended)
Mathematics	3 (1.5 algebra
	1 geometry
	0.5 trigonometry)
Laboratory science	2 (1 physics and 1 chemistry recommended)
Academic electives	4 (2 foreign language recommended)
Free electives	3
Total	15

However, high school programs in mathematics and science vary enormously. Hence most college engineering programs seek the flexibility to accommodate students entering with wide variations in background. For example, some students may have mathematics only to the level of algebra and geometry, while others have received some exposure to trigonometry, analytic geometry, and calculus. Many students enter with only one year of chemistry and physics, while others have taken college-level courses in these subjects while still in high school. Although the path to an engineering degree may be slightly longer for students with weaker high school backgrounds, engineering programs usually provide the student with an opportunity to make up any deficiencies and catch up with the better prepared students.

1.4.2.   The Undergraduate Engineering Curriculum

The role of engineers in society and their responsibility for applying technology to meet human needs have not changed since the beginnings of the profession. However the nature of the problems faced by our society is continually changing, becoming ever more complex and affecting greater numbers of people. Fortunately, the tools that the engineer can apply to address these problems have kept pace with these changes as scientific knowledge and technology have advanced.

The engineer must be able to adapt to an environment of change. In fact, the pace of technological growth is now so rapid that major changes will occur not just during an engineer's career, but even during his or her education. Therefore the engineering curriculum must provide a broad-based education appropriate to this a world of change. It must provide graduating engineers with scientific knowledge. It must also develop their intellectual skills so that they can continue to learn and adapt to their changing environment long after their formal education has ended. The primary emphasis of an engineering education is the development of individuals who can think, respond, and adapt to change.

At the foundation of an engineering education are basic sciences and mathematics. Mathematics is the language of engineers. It provides them with a compactness and accuracy of expression far beyond that of conversational language. Furthermore, the rules of mathematics aid tremendously in structuring a logical approach to problem solving.

Science courses will also play an important role in the early years of an engineering education. Although the primary focus will be on courses in chemistry and physics, many fields of engineering also require coursework in the biological and earth sciences.

The engineering program builds on this foundation of science and mathematics with a sequence of courses referred to as engineering sciences (Figure 1.18). These include courses in solid and fluid mechanics, thermodynamics, materials

FIGURE 1.18. An engineering education is based upon a pyramid of courses in the basic sciences, humanities, social sciences, and engineering sciences.

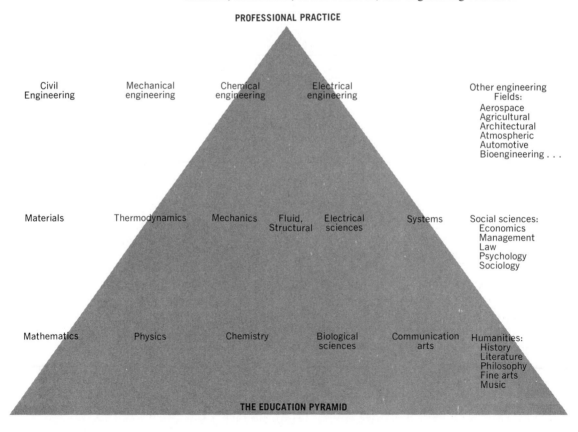

science, and electricity and magnetism that are common to all engineering disciplines. Although such courses have strong roots in basic science and mathematics, they provide the bridge between these fundamental disciplines and engineering application.

During the later years of an engineering education, coursework in engineering design is introduced. These courses serve to orient the student toward engineering practice. In particular, they introduce the student to methods of engineering design, including the solution of open-ended (vaguely-defined) problems under realistic constraints such as economic factors, safety, environmental impact, and social impact.

Another technical area of great importance is modern computing methods. Students may acquire knowledge in this area either by taking formal courses in computer programming and numerical analysis or by acquiring on-the-job experience in other courses that utilize computers. Students should not only learn how to effectively utilize large, centralized computer systems (usually timesharing systems) but also keep abreast of the rapid developments in microcomputers. In fact, at the start of their engineering program, many students will probably find that one of the closest companions is their programmable calculator or desktop computer. The use of mini and microcomputers will enter into many engineering courses, particularly those involving modern laboratory techniques or computer-aided design.

Essential components of any engineering program are courses in the arts, humanities, and social sciences. Traditional subjects in these areas include literature, history, philosophy, fine arts, economics, sociology, and psychology. Not only do such courses complete the education of the engineer, but they provide a point of contact with the rest of society. They help the engineering student understand and develop an appreciation for the potential impact of engineering on society and its natural environment. Of equal importance is the development of the skills to communicate effectively through written, spoken, or graphical means. Engineers cannot be successful without an ability to communicate the results of their labors, their ideas.

Engineers may acquire training in many other specialized areas during their undergraduate studies. For example, many engineers take courses in business, management, or accounting. Others acquire the background in law or medicine necessary for further studies in these fields. Still others specialize in artistic training to assist them in creative design activities.

Sometimes students are confused by the tremendous variety of engineering disciplines and worry that they will be locked into a particular major by a premature choice of courses. However it should be apparent from our discussion that there are only modest differences among engineering curricula during the first two years. In fact the curricula of most engineering majors have more similarities than differences. In most cases the choice of an engineering major can be safely deferred until late in the sophomore year, and in some cases, it is relatively easy to transfer between fields even after this point. In fact, many engineers switch fields (some several times) even after they enter professional practice.

SUMMARY

The emphasis of an engineering education is the development of engineers who can think, respond, and adapt to a world of change. It focuses on the methodology of engineering problem solving and the scientific knowledge that supports this activity. At its foundation are courses in the basic sciences, physics, chemistry, and mathematics (including modern computing methods). But an essential component of the engineer's education includes courses in the arts, humanities, and social sciences, since these subjects round out this education and provide a point of contact with the rest of society. The engineer must also develop the skills to communicate effectively using written, spoken, or graphical means.

1.4.3.   Engineering Technology Programs

Programs in engineering technology differ from those in engineering primarily in the extent to which they emphasize the application rather than the development of engineering knowledge and methods. Graduates of two-year, associate degree programs in engineering technology are called *engineering technicians*. The origin of such programs can be traced back to the early 1960s. At this time increased emphasis of the traditional engineering programs on science and analysis (the "post-Sputnik" era) had left behind an unmet demand in a variety of more applied areas such as drafting and shop methods, machine operations, maintenance of manufacturing systems, and laboratory operations. Two-year associate degree programs in engineering technology were established at a number of junior or community colleges to meet these needs. Graduates of such programs are typically awarded the Associate of Science in Engineering Technology or A.S.E.T. degree.

By the 1970s it had become apparent that many industrial needs were not being satisfied by either the B.S. degree engineering graduate or the engineering technician. Neither of these programs placed enough emphasis on applications in engineering design, testing, and manufacturing. Therefore a number of baccalaureate degree level (four-year) programs in engineering technology were established that awarded the Bachelor of Science in Engineering Technology or B.S.E.T. degree. Graduates of such programs are known as *engineering technologists*. Career opportunities for these graduates lie in the areas of hardware design using proven concepts, product analysis and development, construction and production management. In less technically demanding jobs, engineering technologists and engineers may perform similar functions.

The course requirements for both associate (engineering technician) and baccalaureate (engineering technologist) degree programs are compared against those of the B.S. degree program in engineering in Table 1.3. The somewhat heavier emphasis of these programs on technical applications should be noted. Because of the significant difference in the amount of science, mathematics, and engineering applications required in B.S. (engineering) and B.S.E.T. (engineering technology) programs, it can prove difficult to transfer from one program to another.

SUMMARY

To meet industrial needs for graduates more skilled in specific applications, degree programs in engineering technology have been established. Graduates of a two-year associate degree program (A.S.E.T.) are known as engineering technicians, while those of a four-year baccalaureate program (B.S.E.T.) are referred to as engineering technologists.

TABLE 1.3   Minimum Course Requirements for Engineering and Engineering Technology Degrees[*]

B.S. IN ENGINEERING	B.S. IN ENGINEERING TECHNOLOGY	A.S. IN ENGINEERING TECHNOLOGY
120 SEMESTER HOUR CREDITS OR EQUIVALENT	**120 SEMESTER HOUR CREDITS OR EQUIVALENT**	**60 SEMESTER HOUR CREDITS OR EQUIVALENT**

• 75 semester hour credits in mathematics, science, and engineering:
  • 15 hours of mathematics beyond trigonometry
  • 15 hours of basic sciences
  • 30 hours of engineering sciences
  • 15 hours of engineering design
• 15 hours in humanities and social sciences
	• 23 semester hour credits of basic sciences and mathematics
	• 45 credits of technological courses including technical science, technical specialties, and technical electives
	• 21 credits in communications, humanities, and social sciences
		• 15 semester hour credits of basic sciences and mathematics (about half in mathematics)
		• 30 credits of technical courses including technical science, technical specialties, and technical electives
		• 8 credits in communications, humanities, and social sciences

[*]***SOURCE***   Accreditation Board for Engineering and Technology, 1981.

1.4.4.   Accreditation of Engineering Programs

The formal education received by engineers plays a determining role in their professional competence. For this reason, the Accreditation Board for Engineering and Technology or ABET (formerly known as the Engineer's Council for Professional Development or ECPD) has been given the responsibility for reviewing and accrediting programs in engineering and engineering technology throughout the United States. The ABET determines and monitors the minimum standards that must be met by any accredited program. ABET accreditation is designed to identify to the public, prospective students, potential employers, government agencies, and state boards of examiners for engineering registration, the engineering and engineering technology programs that meet these standards. The ABET also provides guidance for the improvement of existing curricula and for the development of future programs. It attempts to stimulate the general improvement of engineering education in the United States.

The general ABET accreditation criteria have already been detailed in Table 1.3. These criteria are intended to assure an adequate foundation in science and mathematics, the humanities and social sciences, engineering science, and engineering methods, as well as for preparation in an engineering field of specialization. The accreditation process involves a review of the program via questionaires, together with a supplemental report of an on-site visit by a carefully selected team representing the ABET. The ABET is now recognized by the United States Department of Education, the National Bureau of Engineering Registration, and the National Council of Engineering Examiners as the sole agency responsible for the accreditation of engineering programs.

SUMMARY

The accreditation of engineering and engineering technology programs is the responsibility of the Accreditation Board for Engineering and Technology or ABET.

1.4.5.   Postgraduate Education

An engineer's education does not end with graduation. Indeed, an undergraduate program is intended to provide only a broad foundation in science and engineering analysis, to which must be added training on both a formal and informal basis. For instance, all engineers must go through extensive on-the-job training in their particular area of interest. Only after extensive experience does the engineer develop the maturity and judgment necessary for significant contributions.

Engineers with interests in research and development must acquire additional education in both basic and engineering sciences at the graduate level. Most research engineers continue to the doctorate level so that they can obtain research expertise. Many other engineering fields have become sufficiently complex with the result that an undergraduate degree is no longer adequate for professional practice. Indeed, it has even been suggested recently that the engineering profession require the master's degree as the minimum level of education necessary for professional practice.

It is essential that engineering students regard their education as a lifetime commitment. To be relevant to a changing world, engineers must continue their education informally, by reading, discussions with colleagues, attending technical meetings or workshops, or perhaps even returning to college at several points during their career. Without continuing to update their education, engineers would almost certainly find themselves obsolete within several years of graduation.

SUMMARY

An engineer's education does not end with graduation. All engineers require on-the-job training and experience. Additional postgraduate education is needed for many fields such as research. It is essential that engineering students regard their education as a lifetime commitment if they are to adapt to a world of changing demands.

Exercises

Engineering Education

1. Prepare a sample program of courses for your field of interest (consulting your college catalog). Classify each course in your program as: science, mathematics, engineering science, engineering design, and humanities. Compare this program against the ABET criteria listed in Table 1.3.

2. By referring to old college catalogs in your library, compare today's engineering program with the one offered two decades ago.

3. Indicate how courses in humanities and social sciences might prove of value to engineers employed in (a) industry, (b) government, and (c) education.

4. Interview a practicing engineer (or, as a last resort, an engineering professor) to determine the degree to which they use (a) mathematics, (b) physics, and (c) chemistry in their everyday activities.

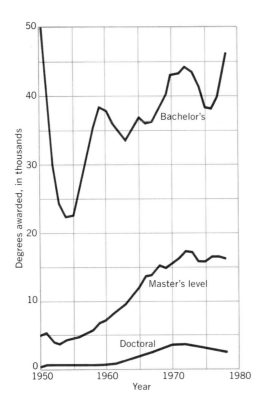

FIGURE 1.19. The number of engineering degrees granted each year in the United States from 1950 through 1980. (Source: Engineering Manpower Commission)

5. One common problem with curricula in science and engineering is that many courses depend on others as prerequisites. Analyze this "pyramid of prerequisites" for your own projected program by tracing the chain of prerequisites for various courses. (Work from your college catalog.)

1.5. ENGINEERING AS A PROFESSION

Engineering is a learned profession, just as law, medicine, or theology are. Engineers must acquire certain skills, through both formal education and experience, to practice their profession. Like members of other professions, engineers assume a responsibility to apply their specialized knowledge and skills for the benefit of society. As professionals they must recognize how profoundly this conduct affects public welfare and act accordingly.

A profession can be distinguished by several important characteristics: (1) It satisfies an indispensable and beneficial social need. (2) It requires the exercise of discretion and judgment and is not subject to standardization. (3) It involves an activity conducted on a high intellectual plane with knowledge and skills not common to the general public. (4) It has as an objective the promotion of knowledge and professional ideals for rendering social services. (5) It has a legal status and requires well-formulated standards of admission.

The engineering profession differs somewhat from the professions of law, medicine, divinity, and philosophy in its origin. It more closely resembles professions such as architecture and journalism that can be traced back to the craft and merchant guilds of medieval times. The engineering profession has struggled on occasion with its desire to retain the voluntary guild pattern of its heritage and its desire to move toward the status of professions such as medicine and law with more rigid laws governing entrance and practice.

The privileges and obligations governing the relationship between a profession and society are reciprocal. A profession offers the technical competence to meet an important social need, in return for which it is granted a monopoly on its activities and self-governing privileges. An essential aspect in the relationship between any profession and society is trust. Society must accept the profession's authority in the field of its competence if its professionals are to discharge their responsibilities. To establish this atmosphere of public trust and respect, the engineering profession has relied on several mechanisms. First, and perhaps most important, it has devised and adopted a code of ethics to guide the activities of the engineer. Second, the profession has assisted government in establishing registration procedures to certify the competence of engineers. Through registration, the public grants engineers certain rights in the practice of their profession. In turn registered engineers accept both the legal and ethical obligation to act with integrity and good judgment. Finally, engineers have joined together to form professional organizations or societies to assist them in carrying out their profession activities.

1.5.1. Professional Ethics

The relationship between the engineer and society must be based on respect, trust, and integrity. Engineers possess specialized knowledge well beyond that of their clients, employers, or the general public. While this knowledge represents the basis for the contributions the engineer can make to society, it also contains the potential for harmful effects on society. To guide this delicate relationship between the engineer and society, the engineering profession has developed and adopted a code of ethics. This code takes the form of a set of rules or canons designed to prescribe the obligations and responsibilities of members of the profession toward other engineers, clients and employers, and the public at large.

Codes of professional ethics have strong roots in history. Perhaps the most well-known statement of professional ethics is the Hippocratic Oath, governing the medical profession, that dates back almost 2000 years. Similar codes have been developed for various engineering groups over the past century.

Most engineering organizations have now adopted a standard code of ethics, first developed by the Engineer's Council for Professional Development and now adopted by the American Association of Engineering Societies. This Code of Ethics for Engineers is given in Figure 1.20. The essence of the code is that engineers must act with complete responsibility toward their clients and with regard to the public interest. To be a member of the engineering profession, the engineer must be worthy of the trust and confidence of both the public and professional colleagues.

Although the general statement of the Code of Ethics is quite simple, the interpretation and application of this code can become more complex. The American Association of Engineering Societies has therefore adopted a set of guidelines for use with the fundamental canons of ethics provided in Appendix A. We strongly encourage the student to read these guidelines carefully and refer to them whenever the potential for an ethical problem arises in professional practice (whether in the classroom or on the job).

Professional ethics are of paramount importance to engineering. Of the many ways in which ethical problems may arise in practice the most familiar examples concern the relationship between consulting engineers and their clients, since this relationship is similar to the one that exists between doctor and patient or lawyer and client. Engineers in industry must also use the code as a guide to their professional responsibilities toward both employers and the public. They may be forced to balance their loyalty toward their company (and perhaps their financial welfare as well) against their professional responsibility toward society at large. The construction engineer, who functions as both engineer and business executive, may be faced with the conflicting demands of the engineering code and business practice. Engineers in government also should keep their professional responsibilities uppermost in mind, since they are frequently subject to intense political pressures.

It is important to remember that a profession such as engineering can function only if it has the trust and respect of the public it serves. If engineers are to earn and deserve this trust and respect, they must adhere closely to the Code of Ethics for Engineers in their professional activities.

1.5.2.   Professional Registration

To establish some degree of control over the qualifications required for the practice of engineering, laws governing its recognition and regulation are now widespread. Beginning with Wyoming in 1907, all states have passed laws that require engineers to obtain a certificate of registration or license as a prerequisite to certain types of professional practice, similar to those laws governing the legal and medical professions. The legal ground for public registration is usually based on the constitutional mandate "to safeguard life, health, and property, and to promote the public welfare."

Registration is required of engineers carrying out professional activities either in an individual capacity (e.g., a consulting engineer) or as the person in charge of an engineering organization. Generally engineers in direct charge of the engineering activities of a company are registered. Other engineers in subordinate posi-

CODE OF ETHICS OF ENGINEERS

THE FUNDAMENTAL PRINCIPLES

Engineers uphold and advance the integrity, honor and dignity of the engineering profession by:

  I. using their knowledge and skill for the enhancement of human welfare;

 II. being honest and impartial, and serving with fidelity the public, their employers and clients;

III. striving to increase the competence and prestige of the engineering profession; and

IV. supporting the professional and technical societies of their disciplines.

THE FUNDAMENTAL CANONS

1. Engineers shall hold paramount the safety, health and welfare of the public in the performance of their professional duties.

2. Engineers shall perform services only in the areas of their competence.

3. Engineers shall issue public statements only in an objective and truthful manner.

4. Engineers shall act in professional matters for each employer or client as faithful agents or trustees, and shall avoid conflicts of interest.

5. Engineers shall build their professional reputation on the merit of their services and shall not compete unfairly with others.

6. Engineers shall act in such a manner as to uphold and enhance the honor, integrity and dignity of the profession.

7. Engineers shall continue their professional development throughout their careers and shall provide opportunities for the professional development of those engineers under their supervision.

Approved by the Board of Directors, October 5, 1977

FIGURE 1.20. The Code of Ethics of Engineers. (Reprinted with permission from the Accreditation Board for Engineering and Technology)

tions need not be registered if the registered engineer in charge signs all plans or has them issued under his or her seal.

Engineering registration carries certain privileges in addition to sanctioning professional practice. Registered engineers are allowed to use the title "Professional Engineer," denoted by the letters P.E. following their name (much as M.D. is used in the medical profession). Furthermore they may advertise themselves as an engineer. However registration also carries with it certain obligations and legal liabilities. For example, registered engineers may be held liable if the plans they prepare prove grossly defective. They are also potentially responsible for damage stemming from their failure to supervise construction work, if this supervision is required, or for handling their duties in a negligent manner. Registered engineers must take care in placing their seal on designs made by other engineers, since they will be held liable for any inadequacies that occur in these plans.

Although all states have adopted separate laws governing registration and established state boards of engineering examiners, the general requirements for registration are quite similar. Typical minimum requirements include an engineering degree from an accredited college (where the accreditation is provided by the Accreditation Board for Engineering and Technology). Generally four or five years of practical experience are also required subsequent to graduation from college. The engineer must pass several examinations covering both the basic theory and practice of engineering and provide references from other registered engineers indicating their approval of the applicant's competency and character.

Engineering graduates are permitted to take the two-day examination on theoretical subjects as soon as they graduate. Successful performance on this examination allows registration as an "Engineer-in-Training" and provides eligibility for membership in the National Society of Professional Engineers. The first day of the EIT examination covers general material in mathematics, chemistry, physics, statics, dynamics, strength of materials, electricity, and engineering economics. The second day of the examination covers the applicant's field of specialization such as civil or mechanical engineering. Since the EIT examination is based upon material covered in the undergraduate engineering curriculum, it is desirable to take this examination as soon as possible following graduation.

After completion of the required four or five years of professional experience, the engineer may take the remaining portion of the registration examination. This is a one-day test consisting of one or more general problems requiring the exercise of judgment expected of persons at that stage of their careers. An oral examination is also usually required. An engineer who has satisfied the registration requirements to practice in a particular state can frequently transfer this registration to another state using reciprocity agreements.

Should an engineering graduate become registered? It is true that professional registration is still regarded as optional, and not all organizations have made it a prior condition for employment. However the rapid growth in engineering registration makes it likely that this credential will be required for increasing numbers of engineering activities in the future. By becoming registered as rapidly as possible, engineering graduates can protect their investment in an engineering education and preserve their future options. In fact, some thought should be given to professional registration even prior to graduation since many universities offer special preparation for the Engineer-in-Training examination. These examinations may even be administered on campus. Information concerning professional registration can be obtained either from your college engineering placement office or directly from your State Board of Engineering Registration.

1.5.3. Professional Societies

Engineers tend to be collaborators, both by instinct and tradition. They are most successful in professional activities when working as teams. Indeed, the body of scientific and engineering knowledge used by the modern engineer has been acquired and distilled from the efforts of thousands of scholars and scientists, inventors and engineers, throughout history. For centuries this body of knowledge was passed on to new members of the profession by trade groups or guilds. Today these groups have evolved into the professional societies of engineering.

There are roughly 20 major engineering societies to serve the technical and professional needs of the engineering profession. These societies generally reflect the major engineering disciplines and are listed in Figure 1.21. In 1980 the major societies joined together to form the American Association of Engineering Societies or AAES. Affiliated with these societies are several hundred more specialized groups.

Engineering societies publish technical journals to distribute engineering information, organize technical meetings, and establish standards and regulations governing both engineering practice and equipment. They provide excellent opportunities for engineers to interact with their professional colleagues. Many such societies have student branches. Engineering students will find it invaluable to become active in one or more of these societies since they assist in developing a professional identity. It is even more critical for engineers to continue this affiliation and active involvement with the professional engineering societies that embrace their technical specialities. In a real sense, these societies represent the corporate bodies of the profession. Engineers have an obligation, both to their profession and to their professional careers, to participate actively in their professional engineering societies.

SUMMARY

Engineering is a learned profession with its own set of standards. Engineers must acquire certain skills through formal education and experience to practice their profession. For many activities they must also

PARTICIPATING BODIES

American Congress on Surveying and Mapping (ACSM)
Founded: 1941. Membership (August 1980) 10,215 incl. students
William A. Radlinski, Executive Director and Treasurer, 210 Little Falls Street, Falls Church, VA 22046 (703) 241-2446

American Institute of Aeronautics and Astronautics, Inc. (AIAA)
Consolidated in 1963 from the American Rocket Society (founded 1931) and the Institute of the Aerospace Science (founded 1932). Membership (August 1980) 30,500 incl. students
James J. Harford, Executive Secretary, 1290 Ave. of the Americas, New York, N.Y. 10019 (212) 581-4300

American Institute of Chemical Engineers (AIChE)
Founded 1908. Membership (August 1980) 53,318 incl. students
J. Charles Forman, Executive Director and Secretary, 345 East 47th St., New York, N.Y 10017 (212) 644-8015

American Institute of Industrial Engineers, Inc. (AIIE)
Founded 1948. Membership (August 1980) 35,100 incl. students
David L. Belden, Executive Director and Secretary, 25 Technology Park/Atlanta, Norcross, Ga 30092 (404) 449-0460

American Institute of Mining, Metallurgical and Petroleum Engineers (AIME)
Founded 1871. Membership (August 1980) 78,811 incl. students
Joe B. Alford, Executive Director and Secretary, 345 East 47th St., New York, N.Y. 10017 (212) 644-7695

American Nuclear Society (ANS)
Founded 1954. Membership (August 1980) 11,034 incl. students
Octave J. Du Temple, Executive Director, 555 North Kensington Ave., La Grange Park, IL 60525 (312) 352-6611

American Society of Agricultural Engineers (ASAE)
Founded 1907. Membership (August 1980) 10,800 incl. students
J.L. Butt, Executive Vice President, 2950 Niles Road, Joseph, MI 49085 (616) 429-0300

American Society of Civil Engineers (ASCE)
Founded 1852. Membership (August 1980) 76,343
Eugene Zwoyer, Executive Director, 345 East 47th St., New York, N.Y. 10017 (212) 644-7490

American Society for Engineering Education (ASEE)
Founded 1893. Membership (August 1980) Engineering Colleges & Affiliates 311; Technical Colleges & Affiliates 136; Industrial 87; Gov't. Assoc. 32; 10,010 incl. students
Donald E. Marlowe, Executive Director, One Dupont Circle, Washington, D.C. 20036 (202) 293-7080

American Society of Heating, Refrigerating and Air-Conditioning Engineers, Inc. (ASHRAE)
Consolidated in 1959 from the American Society of Heating and Air-Conditioning Engineers (founded in 1894 as the American Society of Heating and Ventilating Engineers) and the American Society of Refrigerating Engineers (founded in 1904). Membership (August 1980) 39,131 incl. students
Andrew T. Boggs, III, Executive Vice President and Secretary, 345 East 47th St., New York, N.Y. 10017 (212) 644-7940

The American Society of Mechanical Engineers (ASME)
Founded 1880. Membership (August 1980) 96,440 incl. students
Peter Chiarulli, Deputy Executive Director, 345 East 47th St., New York, N.Y. 10017 (212) 644-7730

The Institute of Electrical and Electronics Engineers, Inc. (IEEE)
Founded 1884. Consolidated in 1963 from AIEE and IRE. Membership (August 1980) 201,673 incl. students
Eric Herz, Executive Director and General Manager, 345 East 47th St., New York, N.Y. 10017 (212) 644-7910

National Council of Engineering Examiners (NCEE)
Founded 1920. Membership (August 1980) approximately 600 Members of 54 Member Boards. Number of Registered Professional Engineers 516,000
Morton S. Fine, Executive Director, Box 5000, Seneca, SC 29678 (803) 882-5230

National Institute of Ceramic Engineers (NICE)
Founded 1938. Membership (August 1980) 1769 incl. students
Arthur L. Friedberg, Executive Director, 65 Ceramic Drive, Columbus, OH 43214 (614) 268-8645

National Society of Professional Engineers (NSPE)
Founded 1934, Membership (August 1980) 80,245 incl. students
Donald G. Weinert, Executive Director, 2029 K. St., N.W., Washington, D.C. 20006 (202) 463-2300

Society of Automotive Engineers (SAE)
Founded 1905. Membership (August 1980) 39,300 incl. students
Joseph Gilbert, Secretary and General Manager, 400 Commonwealth Dr., Warrendale, PA 15096 (412) 776-4841

Society of Manufacturing Engineers (SME)
Founded 1932. Membership (August 1980) 53,059 incl. students
R. William Taylor, Executive Vice President and General Manager, One SME Drive, Box 930, Dearborn, MI 48128 (313) 271-1500

MEMBER BODIES

American Academy of Environmental Engineers (AAEE)
Founded 1955. Membership (August 1980) 1681
Stanley E. Kappe, P.E., Executive Director, Box 1278, Rockville, Md. 20850 (301) 762-7797

American Society for Metals (ASM)
Founded 1913. Membership (August 1980) 50,878 incl. students
Allan Ray Putnam, Managing Director, Metals Park, OH 44073 (216) 338-5151

FIGURE 1.21. A list of engineering societies participating in the Accreditation Board of Engineering and Technology. (Reprinted with permission from the Accreditation Board for Engineering and Technology)

become registered as Professional Engineers. The engineering profession has adopted a Code of Ethics to establish a degree of integrity, trust, and respect for engineering. Engineers have organized themselves into professional societies to meet their professional needs.

Exercises

Professional Aspects

1. Find out the registration laws in your state.

2. What fields of specialization are recognized by your state board of engineering registration?

3. Name several activities that require professional registration in your state.

4. Identify the primary professional societies (and their student affiliate branches) in your field of interest.

Ethics

5. An engineer is hired by a law firm to determine the cause of a propane tankcar explosion that caused extensive public injury and property damage. The engineer finds that the primary cause of the explosion was the negligence of the owner of the tankcar, who also happens to be the law firm's client. What should the engineer report to the law firm? Should this report be provided to anyone else?

6. A company is in the process of preparing a sealed bid for the construction of a dam. An engineer in this company is personal friends with an employee of a competing firm that is also bidding on this project. To what extent would it be proper for the engineer to discuss the project with his friend in the other firm?

7. A government agency issues a request for a proposal (RFP) to determine the strength of fiberglass at high temperatures. Under what conditions can a prospective bidder discuss this request with the issuing agency?

8. Your company has accepted a contract to build a bridge for a fixed fee. During construction the company realizes that the cost of the construction will exceed the agreed-upon fee. The only way to avoid this cost overrun is by lowering the quality of the construction materials. As president of the company, what should you do in this circumstance?

9. Your company has contracted to surface a highway. After the contract is signed, your supervisor directs you to use lower grade con-

crete than was called for in the contract. What should you do in this situation?

10. You have a novel idea regarding the design of an electronic switch. You discuss your idea with your supervisor. A year later you discover that your supervisor has filed for a patent on this idea. What should you do in this situation?

11. A military officer with an engineering background retires after 20 years duty, during which he supervised various aspects of advanced aircraft development and procurement. Upon his retirement, he is immediately approached by several aircraft companies with lucrative offers of employment. What ethical considerations arise in this situation?

12. You are an engineer working for a large company. A newspaper reporter approaches you and asks for an interview concerning recent defects discovered in one of the products of your company. Under what circumstances should you grant the interview?

13. As an engineer working in one of the national laboratories, you are asked by a government agency to evaluate a proposal they have received. Some of the ideas in this proposal are similar to those you have been working on. What should your response be to the government agency's request?

14. You discover that one of your company's products has a defect that could result in serious accidents. You report your finding to your supervisor. Somewhat later, however, you discover that no effort is being made to correct this defect. What should you do in this situation?

1.6.   CHALLENGES FOR THE FUTURE

The modern engineer must be prepared to face a world of change. Scientific knowledge and technology continue to expand at a rapid rate, as do the complexities of the problems faced by our society. Engineers must keep abreast with developments in science and technology if they are to meet the challenges that now lie before our civilization.

The most serious of our immediate dilemmas is due to our rapidly increasing population and the threatening nature of our activities. These activities could alter permanently the very planet on which we live. International conflict using modern weapons of mass destruction now present the very real risk of making large portions of the planet uninhabitable. Equally serious is the slower but nevertheless significant erosion of our natural environment caused by the pollution resulting from human activities. Combustion processes, such as the burning of petroleum or coal, threaten to alter the atmosphere. The discharge of wastes into water supplies has had a serious impact on water quality. Solid waste disposal poses serious problems for our environment, particularly when these wastes are toxic or radioactive.

An equally serious consequence of our growing numbers and activities involves the very real possibility that we shall exhaust many of this planet's natural resources in the near future. Meeting basic human needs for food and energy represent a major challenge in a world of diminishing resources.

For example, there is a serious imbalance between our ever-growing energy consumption and our capacity for supplying this energy. This imbalance is due to many factors, including the energy-intensive nature of our society and way of life, the depletion of existing energy sources (oil and natural gas), and the slow development of new energy sources (synthetic fuels, nuclear power, solar power). This imbalance poses a very serious threat to our society; the ''energy crisis'' is real. A new balance between energy use and supply must be achieved, and this can be accomplished only by simultaneously stressing energy conservation while developing new sources of energy. The engineer must play an essential role in both approaches by developing new energy supply technologies while improving the efficiency of energy utilization.

But a growing population causes still other serious problems. The concentration of large populations in urban areas has given rise to new challenges of both a

FIGURE 1.22.   The imbalance between our ever-growing energy consumption and our capacity for meeting this demand is an example of the challenge facing engineers. (Joe Munroe/Photo Researchers)

social and technological nature. Deteriorating urban environments coupled with inadequate public transportation trap large segments of our society in substandard living conditions. Social problems such as the increasing incidence of crime can be traced both to the deteriorating environment of our cities as well as to the dehumanizing manner in which we are forced to deal with growing populations.

Modern technology may be the answer to many of society's problems, but it can also generate new concerns. For example, the massive implementation of automatic data processing often tends to reduce individuals to numbers. It can threaten privacy if abused. The dramatic developments in modern biology raise the possibility of gene manipulation and genetic engineering, thereby threatening to intensify this dehumanizing trend.

These are but a few of the many challenges to be faced and surmounted by engineers. Engineering is a vital and exciting profession that has played and will continue to play an essential role in the development of civilization. The complexity of these problems will make strong demands on engineers. Evermore sophisticated technological tools require the continued development of engineers' skills. Engineers must also achieve the breadth to allow them to integrate these technical abilities with social, political, and economic factors. In the past engineers have always adapted their skills to meet the changing needs of society, and will continue to do so in the future. They will remain the pathfinders for our society, as they apply science and technology to lead the way for the progress of civilization.

FIGURE 1.23. Modern technology can stimulate new public concerns, a case in point being the reentry of SkyLab in 1979. (Courtesy National Aeronautics and Space Administration)

SUMMARY

Although scientific knowledge and technology continue to expand at a rapid rate, so too do the complexities of the problems faced by modern society. The most serious of our immediate dilemmas are due to our rapidly increasing population and the threatening nature of our activities. Of particular concern is the possibility that we shall exhaust many of this planet's natural resources. Meeting basic human needs for food and energy represent a major challenge in a world of diminishing resources. The concentration of large populations in urban areas has given rise to new challenges of both a social and technological nature. New technological developments such as automatic data processing and genetic engineering have raised new concerns for human diversity and dignity. Since these challenges involve social, political, and economic factors as well as technological concerns, engineers must be continually aware of various dimensions of the complex problems they address.

Exercises

Challenges of the Future

1. Identify several social and political issues associated with each of the following technological developments: (a) artificial satellites, (b) supersonic transports, (c) the artificial kidney, (d) chemical fertilizers.

2. Plot the population of the United States and the world from 1800 to the present. Then extrapolate these curves to project populations by 2000 and 2100.

3. Give several examples of human activities that have inadvertently caused great environmental damage.

4. One serious concern about the continued use of fossil fuels involves the buildup of carbon dioxide in the atmosphere leading to the so-called ''greenhouse effect.'' What is this effect and what might be its consequences?

5. Compare the relative hazards of disposing of waste from coal-fired and nuclear-powered electrical generating plants.

6. What method of solid waste disposal is currently being used in your community? How might this be accomplished with less environmental impact?

7. Give examples of dehumanizing aspects of the computer revolution.

8. What would you project to be the 10 most significant challenges to engineering during your professional lifetime? List them in order of importance.

REFERENCES

Introductory Textbooks on Engineering

1. George C. Beakley and H. W. Leach (with contributions by J. Karl Hedrick), *Engineering: An Introduction to a Creative Profession,* Third Edition (Macmillan, New York, 1977).

2. Arvid R. Eide, Roland D. Jenison, Lane H. Mashaw, and Larry L. Northrup, *Engineering Fundamentals and Problem Solving* (McGraw-Hill, New York, 1979).

3. Edward Krick, *An Introduction to Engineering: Concepts, Methods, and Issues* (Wiley, New York, 1976).

4. Leroy S. Fletcher and Terry E. Shoup, *Introduction to Engineering Including FORTRAN Programming* (Prentice-Hall, Englewood Cliffs, N.J., 1978).

5. R. M. Glorioso and F. S. Hill, *Introduction to Engineering* (Prentice-Hall, Englewood Cliffs, N.J., 1975).

6. D. L. Katz, R. O. Goetz, E. R. Lady, and D. C. Ray, *Engineering Concepts and Perspectives,* Second Edition (Wiley, New York, 1975).

7. L. H. Johnson, *Engineering: Principles and Problems* (McGraw-Hill, New York, 1960).

8. A. S. Weinstein and S. W. Angrist, *The Art of Engineering* (Allyn and Bacon, Boston, 1970).

Introduction to the Profession of Engineering

1. Annual Reports of the Accreditation Board for Engineering and Technology (formerly the Engineers' Council for Professional Development) (345 East 47th Street, New York, New York 10017).

2. *The Young Engineer: A Professional Guide*, prepared by the Development of Young Engineers Committee of the Engineers' Council for Professional Development, 1976, available from the Accreditation Board for Engineering and Technology.

3. William E. Wickenden, *A Professional Guide for Young Engineers* (Engineers' Council for Professional Development, New York, 1975).

4. W. J. King, *The Unwritten Laws of Engineering* (The American Society of Mechanical Engineers, 345 East 47th Street, New York, N.Y., 10017, 1955).

5. A. Pemberton Johnson, *Engineers, Scientists, Technologists, Craftsmen, and Technicians,* (Engineers' Council for Professional Development, New York, 1977).

6. C. C. Furans and J. McCarthy, *The Engineer,* LIFE Science Library (Time, New York, 1966).

Technology and Society

1. J. Bronowski, *The Ascent of Man* (Little, Brown and Company, Boston, 1973).

2. James Burke, *Connections* (Little, Brown and Company, Boston, 1978).

PART II
The Approach

The engineer is a problem solver. Indeed, we might even define engineering as that profession concerned with solving problems of a technological nature that confront our society. All of the major activities of the engineering profession, whether they involve research, development, design, or sales and management, are intimately concerned with solving problems.

By problem solving, we mean a far more general type of intellectual endeavor than simply substituting numbers into familiar formulas or proving abstract mathematical theorems. Engineering problems usually bear little resemblance to the tidy problems encountered in course homework assignments or examinations. In general they are characterized by a degree of complexity that makes stringent demands on imagination and creativity. Such problems will rarely be presented in a carefully stated manner. Rather, engineers will be confronted with situations in which they themselves must define the problem to be solved. The proper formulation of engineering problems is probably the most important—and yet also the most difficult—task faced by engineers. The problem solving role of engineers therefore has two dimensions. Engineers must first identify the problems and the needs of the often complex situation presented to them. Then they must apply science and technology to seek a suitable solution to meet these needs.

The solution of an engineering problem may take several forms. It may be a particular device or product, such as a jet engine, a high speed computer, or a bird feeder. The solution could be a process, such as a technique for producing instant photographs or refining crude oil. It might also be a plan or procedure, for example, to better train airline pilots or carry out maintenance operations on an assembly line.

Engineering problems will rarely possess a unique solution. There may be many possible answers to the problem. Moreover, if the engineer has been too demanding in the specification of the problem, there may be no solutions at all. It is the engineer's job to synthesize as many solutions as possible and then to determine which is most suitable for the particular situation of interest.

Engineers will rarely know in advance if the problem will yield to a particular line of attack. Hence the specific method most suitable for solving the problem must usually be determined by trial and error. Furthermore, because a great deal is at stake in most engineering problem solving, one can ill-afford careless errors. Therefore checking and verification of solutions play a much more important role in engineering than in many other areas of problem solving.

It should be apparent that engineering problem solving requires a careful and disciplined approach. This approach, which forms the basis for engineering practice, will be our concern in this part of the text.

The complexity of engineering problems requires discipline and imagination. (© 1976 by B. Kliban. Reprinted from *Never Eat Anything Bigger than Your Head* by permission of Workman Publishing Company, New York.)

CHAPTER 2

Problem Solving in Engineering

Engineering problems are characterized by a degree of complexity that makes stringent demands on skill, experience, knowledge, and creativity. To be successful in dealing with the array of complex problems that arise in practice, engineers must have a thorough understanding of the scientific principles basic to their field. In particular they must possess a solid grounding in mathematics, physics, and chemistry, along with engineering sciences such as solid and fluid mechanics, thermodynamics, and electrical science. Of course such an educational foundation is one of the primary objectives of a formal engineering education.

Knowledge alone will rarely suffice in engineering problem solving. An essential element of engineering analysis is the ability to abstract. Engineers must be able to identify and isolate the essential elements of a complex engineering problem. These elements can then be used to construct a simplified representation of the original problem, a "model," more suited to engineering analysis. For example, mechanical engineers might approach a complex problem in automobile shock absorber design by identifying and analyzing the mathematical equations characterizing a simple spring system. Or they might even build a simple physical model of the shock absorber using springs and weights. The ability to abstract, to develop simple models, is a critical aspect of engineering.

Success in engineering problem solving also requires a careful and disciplined approach to the problem. Although there is usually no assurance that a specific method of attack will prove successful, the following general procedure has proven useful in engineering problem solving:

1. The problem must be understood and defined, and the information necessary for its solution obtained. This step will generally involve abstraction and model development.

2. The various solutions to the model problem must then be determined or synthesized.

3. The validity of these solutions are examined and evaluated for their

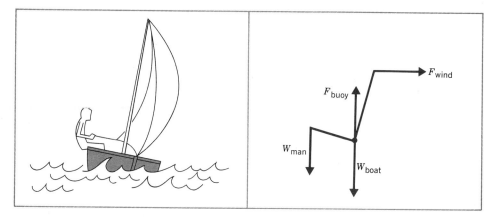

FIGURE 2.1.   An example of the role of abstraction in problem solving: the forces acting on a sailboat are represented by a free-body diagram.

suitability; from these solutions information relevant to the original problem statement is inferred.

4. The solution must be communicated to others in a form suitable for implementation.

Clearly the solution of engineering problems requires a combination of skill, knowledge, and experience. Engineers must extract the essential features of complex problems and apply the basic principles of mathematics and science to their analysis. The same qualities of imagination and creativity so essential to art, music, and literature are also required for synthesizing novel solutions to complex engineering problems. But the complexity of modern engineering problems requires a highly disciplined approach to problem solving that is generally not characteristic of the arts. It is on this methodology of engineering problem solving that we shall focus in this chapter.

SUMMARY

Engineering problems are characterized by a degree of complexity that makes stringent demands on the skill, experience, knowledge, and creativity of the engineer. Engineering problem solving is both a science and an art. It requires a careful and disciplined approach that can be summarized as: (1) understanding and defining the problem and obtaining the information necessary for its solution; (2) determining possible solutions to an abstraction or model of the original problem; (3) assessing the validity of these solutions, evaluating them for their suitability, and inferring information from them about the original problem; and (4) presenting the solution in a form suitable for implementation.

Example An engineer is presented with the problem of designing a high speed printer for a computer. The first task is to define the problem as fully as possible by finding answers to questions such as the following: What is the minimum acceptable speed of operation that will meet the needs of the application? What is the maximum cost that can be tolerated in a single unit? Is the unit to be mass produced or are only a limited number of units to be supplied? Are there size or weight restrictions that must be observed? Must the unit be compatible with other existing equipment? What are the requirements on reliability and in-field servicing?

In the second step, the engineer develops an abstraction or model of the problem that will incorporate the information obtained in the first step into a set of specifications and performance criteria. Possible solutions are then outlined that may consist of adaptations of existing units such as electric typewriters or line printers, or of totally new approaches such as electrostatically guided "ink jet" or laser printers.

Each of these possibilities is then closely examined for its suitability in the third step. The typewriter-based solutions are rejected because they cannot meet the speed requirements. The ink jet systems are very fast, but too expensive for the intended application. A design based on the development of the line printer principle is chosen as the best compromise among all the performance criteria.

As the final step, the engineer is likely to first document the elements of the suggested solution on paper. After review and refinement through consultation with others, a working model is assembled for more thorough evaluation, and detailed technical specifications are prepared.

Exercises

(Introduction to Problem Solving)

1. Identify each of the four problem solving steps involved in approaching the following problems:

    (a) Developing a human-powered aircraft to cross the English Channel.

      (b)   Improving air quality in the Los Angeles basin.

      (c)   Producing a 50 mpg automobile.

      (d)   Redesigning residential dwellings to reduce energy consumption.

2.   Identify the problem solving steps in the following nontechnical problems:

      (a)   Reducing inflation in the national economy while maintaining full employment.

      (b)   Securing the release of hostages in a bank robbery without endangering them or conceding to the demands of their captors.

      (c)   Reducing the cost of a college education while maintaining its quality.

      (d)   Reducing the cost of medical care without a significant decline in public health.

      (e)   Reducing the rate of population growth in underdeveloped countries.

3.   Each of your college courses can be interpreted as a solution to a particular problem that arises in a career in engineering. Analyze one of your courses from this perspective.

4.   Analyze a problem on a typical homework assignment in mathematics, physics, or chemistry according to the problem solving steps.

5.   Contrast the type of problems typically assigned for homework in your other courses with those listed in Exercises 1 and 2.

2.1.   THE METHODOLOGY OF ENGINEERING PROBLEM SOLVING

Perhaps it is appropriate to begin by asking just what engineers mean when they refer to a problem. In its most abstract form, a *problem* is merely an expression of a desire to achieve a transformation from one situation to another, to change "what is" into "what is desired." For example, we might wish to change the situation in which automobiles achieve a fuel economy of 25 mpg to one in which they achieve at least 50 mpg. The *solution* would then be the means for realizing this transformation, such as by the use of lightweight materials or new engine designs. Since there is generally more than one possible solution, we must also utilize various *criteria* that provide a basis for a preference among solutions. In redesigning automobiles to achieve higher fuel economy, one would be concerned with solution criteria such as fuel consumption, cost, and reliability. Problem solving will also be accompanied by various *constraints* or restrictions that represent absolute requirements on the solution. Because of such constraints certain solutions will be unacceptable. Typical constraints in the automobile fuel economy problem might include exhaust emission or safety standards.

FIGURE 2.2 A problem, possible solutions, constraints, and criteria.

Example An engineering student awakens after a particularly active night of partying to find himself marooned on a small tropical island. After some exploration, he finds that he is not alone on the island. Snoring peacefully away under a nearby palm tree is a large gorilla. The *problem* facing the engineer is to get off the island before the gorilla wakes up. That is, the engineer desires to change the situation in which he and the gorilla coinhabit the island to one in which he escapes back to the mainland (and his college studies). But there are *constraints*. The danger posed by the shark-infested waters surrounding the island represents a constraint that will rule out a *solution* such as swimming. The lack of suitable materials is also a constraint that eliminates the possibility of building a small helicopter to fly off the island. *Criteria* might be applied to evaluate various solutions, such as the length of time they require for implementation (hollowing out a log to make a dugout canoe will take a long time) or the risk they pose. In this particular case, the engineer arrives at a clever solution. He simply waits until low tide and sneaks across an exposed sandbar connecting the island with the mainland. Unfortunately, the engineer failed to evaluate this solution adequately. The gorilla wakes up, spots the engineer sneaking away, and chases him across the sandbar and back to his college. Ah, but not to worry. As it happens, the gorilla is an excellent student, enrolls in mechanical engineering, graduates with honors, and now is a professor (moonlighting on the side as a linebacker on the football team).

The identification and proper definition of engineering problems, the search for solutions subject to constraints, and the evaluation of these solutions according to precisely defined criteria necessitates a carefully planned approach to problem solving. Our approach is to group the various general activities involved in solving engineering problems into four steps, as shown in Figure 2.3. In the first step an effort is made to understand the problem, to determine what is perplexing in the situation, to determine what transformation in circumstances is desired. After a thorough understanding of the problem is acquired, an explicit statement of this problem is then formulated along with the goals to be pursued. As mentioned previously, the precise statement of the problem usually involves the process of abstraction in which the essential elements of the general problem are identified, and a simplified model of it proposed.

The second step of problem solving involves determining possible solutions to the problem. This stage is probably most familiar to the engineering student since it involves taking a well-defined, precise problem and determining its solution. This is just what the student is expected to do on homework assignments in most engineering courses. But in practical engineering problems, there are generally many possible solutions that satisfy the constraints of the problem. The engineer employs the tools of engineering analysis to determine these solutions.

The third step consists of establishing the validity of solutions. The engineer must check both the logic and the mathematical manipulations involved in obtain-

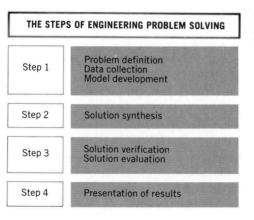

THE STEPS OF ENGINEERING PROBLEM SOLVING

Step 1 — Problem definition / Data collection / Model development

Step 2 — Solution synthesis

Step 3 — Solution verification / Solution evaluation

Step 4 — Presentation of results

FIGURE 2.3. The steps of engineering problem solving.

ing the solution. The solutions are evaluated and compared against various criteria to determine the solution that is most suitable to the original problem statement. The implications of this optimum solution must then be studied in detail.

The fourth and final step requires the presentation of the solution in a form that is both clear and convenient for implementation. Although the presentation of the results of engineering problem solving usually is in written form, it frequently also requires oral and graphic presentation.

Although these four steps of problem definition, solution synthesis, solution verification and evaluation, and presentation comprise the general procedure of engineering problem solving, we can identify additional activities. For example, the process of achieving the optimum solution may involve returning after the third step to revise the original problem statement (or model) formulated in step one so that new solutions may be obtained that more closely satisfy the original problem goals. Furthermore, the engineer must frequently augment the solution of the problem with plans for its implementation. These additional aspects of problem solving become more apparent as we proceed with a detailed discussion of each step.

Our present discussion of a procedure for problem solving is intended more to establish an attitude toward this activity than to provide the student with a precise recipe. Rarely can engineers approach problems in a step-by-step, mechanical fashion, by merely turning the crank. Instead they should acquire an attitude toward the solution of complex problems, a knowledge of the general structure of problem-solving methods, which can be adapted to the constraints of the situation at hand and allow for creativity and imagination.

SUMMARY

A **problem** is an expression of a desire to achieve a transformation from one situation to another.

A **solution** is the means for achieving this transformation.

Solution criteria are goals that provide a basis for a preference among solutions.

Solution constraints represent absolute requirements that the solution must satisfy to be acceptable.

The four-step approach to problem solving includes problem definition, solution determination, solution verification and evaluation, and presentation of results.

Example In the quest for perfect sound reproduction, a number of "solutions" have been introduced at various times over the past several decades, only to be replaced by better solutions made possible by the newer technology. The fidelity permitted by playback of the standard phonograph record was improved with the development of sophisticated magnetic tape units. Noise suppression electronics further improved playback quality.

The latest development in this field is digital recording, an excellent example of innovative engineering. The problem definition has remained the same for many years: reproduction of sound that resembles the original source as closely as possible. The breakthrough has taken place in the second step in problem solving—the definition of possible solutions. Whereas all previous methods of reproduction dealt with the audio signal in analog form (i.e., the degree of displacement of a phonograph stylus or the magnetization strength on tape), an entirely new solution is introduced by converting the signal to digital form. Here the signal is represented by a sequence of numbers, and many of the problems of distortion (e.g., noise, dynamic range) are completely avoided by processing the signal through circuitry that is entirely digital in nature.

Practical implementation of digital recording required the solution of several related problems. One of these was finding a way to store the digital signal in a form that would permit ready playback in the home. The solution in this case has been the digital disk that is played back by means of a laser beam. The digital record has no grooves. Rather it has a smooth plastic surface that protects a layer of microscopic pits that are used to code millions of binary numbers representing the sound. A beam of laser light scans the pits, converting them into signals that a tiny computer (microprocessor) can understand. Dirt, scratches, and fingerprints are read as out-of-focus information and rejected. Therefore not only is the sound reproduction perfect, but the disks are virtually indestructible. You can drop them, step on them, play frisbee with them, but when you put them in the player, they will sound as good as new. The digital record is an example of an imaginative solution that met the needs of high information density and reliable operation. The most difficult task in problem solving is maintaining a broad enough perspective from the start to include such innovative approaches that may turn out to be the best solution.

2.1.1.   Step One: Problem Definition

Engineering problems are usually characterized by a vagueness, a lack of definition, which demands very careful attention by the engineer. In many instances the engineer knows that something is wrong with a given situation but does not know just what the source of the difficulty is. A laser communication system may be subject to spontaneous failure; the efficiency of an automotive assembly line may suddenly deteriorate with the addition of new equipment; the price of a solar photocell may be prohibitively high. It is one thing to spot such difficulties, and quite another to identify their causes. Therefore the first job of the engineer is to determine just what the problem is—to define as precisely as possible the problem to be solved.

The most common mistake made by inexperienced engineers is to leap immediately into analysis before adequately defining the problem to be solved. In fact, many engineers avoid defining the problem altogether and begin instead by searching for solutions that others have provided to similar problems. These unfortunate tendencies are understandable since one of the most frustrating aspects of engineering is the vague nature of many problems. Indeed, if these problems were cut-and-dried, they probably would have already been solved, and the engineer would not be necessary in the first place.

Effective problem definition usually begins with a thorough study of the situation to determine its essential elements, that is, what is known and unknown and what is desired. For example, the communications engineer might examine the detailed performance record of a microwave transmission system to classify the precise nature of each failure. The experience of other systems would also be studied in an effort to identify possible defects of a more general nature. At this stage one should attempt to formulate each problem as generally as possible, at least to the extent that the constraints on the problem permit a broad formulation. This counteracts the natural tendency to become enmeshed in details, for once that happens, it is difficult to reestablish a broad perspective.

Novice engineers frequently choose initial problem definitions that are far too narrow. An overly restricted problem definition at an early stage will needlessly inhibit an imaginative approach to solving the problem. It may exclude whole realms of suitable possibilities. An effort to concentrate on improving the light collector efficiency to reduce the price of a photovoltaic solar power source might well miss the fact that the photocell materials themselves may be the overriding cost factor. Narrow problem definitions usually lead down the same old rutted paths that others have followed without success. Creativity demands freedom and, therefore, broad problem statements.

One can usually assemble a broad, detail-free problem formulation quite rapidly. This will serve to orient the engineer and will enlarge the perspective of the problem and its possible solutions. It will also suggest the general types of information required to attack the problem. It is only after this broad formulation that one turns to the more laborious task of translating the general problem statement into specific problem characteristics. This latter activity will usually involve considerable effort, as well as extensive consultation with reference materials and colleagues, fact gathering, observation, and careful thought.

SUMMARY

Problem definition begins with a thorough study of the situation to determine the essential elements of the problem, what is known and unknown, and what is desired. One should then attempt to formulate a broad, detail-free statement of the problem.

Example   An engineer is faced with the problem of predicting the turbulent air patterns that arise in the neighborhood of an airport. Her detailed measurements of the air currents indicate that the turbulent air appears in a very odd pattern of polygon-shaped cells. She generalizes her study to search for other examples of a cell structure in a turbulent fluid. This search quickly reveals many other similar situations, ranging from ocean current patterns to the atmospheric patterns photographed on Jupiter by the Voyager space probes. The general phenomenon is known as the Bénard fluid instability, and it arises when a column of fluid is heated from below. Returning to the airport problem, the engineer realizes that a neighboring city rejects heat on certain days in such a way as to establish the Bénard fluid instability pattern about the airport. Thus the engineer can simply monitor the heat rejection from the city to infer the atmospheric patterns.

What then are the essential elements of problem definition? First, it should be evident that we cannot pose problems until we have uncovered the difficulties that have led to them, and we cannot uncover these difficulties until we have clarified what our goals or criteria for solving the problem are in relationship to the situation of interest. These goals or *solution criteria* are used later to evaluate the

FIGURE 2.4. A photograph of turbulence in the Jovian atmosphere taken by the Voyager spacecraft. (Courtesy National Aeronautics and Space Administration)

situation of interest, for example, the cost and fuel consumption of new automobile engine designs. Personal judgments enter at this stage. What may be an acceptable goal or criterion for one individual may prove quite unacceptable to someone else.

Solution criteria are not sufficient in themselves to allow us to evaluate a solution because generally the *constraints* on the problem restrict our available options, that is, the set of possible solutions. For example, does a new automobile engine design meet federal exhaust emission standards? If it does not, it may have to be redesigned or even abandoned.

There exists important distinction between solution constraints and criteria. A constraint is an either/or proposition. That is, the solution either satisfies the constraint or it does not. An automobile engine design either meets federal emission standards or it does not. A criterion, on the other hand, is a performance index. It is used to determine which solutions are preferable. Both the cost and fuel efficiency of an automobile design may vary considerably. One would choose a design which optimizes these parameters.

Example   An engineer wishes to design a television broadcast antenna mast. He identifies the following solution constraints and criteria:

Constraints

(1) The mast must be able to qualify for a construction permit. (2) It must be tall enough to reach fringe broadcast areas. (3) It must exceed minimum strength requirements for safety. (4) The mast supports must be confined to the available land.

Criteria

(1) Construction cost. (2) Ease of maintenance. (3) Aesthetic appeal. (4) Minimum hazard to low-flying aircraft.

Example   One interesting example of the role of constraints in influencing the number of possible solutions is the siting of a large industrial facility. If there were no constraints on such facilities, then any site might be suitable—even in the center of a park or residential area. However there are usually many constraints on the site: availability of transportation, power, labor, and water; air and water quality standards; local zoning ordinances; land costs and taxes; and so on. As the number of constraints increase, the number of suitable sites (solutions) decrease, until eventually (and not infrequently), no sites are available that meet all of the constraints. The siting problem thus becomes overconstrained.

After a broad problem formulation and identification of relevant constraints and solution criteria, one now formulates a more detailed problem statement in which only the essential elements of the problem are included. This is the stage at which abstraction plays an essential role. That is, the engineer must sift out the essential features of the problem from all of the extraneous (and usually irrelevant) details that are present in the original problem statement.

An important tool in engineering problem solving is the use of a *model*. Models are simplified descriptions of reality in which only the essential features of the problem are represented. Models come in a variety of forms or representations. For example, the determination of the optimum timing of a sequence of traffic lights can be represented abstractly by a mathematical equation. The design of a jet aircraft may be represented by a simple wooden model suitable for study in a wind tunnel. In both cases, the models serve as an aid to abstraction that enables us to focus on only the essential features of a problem.

A carefully worded problem statement of only these essential elements is an important feature of problem definition. Such a statement usually suggests an appropriate model of the problem (e.g., a set of mathematical equations or a symbolic diagram of a physical model) to simplify further analysis. The engineer should determine just what *solution variables* are involved in the problem, that is, which characteristics of the problem can be altered in a search for acceptable solutions. After formulating a precise verbal statement of the problem, the solution criteria, constraints, and solution variables, the engineer then develops an appropriate model that can be used for further analysis in an attempt to arrive at a set of solutions in the next stage of the problem solving procedure.

In the process of developing the model for the problem, assumptions must be made and carefully considered for validity. It is also at this stage that the search for relevant data or information is begun. In particular, the engineer must determine just what data are necessary to facilitate the solution and then determine how to obtain these data. If sufficient data are not available, it may be necessary to fill in the gaps by making reasonable assumptions. Of course, there may also be situations in which there is too much data, some of which is of a conflicting nature. Poor or redundant data must be evaluated and weeded out to arrive at the most appropriate set of information.

Although problem definition is certainly the most important and complex aspect of engineering problem solving, it is rarely emphasized sufficiently in the formal education of the engineer. Furthermore it may represent only a small fraction of the effort expended in arriving at a solution. Nevertheless, the importance of adequate problem definition cannot be overemphasized, for without it all further analysis and evaluation of the problem will be futile. Problem formulation is the crucial stage of an engineering project. It is here that the engineer breaks down a complex problem into its essential elements and examines each of these elements to see what it entails. The approach taken in solving the problem is largely determined at this stage.

SUMMARY

The next facet of the problem definition step is to identify the relevant solution criteria and constraints that will be used to evaluate the solution. This step leads to a more precise statement of the problem. At this stage engineers usually develop an abstract model of the actual problem by identifying its essential features and introducing assumptions to simplify the model. They break down complex problems into their essential elements and examine each of these elements to see what they involve. They begin the search for relevant data or information to facilitate a solution of the problem.

Example An interesting example of the abstraction and modeling process is provided by the analysis of a railroad tankcar explosion that occurred in the Chicago freight yards in the mid-1970s. The explosion was apparently caused by the collision of another railcar with the tankcar containing flammable liquids. The motion of each railcar, including the vertical motion on the shock absorber springs of their wheel trucks, was modeled using well-known equations of mechanics. To more easily visualize the accident, the solution of these equations was represented using computer graphics to display the location and configuration of each car (Figure 2.5). This analysis revealed that the swaying motion in the moving car caused its coupler to ride up and penetrate into the tank car, causing the explosion.

2.1.2. Step Two: Determination of Solutions

Suppose we have formulated a specific problem statement and have simplified it with assumptions and approximations to arrive at an abstracted form (a model) suitable for analysis. We are now ready to begin our search for solutions that satisfy the constraints of the problem and yield acceptable goals. At this stage we draw on our arsenal of methods of engineering analysis. These include not only scientific concepts and mathematical techniques, but also methods of experimental investigation. We begin by choosing a preliminary method of attack to generate or synthesize these solutions. This initial approach is still tentative since one is rarely assured that it will yield acceptable solutions.

For the most part the solutions to engineering problems need not be entirely original. Indeed, most components of engineering solutions are already present in

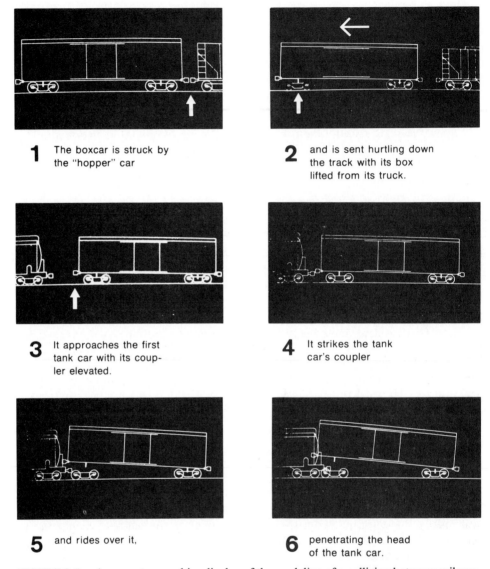

1 The boxcar is struck by the "hopper" car

2 and is sent hurtling down the track with its box lifted from its truck.

3 It approaches the first tank car with its coupler elevated.

4 It strikes the tank car's coupler

5 and rides over it,

6 penetrating the head of the tank car.

FIGURE 2.5.   A computer graphics display of the modeling of a collision between railcars. (Courtesy M. Chace, University of Michigan College of Engineering)

the engineer's knowledge and experience. Additional information can be found in technical libraries or data sources. The engineer will also benefit enormously from interactions with colleagues. It is rare that engineers are as effective working in isolation as when interacting with one another. Most modern engineering activities are performed as team projects to encourage this interaction.

The tools used for drawing this information together, for synthesizing it into possible solutions, form the discipline of engineering analysis that characterizes each engineering field. For example, the mechanical engineer relies on methods of thermodynamics, fluid dynamics, mechanics, and stress analysis. The electrical engineer employs methods of circuit analysis, electromagnetic theory, and control engineering. The civil engineer applies structural mechanics, materials science, and soil mechanics. In each instance, these subjects form the core of the engineer's formal education.

There are many useful methods available for searching for solutions. We shall return to consider this aspect of problem solving in more detail in the next section.

SUMMARY

The determination or synthesis of possible solutions to the problem of interest relies heavily on the engineer's knowledge and experience. Since the individual components of these solutions are rarely novel, the engineer makes extensive use of reference material or information supplied by colleagues. This information is then drawn together and applied using the tools of engineering analysis, mathematics, and science.

Example   A lecturer in physics for a class of freshmen engineering students had just completed a discussion of atmospheric pressure and decided to test the class's comprehension by including the following question on the next exam: "How could you determine the height of a tall building with the aid of a barometer?" One of the engineers, who later went on to great fame and riches, responded as follows: "Well, there are any number of ways I might go about accomplishing this task. Let me list just a few. For example, I could take the elevator to the top floor of the building, find an open window, hold the barometer a little away from the wall, and let it go. By timing the interval before the barometer crashes to the pavement below and applying well-known formulas for free-falling bodies (with small corrections for air resistance), the height of the building could easily be calculated. Or, I could tie a rope to the barometer, lower it carefully from the open window until it reached the ground, then mark the rope at the window sill and pull the barometer back up. The length of the rope between the barometer and the mark then gives the height of the building. Better still, I could save some effort by going directly to the superintendent's office in the building and telling him the following: 'Sir, I have here a fine new barometer. I will be happy to give it to you in exchange for a look at the building plans on which dimensions are recorded.' Of course, the building height could also be deduced from measurements of the barometric pressure at its base and top, but this method seems to be utterly lacking in imagination."

The moral of this story: Do not limit possible solutions to those indicated by prior conventional wisdom.

2.1.3.   Validation and Evaluation of Solutions

Once the engineer has obtained a number of solutions that appear suitable, that satisfy the required constraints and exhibit to some degree favorable solution characteristics, then these solutions must be carefully verified to determine their validity. It would be useless to continue with a detailed evaluation of these solutions without the assurance that they are the result of correct logic and analysis.

The verification or checking of solutions takes on a very important role in engineering problem solving. This is due in part to the massive scale of many engineering projects and their impact on society. This impact is frequently so profound that the slightest error in design can have the most disastrous consequences. We have only to think of the implications of serious failures in the engineering designs of aircraft or automobiles. The complexity of most engineering problems makes careful verification all the more important. Numerous assumptions and approximations are often required to simplify such problems so that they yield to the available tools of analysis, and the validity of these assumptions must be checked.

Careful thought should be given to developing systematic methods to verify solutions. The most desirable method of verification involves a comparison of the solution with experimental observations of the behavior of the real system. Sometimes it is impractical to run actual full-scale tests because of cost, inaccessibility, or other restrictions. An alternative is to apply the models used in the analysis to study operating data obtained from previous designs or to analyze the results of tests performed on individual system components. Both approaches are commonly employed in the design of very large and expensive systems for which detailed testing of the entire system is impractical. An excellent example is the verification of the design of a nuclear power plant. Since such a plant costs roughly a billion dollars to construct, the direct testing of a prototype is usually not feasible. Instead nuclear engineers rely on tests performed on subsystems, or components, or upon limited experiments performed on similar nuclear plants already in operation.

Various analytical methods can be used to check the mathematical analysis utilized in obtaining solutions. One should always check the equations comprising the mathematical model of the problem to see that the dimensions or units characterizing physical quantities are consistent. (Refer to Section 3.2 for a detailed discussion of dimensions and units.) For example, if the minimum time calculated for a spacecraft journey is expressed in kilometers (rather than years), we know a mistake has been made. The study of various limiting cases can also be used to verify that the mathematical solution yields the proper physical behavior. If the travel time is calculated based on the average speed of the spacecraft, we might examine whether this time approaches zero as the speed becomes very large. If not, then we might once again suspect that an error has occurred somewhere in the analysis.

Of most importance is a simple check to see if the answer makes good sense. Are the results what one might expect; are they in the right ballpark? If the answer does not seem reasonable, odds are that there has been an error somewhere along the way. If we find that the transit time for a spacecraft journey from Earth to Jupiter is several hours rather than several years, we should once again be skeptical of the solution.

Occasionally it may be possible to repeat the solution of the problem using an entirely different procedure. This may prove to be an excellent check of the method of analysis. We shall return to discuss these and other aspects of verifying mathematical calculations in the next chapter.

Although we have chosen to discuss the verification process in the third stage of the problem solving method, checking and verification are ongoing activities of the engineer that occur at all stages of the analysis. Even during the problem definition stage, the engineer should continually be questioning the validity of any assumptions and determining ways to verify any approximations introduced along the way.

It is usually much more difficult to check the original assumptions that are used in reducing the general problem statement to a workable form suitable for analysis. Since these assumptions usually involve factors that lie outside the

FIGURE 2.6. Engineers should always verify their solutions before implementing them. (© 1978 by B. Kliban. Reprinted from *Tiny Footprints* by permission of Workman Publishing Company, New York)

analysis used in obtaining the solutions, particular care should be taken in their evaluation.

The will to doubt, to question your efforts, is one of the most important aspects of successful problem solving. Verification and evaluation of solutions requires time and effort. You must avoid the temptation to seize upon a "quick and dirty" solution just to get the problem off your back, since this approach may later create more problems that it solves.

SUMMARY

In engineering practice all solutions must be carefully verified. Not only must the mathematical analysis and logic applied in generating solutions be thoroughly checked, but the assumptions and approximations used to simplify the original problem statement should also be evaluated. Although comparison with experimental data or alternative methods of solution usually provide the best verification, other methods of checking solutions such as dimensional consistency or ballpark estimates are useful.

Example   An engineer is asked to estimate the lifetime of domestic petroleum reserves, assuming that recoverable reserves in the United States are some 300 trillion barrels (bbl). The present consumption of domestic oil is determined to be roughly 15 million bbl/day. Hence the engineer infers how long our domestic reserves will last:

$$T = \text{time} = \frac{300,000,000,000 \text{ bbl}}{15,000,000 \text{ bbl/day} \cdot 365 \text{ day/year}} = 54.8 \text{ years}$$

Since this is an important result, the analysis is checked very carefully. Certainly the units of years are appropriate. Furthermore the magnitude of about 50 years seems reasonable. Therefore the engineer concludes that the calculation is correct, and that petroleum reserves will certainly last for the remainder of his lifetime (Figure 2.7).

But the engineer forgot to check the assumptions. By taking the oil consumption rate over the next 50 years as equal to the present rate of 15 million bbl/day, it has implicitly assumed that this consumption rate will not change. Yet past experience has indicated that this rate increases by as much as 3% each year. If we assumed that this 3% annual increase in the consumption rate were to continue (still another questionable assumption), we would calculate a new time for exhaustion of domestic petroleum reserves by solving:

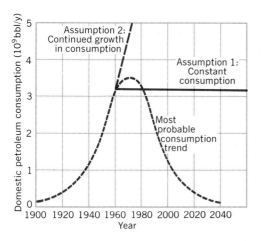

Domestic petroleum consumption (10⁹ bbl/y)

Assumption 2:
Continued growth
in consumption

Assumption 1:
Constant
consumption

Most
probable
consumption
trend

1900 1920 1940 1960 1980 2000 2020 2040

Year

FIGURE 2.7. Petroleum production in the United States as described by various models.

$$(15{,}000{,}000 \text{ bbl/day}) \ (365 \text{ day/year}) \ T(\text{years}) \ (1 + 0.03)^T = 300{,}000{,}000{,}000$$

If we solve this, we find $T = 18$ years. Actually, even this second assumption is unrealistic, since it fails to account for the falloff in demand as petroleum becomes more expensive due to its increasing scarcity. The actual demand curve should probably resemble the one shown in Figure 2.7.

After the verifying or checking step, the engineer next evaluates each of the solutions from the perspective of the problem constraints and solution criteria. Any that fail to satisfy the constraints are unsuitable. These remaining solutions that satisfy all the constraints of the problem are then evaluated to determine which yields the most favorable solution criteria. This optimum or most desirable solution is then analyzed in detail to determine its implications for the problem. We must remember that usually these solutions have been obtained for a simplified model. If we have chosen to describe the problem by a mathematical model, then our solution will be in the form of a mathematical expression (i.e., an equation) that must be analyzed and translated back into physical terms. It is a bit difficult to stare at a complicated algebraic or trigometric expression and know immediately whether the force loads on the beam or the trajectory of a space vehicle are acceptable.

The third stage of problem solving also includes related activities. One usually would like to modify the solution to achieve the most favorable result, to optimize the solution. Furthermore, if this solution is to be translated into an actual engineering design, then one must study the solution and its implications in detail, documenting it for further use. This stage of problem solving may be accompanied by iteration, in the sense that the engineer may return to modify the initial assumptions (or model) to allow for a more acceptable solution. Also, it may happen that the initial assumptions were unnecessarily restrictive and can be relaxed to yield a more general solution. In this sense, the problem definition, solution synthesis, and solution verification and evaluation stages are tightly coupled.

SUMMARY

The engineer must evaluate each of the solutions in light of the problem constraints and solution criteria to determine which solution most adequately achieves the goals of the problem statement. At this stage the solution may be adjusted to optimize its suitability. Finally one studies the solution and its implications in detail, documenting it for further use.

Example   Optimization plays an important role in the routing of highways. The highway engineer must work with a number of solution criteria such as the cost and time required for right-of-way acquisition, the length of the route, its proximity to major population centers, the transportation cost of construction materials to the route, land excavation and preparation costs, and so on. Since several of these criteria may be contradictory in nature (e.g., low right-of-way costs and proximity to population centers), it is usually necessary to adjust the route until an optimum mix of cost, convenience, and access is achieved. The final route chosen will rarely be the shortest distance between two points.

2.1.4.   Step Four: Presentation of Results

An important aspect of engineering problem solving involves the presentation of results. Even the most elegant and successful results will be for naught if they are not presented in an understandable fashion to those who will use them. A carefully organized presentation of both the analysis and the solution of the problem is essential, whether the content be homework assignments prepared for an instructor, a professional report prepared in industry, or research notes prepared for the engineer's own records. Careful organization will enhance the clarity of the presentation, which can also provide a logical structure to the solution of the problem itself by helping to focus the analytical skills of the engineer. A careful documentation of the solution of an engineering problem is also important in providing a record of the analysis (possible for later patent or legal use) or in aiding a detailed evaluation of the solution method. A premium is placed on a neat, careful presentation in which all relevant details of the specification and the solution of the problem are included.

The detailed style chosen for the presentation may vary. The choice of style reflects both the nature of the problem and the method of its solution. For example, the format chosen for presentation of a homework problem differs from that characterizing a research paper. So, too, will a report on an experimental study differ from a computer analysis. The style will also depend on the anticipated audience. A survey report intended for a lay audience will differ dramatically from that directed toward an experienced engineer.

Despite these differences, we can provide some guidelines on the format of engineer presentations. The organization should generally follow the pattern adopted in the solution of the problem. An outline for a technical presentation might include:

1. A concise statement (or summary) of the problem in its original form, complete with sketches if appropriate. This statement should include all given data that are relevant.

2. A brief statement of relevant engineering principles.

3. A list of the assumptions and approximations necessary to simplify the problem to a soluble form.

4. A precise statement of the simplified problem (or model) to be studied.

5. The detailed solution, including all relevant steps (with particular care to include the units of physical quantities where appropriate). Often, an estimate of the uncertainty associated with numerical results is also given.

6. A verification of the solution method (e.g., by a dimension check) and the assumptions (e.g., by comparing with experimental data).

7. A discussion of the implications of the results and possible conclusions.

Of course the technical details of problem presentation should be considered. For example, it might be convenient to present results on engineering computation paper that is specifically formatted to assist in technical sketches and calculations. There are many variations in the technical requirements of presentations. Many presentations require a particular type of lettering, and presentations at technical meetings are often organized around a sequence of slides or viewgraph visual aids. The rapid advances in computer-aided composition allow today's engineer to prepare presentations on the screen of a computer display. Such aspects of the preparation of written or graphical reports will be discussed in detail in Chapter 6.

Figures 2.8 to 2.10 are examples of different styles of presentation. Engineering students are advised to practice the presentation of homework problem solutions using these or similar formats as early as possible in their studies.

Example Solution format for short mathematical problems:

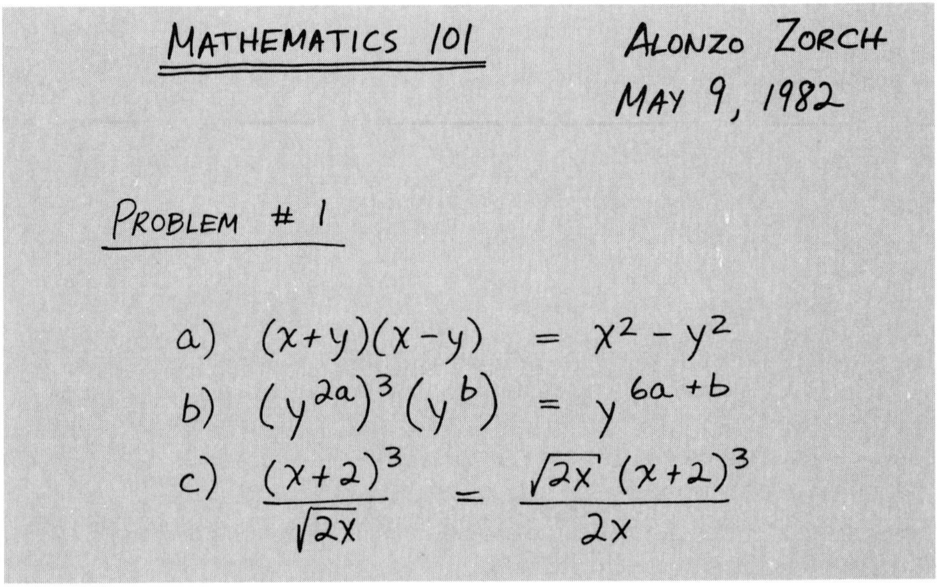

$$\underline{\text{MATHEMATICS } 101} \qquad \text{ALONZO ZORCH}$$
$$\text{MAY } 9, 1982$$

$$\underline{\text{PROBLEM \# 1}}$$

a) $(x+y)(x-y) = x^2 - y^2$

b) $(y^{2a})^3 (y^b) = y^{6a+b}$

c) $\dfrac{(x+2)^3}{\sqrt{2x}} = \dfrac{\sqrt{2x}\,(x+2)^3}{2x}$

FIGURE 2.8. A possible solution format for short mathematical homework problems.

Example Solution format for short problems:

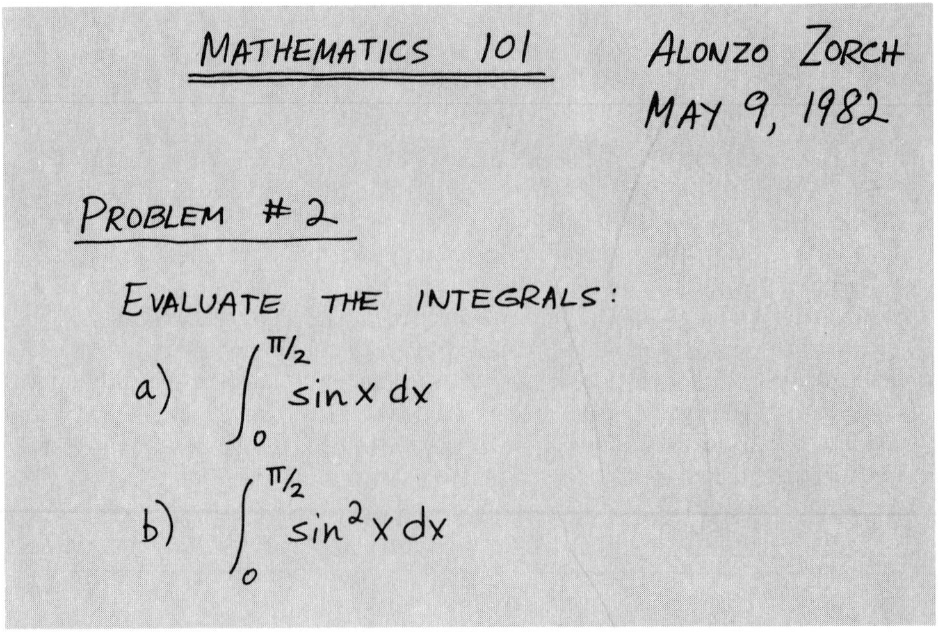

$$\underline{\text{MATHEMATICS } 101} \qquad \text{ALONZO ZORCH}$$
$$\text{MAY } 9, 1982$$

$$\underline{\text{PROBLEM \# 2}}$$

EVALUATE THE INTEGRALS:

a) $\displaystyle\int_0^{\pi/2} \sin x \, dx$

b) $\displaystyle\int_0^{\pi/2} \sin^2 x \, dx$

SOLUTION :

a) $\displaystyle\int_0^{\pi/2} \sin x \, dx \;=\; -\cos x \,\Big]_0^{\pi/2}$

$\qquad = -\cos \dfrac{\pi}{2} + \cos 0 \;=\; -0 + 1$

$\qquad = 1$

b) $\displaystyle\int_0^{\pi/2} \sin^2 x \, dx \;=\; \left[-\tfrac{1}{2}\cos x \sin x + \tfrac{1}{2} x \right]_0^{\pi/2}$

$\qquad = \left[\tfrac{1}{2} x - \tfrac{1}{4} \sin 2x \right]_0^{\pi/2}$

$\qquad = \dfrac{\pi}{4} - \tfrac{1}{4}(0) - \tfrac{1}{2}(0) + \tfrac{1}{4}(0) \;=\; \pi/4$

FIGURE 2.9. A possible solution format for short homework problems.

Example   Solution format for longer problems:

THERMODYNAMICS 235   A. ZORCH
                             MAY 9, 1982

PROBLEM: Gas enclosed in cylinder (dia = 20 mm)
   by piston. Pressure inside cylinder is
  .40 kPa above atm. Find mass of piston.

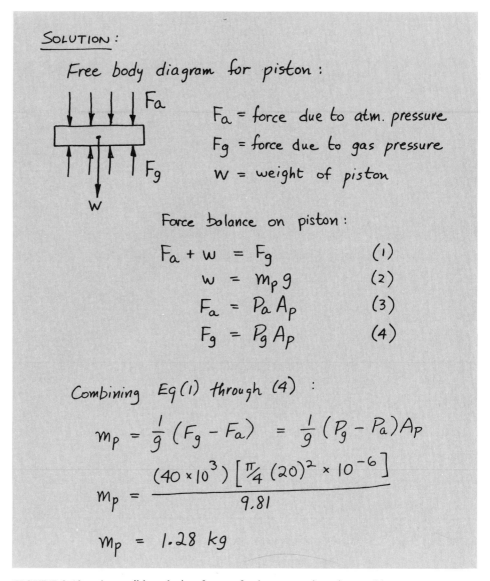

SOLUTION:

Free body diagram for piston:

F_a = force due to atm. pressure

F_g = force due to gas pressure

w = weight of piston

Force balance on piston:

$$F_a + w = F_g \qquad (1)$$
$$w = m_p g \qquad (2)$$
$$F_a = P_a A_p \qquad (3)$$
$$F_g = P_g A_p \qquad (4)$$

Combining Eq (1) through (4):

$$m_p = \frac{1}{g}(F_g - F_a) = \frac{1}{g}(P_g - P_a)A_p$$

$$m_p = \frac{(40 \times 10^3)\left[\frac{\pi}{4}(20)^2 \times 10^{-6}\right]}{9.81}$$

$$m_p = 1.28 \text{ kg}$$

FIGURE 2.10.   A possible solution format for longer engineering problems.

SUMMARY

The engineer should arrange the analysis and solution of the problem into a form suitable for communication of the essential results to others. A logical and carefully organized structure of problem presentation can help to focus the analytical skills of the engineer. It can also provide a record of the solution. The detailed format of the presentation will depend on the nature of the problem, the method used to solve it, and the intended audience. Generally the format will follow the pattern adopted in problem solving:

1. *A concise statement of the problem in its original form (complete with sketches and given data).*
2. *A brief statement of relevant engineering principles.*
3. *A list of simplifying assumptions and approximations.*
4. *A precise statement of the simplified problem (or model).*
5. *The detailed solution, including all relevant steps.*
6. *A verification of the solution method and assumptions.*
7. *A discussion of the implications of the results.*

Exercises

Engineering Problem Solving

1. Describe a problem that arises in each of the following areas by identifying ''what is'' and ''what is desired'' in each case:
   (a) Mass transportation
   (b) Air quality
   (c) Nonreturnable bottles
   (d) Urban renewal

2. Choose a personal problem that you have and analyze it in terms of our definition of a problem.

3. Choose a social problem of major concern and analyze it using our definition of a problem.

4. Fill in the blanks in the problem definition table:

	What is	*What is desired*	*Constraints*	*Criteria*
(a)	A Friday afternoon traffic jam in Manhattan	Free flowing traffic	—	—
(b)	Radioactive waste produced in nuclear power plants	—	Public safety	—

(c)   —   Adequate water   —   —
                  resources in the
                  desert southwest

(d)   Road salt   —   —   —
      corrosion of
      automobiles

Problem Definition

5.   List the *criteria* that should be considered in designing:

     A pocket computer   A vacuum cleaner
     A supersonic transport plane   A nuclear-powered rocket
     A tape dispenser   A solar power station

6.   List the *constraints* that should be considered in designing:

     A highway   A typewriter
     An artificial heart   A lawn sprinkler
     A bicycle lock   A disposable coffee cup

7.   Given a list of what is "known" and what is desired in each of the following problems; then give a broad definition of the problem:

     (a)   Transport coal as a liquid slurry through a pipeline
     (b)   Design an artificial kidney
     (c)   Eliminate a dangerous pedestrian crossing
     (d)   Develop a three-dimensional television

8.   Describe how you would model the following problems:

     (a)   The collision of two automobiles
     (b)   Pumping oil out of a well
     (c)   An explosion in a mine
     (d)   A coal burning furnace

Determination of Solutions

Suggest several possible solutions to each of the following problems:

9.   Measuring the distance between two points along a highway.

10.   Measuring the diameters of precisely manufactured shafts.

11.   Preventing the wearing down of the tips of ball point pens.

12.   Controlling the flame temperature of gas ovens.

13.   Eliminating the noise occurring during the playback of cassette tapes.

14.   Mixing in precise portions two streams of chemicals piped from large tanks.

Validation and Evaluation of Solutions

15. An advertisement claims that the new Muscle Car can accelerate from a standing start to cover 400 m (roughly a quarter mile) in 6 seconds. Is this claim reasonable?

16. An engineer determines that it will take approximately 3000 liters of heating oil to raise the temperature of her swimming pool from 10 to 25°C. Should she be suspicious of this calculation?

17. A strategic oil reserve of 1000 million barrels of oil has been proposed for the United States. It has been assumed that this amount of oil would ensure that there would be no shortage of oil for 12 months, even if all oil imports were suspended. Is this assumption reasonable?

18. It has been suggested that all the human inhabitants of the Earth would fit into a box 1.5 × 1.5 × 1.5 km in dimensions. Do you agree with this?

19. The star on a soccer team brags that he can kick the ball so hard that it travels at a speed of 3600 meters per minute. Could this be a valid claim?

Solution Presentation

Give a one page outline of how you would present the results or solutions to the following problems:

20. A set of simple algebra problems.

21. A proof of the Pythagorean theorem in geometry.

22. The design of a mechanism for a water flow meter.

23. The plan for packing and shipping of a delicate instrument.

24. A plan for collecting toll at a bridge entrance.

25. A plan for feeding a cat at predetermined times.

26. An alarm system for alerting residents of an apartment building in the event of fire.

27. The results of a computer analysis of the 1980 United States Census data to establish possible trends in the migration of urban populations to the suburbs.

General Problem Solving Methodology

Sketch the step-by-step approach you would use to solve the following problems. (Do not carry through any of the details of the solutions.)

28. Reduction of the particulate emission from diesel engine exhausts.

29. Increasing the size of television screens without increasing the weight of the unit.

30. Reduction of the tuition at your university without sacrificing academic quality.

31. Provision of a system to provide public warning in the event of tornadoes.

32. Designing a luggage handling system for an airport that expedites luggage delivery without damage or loss.

33. Reducing the interference between CB radio transmission and television reception.

2.2.   THE SEARCH FOR SOLUTIONS

The search for solutions to engineering problems calls most heavily on the creativity and imagination of the engineer. This phase of the problem-solving process closely approaches the creative arts in the degree to which it depends upon original thought. Yet, even though the synthesis of solutions to engineering problems must of necessity be a free-form, unrestricted endeavor, several guidelines nevertheless can prove of considerable value.

2.2.1.   The Search

When we are stumped with a problem, we usually begin to cast about for additional information. Yet the most common difficulty in problem solving is usually not our lack of information. Rather it is our failure to use the information that we already have at our disposal. There are emotional, cultural, and environmental factors that contribute to our "blindness," our inability to recognize possible solutions that are right before our noses. For example, you may be surrounded by distractions that hinder your concentration. The Nobel prize winning biologist, James Watson, noted in his book, *The Double Helix*, that the solution of difficult problems sometimes seems to require a certain level of boredom, an absence of distraction that keeps one returning time and time again to attempt yet another solution.

Frequently our problem definition is at fault. If this definition is too narrow, we may inadvertently eliminate attractive solutions. Or emotional or cultural factors may constrain us to look at the problem from only one restricted viewpoint. Sometimes our over-familiarity with similar problems preconditions us to keep trying the old solutions repeatedly, even though the new problem may be sufficiently different to prevent these solutions from working. This particular constraint is referred to by psychologists as "Einstellung" (the German word for "set"). A term of more recent vintage for this fixation on old solutions to new problems is "mind set."

Example   The term "mind set" was popularized in an investigation of the nuclear power plant accident that occurred at Three Mile Island in Pennsylvania in spring of 1979. The operators of this plant had been preconditioned, both by prior experience and operating instructions, to place a top priority on the avoidance of injecting too much cooling water into the nuclear reactor of the plant. When the reactor coolant system began to lose water through a safety valve that had stuck in an open position, the operators misread the situation, assumed that the coolant system was filling too rapidly with water, and shut off all water injection systems and cooling pumps (including emergency cooling systems), leading to reactor overheating and severe damage. This response was an excellent example of how Einstellung prevented the proper action (in this case, allowing the cooling pumps to continue to run to keep the reactor from becoming too hot).

Many abstract methods have been formulated to facilitate the search for solutions. These are discussed in many books concerned with problem solving. However we confine ourselves here to general recommendations of a more practical bent.

In approaching problems try to keep an open mind, that is, a general perspective coupled with a flexibility to examine different approaches. Be wary of becoming bogged down in details too early. In particular, in approaching a problem avoid a premature evaluation, which may make it difficult to continue the search for additional solutions. In addition, do not rely too heavily on your previous experience in trying to solve similar problems. Sometimes a little bit of knowledge is a dangerous thing. It can actually be an advantage to approach a problem with only limited experience since this may result in a fresh insight to a problem that has resisted earlier solution attempts. Frequently newcomers to a field make dramatic contributions merely because in their ignorance (or naiveté) they avoid the ruts that have trapped other engineers.

It usually helps to discuss the problem with others. They might be able to provide helpful suggestions or provide a new perspective. Moreover, attempting to explain the problem to others can force you to organize it better in your own mind. Of course, if you are going to ask others to listen to your problem, you owe it to them to listen to their suggestions and criticism with an open mind. Otherwise you might as well explain your problem to a blank wall (or perhaps a computer).

Here again an open-minded approach is of great importance. You must be prepared to doubt your own analysis, to be its harshest critic, to pry yourself loose from unproductive approaches and strike out along new paths.

A common difficulty in problem solving is the limited nature of human memory. Most of us are simply incapable of remembering all of the bits and pieces of information relevant to the problem at hand. We may forget key facts. If we look at the brain as a computer, then we would find that its information processing unit is capable of dealing with only a few items at a time (psychologists suggest about seven), although our longer term memory is capable of storing vast amounts of information that will be available for recall with proper stimulation (cues).

To cope with this human frailty, it is wise never to depend on your memory. Instead, always write down ideas (or facts or hunches). Make frequent use of diagrams or charts or mathematical symbols.

A second aid is subdivision or "chunking." The general idea is to break a problem into bite-sized (or brain-processor-sized) chunks or subproblems that you can deal with. In particular, try to choose these chunks so they have their own subgoals. That is, in order to get from A to Z, you might first solve the problem of going from A to B, then B to C, and so on. At any stage you can take a break from the problem without fear of forgetting just where you are (provided you have written down your analysis of the subproblems). Interestingly enough, this structured approach to problem solving has become the cornerstone of modern computer programming, and we will return to consider it later in Chapter 4.

One of the most important devices in problem solving involves the use of a model. Models can assume any of a number of forms or representations. They might be a diagram, a physical object, or perhaps a mathematical equation. When you get stuck on a problem, it frequently helps to change the form of the model you are using. For example, if you cannot solve the mathematical equations you are using to model the problem, perhaps a diagram would be more useful.

Related to the use of models in problem solving is the use of analogies. You may be able to identify similarities between the problem of interest and other problems you have already solved. Maybe you can recognize your problem to be merely a special case of a more general problem.

There are other useful tricks. Although one usually thinks of problem solving as proceeding from "what is" to "what is desired," it may sometimes be useful to reverse this order. We may know more about where we want to go, our goals, than where we are at present. Then working backwards from the goal can prove useful. Mathematicians have developed other useful approaches. One such approach is the method of induction in which one infers (induces) a general conclusion from a specific example. Another common technique is the method of contradiction in which you make an assumption and then try to disprove it. We leave it to your mathematics courses to develop these approaches in detail.

There will come a time, however, when you finally have exhausted all your ideas and you come up against a brick wall—you are stumped. At this point the best advice is simply to leave the problem for awhile. Do something else. Of course, as you push the problem into the farther reaches of your mind, it is still lurking there, fermenting away. This process of "incubation" may trigger a sudden breakthrough, a bolt of lightning out of the blue. It may also allow you to return to the problem with a fresh approach. There are sound psychological reasons why incubation can assist in problem solving, such as relieving fatigue or letting your subconscious mind work at the problem. But whatever the reason, this is sometimes the best approach to breaking the mental block.

SUMMARY

Difficulties in problem solving frequently arise from not using all the information at our disposal for a variety of reasons including distraction, narrow problem definition, and fixation on old solutions. Useful aids in the search for solutions include keeping an open mind, discussing the problem with others, compensating for the limited nature of human memory by writing down ideas, breaking the problem into subdivisions, and using models and analogies. When all else fails, sometimes the best tactic is to leave the problem alone for awhile and let it incubate in your subconscious mind.

2.2.2. The Search for Technical Information

An important aspect of problem solving is the gathering of information. While some information can usually be found in the engineer's personal collection of textbooks, journals, or technical manuals, at some point a visit to the technical library is usually necessary. A search of the technical literature—of books, journals, reports, and patents—can provide the background necessary for solving problems.

Unfortunately many engineering students as well as practicing engineers are unfamiliar with the resources of technical libraries. Engineering students should learn how to use the technical library and its various resources early in their education. Library research is an important tool in engineering analysis and design.

Since the volume of technical literature is very large, a systematic approach is desirable. It is common to begin a literature search using the card catalog (Figure 2.11). Books and conference proceedings are indexed by author and subject. For example, an engineer interested in energy conversion might search through the card catalog under the main headings of "power," "power resources," or "energy conservation." Titles to relevant books and conferences are listed under these headings.

The next source of information is scientific or technical journals. Most engineers subscribe to several technical journals in their particular field, and routinely browse through other relevant journals in the technical library as they are received. However, an organized approach is required to conduct a systematic search through the thousands of technical journals and conference proceedings concerned with engineering.

Of particular importance are indexes of titles and abstracts (short summaries) in science and engineering arranged by subject and author (Figure 2.12). For

CARD CATALOGUE

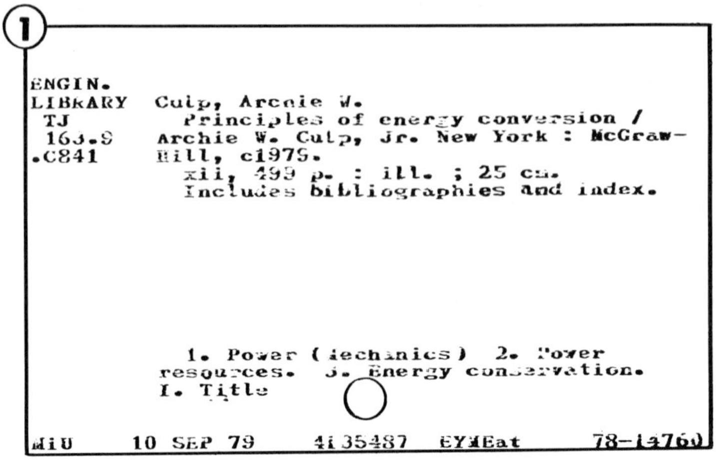

```
ENGIN.
LIBRARY    Culp, Archie W.
TJ            Principles of energy conversion /
163.9       Archie W. Culp, Jr. New York : McGraw-
.C841       Hill, c1979.
              xii, 499 p. : ill. ; 25 cm.
              Includes bibliographies and index.




              1. Power (Mechanics)  2. Power
            resources.  3. Energy conservation.
            I. Title

MiU       10 SEP 79      4135487    EYIEat      78-13760
```

```
Engin. Library
TH
1095
.W93      World Conference on Earthquake Engineering.
            Proceedings.  [1st]-      1956-

             v. in     illus., ports., diagrs.  27 cm.
                              For Complete Record See Shelf List




            1. Earthquakes and building—Congresses.

         TH1095.W6              693.852        57-27345 rev

         Library of Congress         [r61b2]       Engin. Library  MiU
```

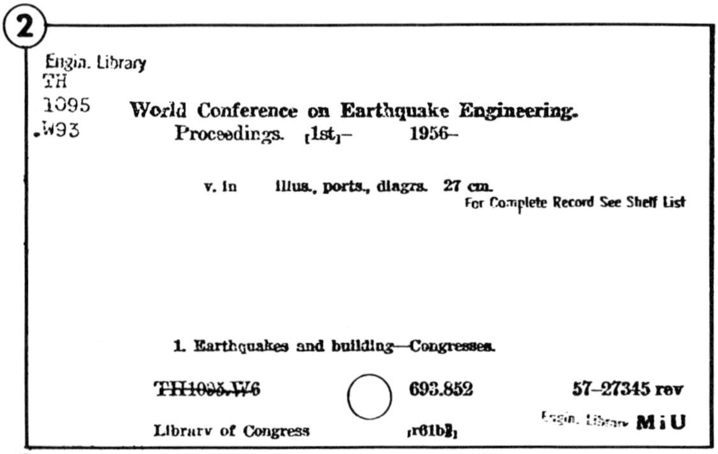

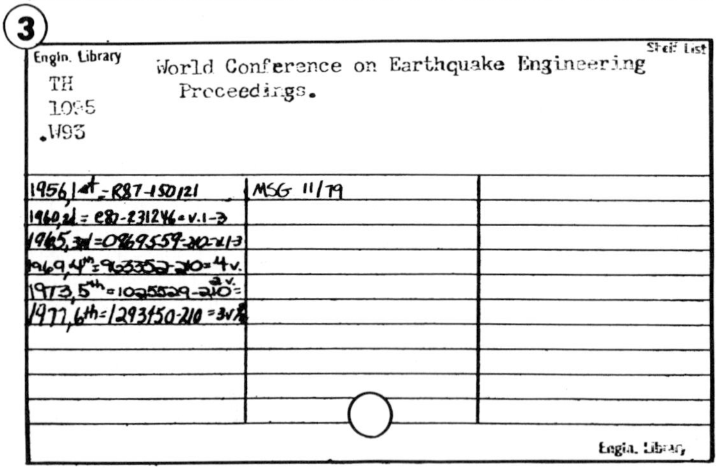

Conferences and symposia may be found by title, as shown to the left, or under the name of the sponsoring organization.

  e.g. Michigan. University. Engineering Summer Conferences.
  U.S. National Bureau of Standards. Symposium on...

A conference's proceedings, papers, or transactions may be found under the title of the conference (card 2), or under the name of the sponsoring organization.

  e.g. Association for Computing Machinery. Proceedings of ...

```
                                         Shelf List
Engin. Library
TH       World Conference on Earthquake Engineering
1095          Proceedings.
.W93


1956,1st = R87-150121   MSG 11/79
1960,2d = C82-231246 = v.1-3
1965,3d = 0869559-210-213
1969,4th = 963352-210 = 4v.
1973,5th = 1095529-210 =
1977,6th = 1293150-210 = 3v.




                                        Engin. Library
```

FIGURE 2.11. Typical references in a card catalog. (Courtesy University of Michigan Libraries)

94

INDUSTRIAL PLANTS—Location —
Contd.

established, and an appropriate generalization of the Weiszfeld iterative approach is given. A convergence proof is supplied for an ϵ-approximation to the original problem, under certain restrictions on p and K. 21 refs.

Morris, James G. (Univ of Wis, Madison). *Oper Res* v 29 n 1 Jan-Feb 1981 p 37-48.

Maintenance See Also HYDRAULIC FLUIDS—Analysis, LUBRICATORS

083155 MASONRY WALL REPAIR. Existing facilities are sometimes in such disrepair that severe curtailment of operations seems imminent. However, owners and operators of industrial buildings are discovering the economic advantages of repairing these buildings for an extended useful life. This article discusses repair and rehabilitation of exterior masonry walls. The key to understanding masonry wall behavior is in recognizing two fundamental principles: That a wall is as strong as its mortar brick or block, and that masonry walls have practically no strength in tension. The causes of masonry wall problems can generally be divided into two categories: man-made and environmental. The article covers mortar joint deterioration and spalling brick, cracking, dislodging bricks, and settlement problems.

Monhait, Michael M. (Schumacher & Svoboda Inc, Chicago, Ill). *Plant Eng (Barrington Ill)* v 35 n 11 May 28 1981 p 229-231

Mathematical Models

083156 MATHEMATICAL MODELING OF MANUFACTURING PROCESSES. Two major concerns in manufacturing processes addressed in this paper are heat flow and material flow. Finite element analyses of an injection-molded plastic part and of ceramic tube extrusion are two examples that illustrate how this technique can be used to improve both productivity and quality in industrial manufacturing processes. 6 refs.

Wang, H.P. (GE, Schenectady, NY); Klint, R.V. *Adv in Comput Technol. Presented at Int Comput Technol Conf, ASME Century 2 - Emerging Technol Conf, v 2, San Francisco, Calif, Aug 12-15 1980* Publ by ASME, New York, NY, 1980 p 329-333.

Water Cooling Systems See Also PUMPS—Control

INDUSTRIAL POISONS

083157 SOSTOYANIE PISHCHEVARITEL'NOI FUNKTSII TONKOI KISHKI U BOL'NYKH S KHRONICHESKOI INTOKSIKATSIEI NITRO-PROIZVODNYMI TOLUOLA. [State of the Digestive Function of the Small Intestine in Patients with Chronic Poisoning by Nitro Derivatives of Toluene]. Clinical and laboratory studies of the state of the intestine in 38 patients with chronic intoxication by toluene nitro derivatives revealed disturbances of a functional nature in most cases. Both cavitary (elevated enterokinase and alkaline phosphatase in duodenal juice and in feces) and parietal (a tendency of intestinal amylase activity to fall and of adsorbed amylase to rise) digestion was found to be disturbed, as was, though to a small degree, the absorption of carbohydrates (d-xylose) and fats. Long-term (up to 5 months) experiments on dogs with a fistula made in the small intestine by Thiry-Vella's method confirmed the etiologic role of trinitrotoluene in causing disturbances of intestinal function. It is concluded that the functional status of the intestine is a factor to be considered in conducting periodical medical checkups of workers handling toluene nitro derivatives and in treating patients with chronic intoxication with these substances. 4 refs. In Russian.

Kleiner, A.I. (Inst of Lab Hyg & Occup Dis, Khar'kov, Ukr SSR); Stovpivskaya, Yu.R. *Gig Tr Prof Zabol* n 2 Feb 1981 p 23-26.

Research

083158 O TAK NAZYVAEMOM SPETSIFICHESKOM I NESPETSIFICHESKOM DEISTVII PROMYSHLENNYKH YADOV. [On the So-called Specific and Nonspecific Actions of Industrial Poisons]. Exposure to poisons showing systemic toxicity results both in direct damage to the target organs (organ-specific effects) and in indirect, mediated damage to other organs (nonspecific effects). Early in the course of intoxication, the specific effects may not be reflected in the clinical picture or may appear as nonspecific. Many of their signs are similar to those of systemic disease. In such cases the etiologic role of the toxic factor in question in a given disorder can be confirmed or excluded only by applying what is called a syndrome-oriented principle of diagnosis. The essence of this principle is discussed in detail. 3 refs. In Russian.

Zislin, D.M. (Inst of Lab Hyg & Occup Dis, Sverdlovsk, USSR). *Gig Tr Prof Zabol* n 3 Mar 1981 p 30-33.

INDUSTRIAL ROBOTS See Also MATERIALS HANDLING—Manipulators; WELDING—Mechanization

083159 SYSTEM FOR THE DYNAMIC MODIFICATION OF A ROBOTIC PATH PROGRAM. Dynamic reprogramming extends the adaptation ability of a stand-alone industrial robot whose path program has been generated off-line. Adaptation information for reprogramming is obtained through pattern recognition of a video signal. Pattern recognition as a method of gaining adaptation information was chosen for the facility of collecting large amounts of environmental information in a generalized form. The dynamic reprogramming of an off-line program for an industrial robot within a computer controlled manufacturing system was shown to be a logical extension allowing robotics to be integrated into existing computer controlled manufacturing environments. 34 refs.

Washburn, Mark A. (DuPont, Wilmington, Del). *Proc Micro Delcon '81, Del Bay Comput Conf, Newark, Del, Mar 10 1981.* Publ by IEEE Comput Soc Press (Cat n 81CH1628-7), Los Alamitos, Calif p 63-70.

083160 GEBRUIK VAN SENSOREN BIJ INDUSTRIELE ROBOTS. [Use of Sensors in Industrial Robots]. The various types of sensors applicable for controlling the operations of industrial robots are reviewed, and the specific applicability fields characteristic of the different sensor types are outlined. The use of force feedback in automatic assemblies has been thoroughly explored and is discussed in detail. 15 refs. In Dutch.

Van Brussel, H. (Kathol Univ Louvain, Belg). *Rev M Mec* v 26 n 2 Jun 1980 p 111-120.

083161 METHOD FOR MODELING OF A ROBOT MOVING IN SPACE. The problems of an economic description of topology and of the organization of a quick access to the external environment model, as well as the computer methods for construction of images in which invisible portions of lines are eliminated, are discussed within the framework of mathematical modeling of a walking robot with the help of a display system. The three-level environment model constructed in this article has a sufficient amount of detail. Its cell structure allows a compact description in terms of the input language, and an effective method of visualization. 12 refs.

Okhotsimskiy, D.Ye.; Platonov, A.K.; Pryanichnikov, V.Ye. *Eng Cybern* v 18 n 1 Jan-Feb 1980 p 40-47.

083162 ADAPTIVE CONTROL ALGORITHM FOR A MANIPULATOR. An adaptive control algorithm is proposed for a manipulator when the weight and the moments of inertia of the object and the dynamic characteristics of the manipulator are all uncertain. A piecewise-constant control is synthesized, using information about the generalized coordinates of the manipulator and their first derivatives. The proposed control, after the completion of the adaptation stage, tracks the specified programmed motion with the required accuracy. 12 refs.

Gusev, S.V.; Yakubovich, V.A. *Autom Remote Control* v 41 n 9 pt 2 Sep 1980 p 1268-1277.

083163 OBJEKTERKENNUNG UND AUTOMATISCHES SPANNEN EMPFINDLICHER WERKSTUECKE. [Pattern Recognition and Automatic Clamping of Sensitive Workpieces]. The article reports on a possibility of automation of the process of clamping sensitive workpieces in which a comfortable pattern recognition system for identifying of three-dimensional objects is employed. 5 refs. In German.

Jung, A. *TZ Metallbearb* v 75 n 3 Mar 1981 p 25-26.

083164 ORIENTING ROBOT FOR FEEDING WORKPIECES STORED IN BINS. A robot system has been tested which can acquire a class of workpieces that are unoriented in bins and transport them, one at a time, to a destination with the proper orientation. This kind of robot has numerous applications in industries which manufacture discrete-part products in batches. A variety of approaches to the design of an orienting robot have been identified. 31 refs.

Birk, John R. (Univ of RI, Kingston); Kelley, Robert B.; Martins, Henrique A.S. *IEEE Trans Syst Man Cybern* v SMC-11 n 2 Feb 1981 p 151-160.

Computer Interfaces

083165 ASSEMBLY ROBOTS AND MACHINE VISION. PUMA, an acronym for Programmable Universal Machine for Assembly, has received widespread attention from the academic and industrial communities during the past two years. This paper describes the rationale of the development, the development of the robot associated with the system, the specifications for the system, as well as some thoughts on future needs and direction. This paper also explains what a machine vision system consists of and describes other developments preceding this system.

Cwycyshyn, W. (GM, Dearborn, Mich); Larson, D.F. *Adv in Comput Technol. Presented at Int Comput Technol Conf, ASME Century 2 - Emerging Technol Conf, v 2, San Francisco, Calif, Aug 12-15 1980* Publ by ASME, New York, NY, 1980 p 94-97.

Control

083166 DIRECT DESIGN METHOD FOR THE CONTROL OF INDUSTRIAL ROBOTS. The models of industrial robots are characterized by highly nonlinear equations with nonlinear couplings between the variables of motion. In the paper three nonlinear methods are presented where two of the approaches are direct design procedures for industrial robots. The design procedures presented simplify the derivation of the algorithm for computer-controlled industrial robots extremely. The methods are applied to two different types of industrial robots. 8 refs.

Freund, E. (Fernuniv, Hagen, Ger). *Adv in Comput Technol. Presented at Int Comput Technol Conf, ASME Century 2 - Emerging Technol Conf, v 1, San Francisco, Calif, Aug 12-15 1980* Publ by ASME, New York, NY, 1980 p 175-184.

083167 KINESTHETIC COUPLING BETWEEN OPERATOR AND REMOTE MANIPULATOR. The basic mechanism and an application of a universal force-reflecting hand controller are described. The hand controller measures the three position (x, y, z) and three orientation (pitch, yaw, roll) coordinates of the hand grip as the operator moves it around. Forces and torques can be generated on the three positional and three rotational axes of the hand controller permitting the operator to "feel" the task he is controlling. 12 refs.

Bejczy, A.K. (JPL, Pasadena, Calif); Salisbury, J.K. Jr. *Adv in Comput Technol. Presented at Int Comput Technol Conf, ASME Century 2 - Emerging Technol Conf, v 1, San Francisco, Calif, Aug 12-15 1980* Publ by ASME, New York, NY, 1980 p 197-211.

FIGURE 2.12. A sample page from the Engineering Index. (Courtesy University of Michigan Libraries)

example, the *Engineering Index* contains the abstracts from 300 scientific and technical journals and selected conferences and meetings; *Chemical Abstracts* covers 13,000 chemical related journals, series, books, and patents, (Figure 2.13); and the *Government Reports Announcements and Index*, published biweekly by the National Technical Information Service, indexes and abstracts United States government sponsored research (Figure 2.14). Most libraries also maintain an up-to-date list of all journals and serials that are received on a regular basis.

Other sources of useful information include standards and patents (Figure 2.15). Most technical libraries will subscribe to an index of the major sets of standards such as those of the American National Standards Institute (ANSI), the Occupational Safety and Health Administration (OSHA), and the Society of Automotive Engineers (SAE). Many libraries will also contain materials related to patents, including the *Index of Classification, Manual of Classification, List of Patents by Class and Subclass*, and the *Official Gazette*. Patent searches can become quite complex, so that library assistance is usually desirable (Figure 2.16).

CHEMICAL ABSTRACTS

Abstracts approx. 13,000 chemical related journals, series, books and patents. Broad in scope.
Indexes = Author, subject, patent number, formula and ring system, chemical substance, and registry number.

Available online 1970–

VOL. 89, 1978 - GENERAL SUBJECT INDEX

Tires

abrasion resistance of, R 216545s
adhesion in recapped, peptizer-contg. adhesive
   sheets for, P 25838e
adhesion of brass-coated steel cords in, to rubber,
   181057e
adhesion of polyamide fibers and rubber, adhesive
   systems for improvement of, 25759e
adhesion of rubber, to brass-plated steel cords,
   optimization of, 181058f
adhesion of steel cords of, to rubber, R 164695u

air pollution by wear of 64248g

ASTM stds. for, B 147928e
bias–belted
   crit. rolling velocity of, aspect ratio effect on,
      25791j
   for trucks, quality index of, calcn. of, equation
      for, 147920w
bicycle, unpressurized spare, from skinned
   polyurethane foam, P 130829a

89: 64248g Gas and particle emissions from automobile tires in laboratory and field studies. Cadle, S. H.; Williams, R L. (Environ. Sci. Dep., Gen. Motors Res. Lab., Warren, Mich.). *J. Air Pollut. Control Assoc.* 1978, 28(5), 502 7 (Eng). The gaseous hydrocarbon and S compds. emitted in lab. tests were identified. Although these hydrocarbons can participate in smog reactions, their mass emission rate is <0.1% current exhaust hydrocarbon emission rate. Hydrocarbons from tires are not measurable near a freeway. The particulate emitted from tires have sizes $0.01 \rightarrow 30\mu$, with the larger particles dominating the total mass. Measurements along a California freeway showed that most of the tire debris had settled within 5 m of the pavement edge. Airborne rubber concns. were <0.5 $\mu g/m^3$, or <5% total tire wear. These field measurements confirm the indoor emission pattern and verify that tire wear products are not a significant air pollution problem.

89: 64249h Chemical-kinetic criteria of the effect of the substances of natural and anthropogenic origin on the ozonosphere. Tal'roze, V. L.; Poroikova, A. I.; Larin, I. K.; Vinogradov, P. S.; Kasimovskaya, E. E. (Inst. Khim. Fiz., Moscow, USSR). *Izv. Akad. Nauk SSSR, Fiz. Atmos. Okeana* 1978, 14(4), 355–65 (Russ). The basic cycles of O_3 destruction are analyzed including N, H, halogen, and O. The role of each of these cycles was detd. by 1 or 2 main processes with participation of O atoms or O_3 mols. Expressions are obtained to est. contribution of each cycle to the total rate of O_3 destruction.

89: 64250b Control of ammonium nitrate prill tower emissions. Stover, John C. (Coop. Farm Chem. Assoc., Lawrence, Kans.). *Environ. Symp., [Proc.]* 1976, 251–85 (Eng). Fert. Inst.: Washington, D. C. With a collection duct and a glass-fiber filter, particle and small prill emissions are reduced to <1 lb/ton NH_4NO_3. This meets government regulations for opacity of the emission. The particulate loading of the tower effluent is ~0.006 grains/ft³ which is ~10% opacity. A model was developed to calc. the particulate NH_4NO_3 formed by dissocn. and the heat loss of prills as a function of distance.

FIGURE 2.13.   Examples of Chemical Abstracts and the Union List of Serials. (Courtesy University of Michigan Libraries)

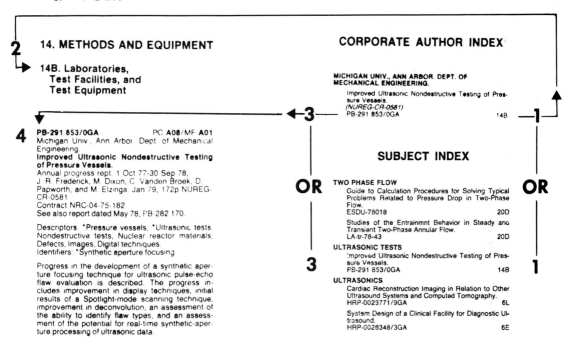

2

14. METHODS AND EQUIPMENT

**14B. Laboratories,
Test Facilities, and
Test Equipment**

4 PB-291 853/0GA   PC A08/MF A01
Michigan Univ., Ann Arbor. Dept. of Mechanical
Engineering.
**Improved Ultrasonic Nondestructive Testing
of Pressure Vessels.**
Annual progress rept. 1 Oct 77-30 Sep 78,
J. R. Frederick, M. Dixon, C. Vanden Broek, D.
Papworth, and M. Elzinga. Jan 79, 172p NUREG-
CR-0581
Contract NRC-04-75-182
See also report dated May 78, PB-282 170.

Descriptors. *Pressure vessels, *Ultrasonic tests,
Nondestructive tests, Nuclear reactor materials,
Defects, Images, Digital techniques.
Identifiers: *Synthetic aperture focusing

Progress in the development of a synthetic aperture focusing technique for ultrasonic pulse-echo
flaw evaluation is described. The progress includes improvement in display techniques, initial
results of a Spotlight-mode scanning technique,
improvement in deconvolution, an assessment of
the ability to identify flaw types, and an assessment of the potential for real-time synthetic-aper-
ture processing of ultrasonic data.

CORPORATE AUTHOR INDEX

MICHIGAN UNIV., ANN ARBOR. DEPT. OF
MECHANICAL ENGINEERING.
Improved Ultrasonic Nondestructive Testing of Pres-
sure Vessels.
(NUREG-CR-0581)
PB-291 853/0GA   14B

3

SUBJECT INDEX

OR

TWO PHASE FLOW
Guide to Calculation Procedures for Solving Typical
Problems Related to Pressure Drop in Two-Phase
Flow.
ESDU-78018   20D
Studies of the Entrainmnt Behavior in Steady and
Transient Two-Phase Annular Flow.
LA-tr-78-43   20D
ULTRASONIC TESTS
Improved Ultrasonic Nondestructive Testing of Pres-
sure Vessels.
PB-291 853/0GA   14B
ULTRASONICS
Cardiac Reconstruction Imaging in Relation to Other
Ultrasound Systems and Computed Tomography.
HRP-0023771/9GA   6L
System Design of a Clinical Facility for Diagnostic Ul-
trasound.
HRP-0028348/3GA   6E

OR

1

Government Reports, Announcements, & Index, published by NTIS
(National Technical Information Service) indexes and
abstracts U.S. government sponsored research. Information
may be looked up by subject, personal author, corporate
author, contract grant number, or accession/report number.

There are numerous other publications which index government funded
research in specialized areas. These include:

(DOE)   Energy Research Abstracts
          Energy Abstracts for Policy Analysis
          Geothermal Energy Update
          Solar Energy Update

(EPA)   EPA Cumulative Bibliography
          EPA Publications Bibliography

(NASA)   Scientific and Technical Aerospace
          Energy: A Continuing Bibliography
          International Aerospace Abstracts

(RAND)   Selected RAND Abstracts

FIGURE 2.14. Examples of government reports and announcements. (Courtesy University of Michigan Libraries)

STANDARDS

Major standards sets include:
American National Standards Institute (ANSI)
ASHRAE
ASTM
IEEE
NEMA
OSHA
SAE
and many others

The primary index is from Information Handling
Service
1. <u>Product/Subject Master Index</u>
2. <u>VSMF Standards Locator</u> (on microfilm)

(Also available online as Technet)

① NOSE TREATMENT UNIT C-90-51

NOSEBLEED BALLOON P-51-07

NOSEPIECE,
Microscope W-17-35
Riveting Gun P-11-10

NOSING,
Counter C-55-21
Stair,
(Except Molding Type)--See
Specific Type TREAD
Molding Type C-55-21

NOTCHER,
Address Plate S-07-03
Fiber Tube/Core M-16-15
Test Specimen F-73-23

NOTCHING (adj.)
Press--See Specific Type PRESS,
Industrial, Punch
Tool, Hand Held Cutting Plier
Type, Nonpowered P-09-15

NOVAL TUBE SOCKET A-23-13

NOZZLE--See Specific Type
Equipment Except As Listed Below:

NOZZLE,
Abrasive Blast Cleaning N-14-05
Atomizing (Except As Otherwise
Listed) E-21-05
Blow Gun, Air (See Also
BLOWGUN, Air) E-21-05
Douche C-85-55
Electric Welding Torch (See Also
Specific Type WELDING
EQUIPMENT) N-03-27
Fire Fighting Equipment S-13-53
Flow Metering (Primary Sensing
Element) F-06-05
Fog,
Fire Fighting S-13-53

```
1    3333333
1   3       33
1           33
1        33333
1        33333
1           33
1111 3      33
1111    3333333
```

② Section S-13, part 53

FIRE, FIRE PREVENTION
& FIRE FIGHTING

SOCIETY	DOC N°	DOCUMENT DESCRIPTION
**** 50	FIRE FIGHTING EQUIPMENT IN GENERAL (CONT)	
U.L.	U 405-75	FIRE DEPARTMENT CONNECTIONS FEBRUARY 12, 1979
**** 53	NOZZLES & PLAY PIPES	
ANSI	B111.1-75	PLAY PIPES FOR WATER SUPPLY TESTING IN FIRE-PROTECTION SERVICE (UL 385-75) MARCH 28, 1975
NFPA	GOFPRA198	CARE OF FIRE HOSE (COUPLING AND NOZZLES) 1969 EDITION OF NO. 198
U.L.	ULA385-75	PLAY PIPES FOR WATER SUPPLY TESTING IN FIRE-PROTECTION SERVICE (ANSI B111.1-75) MARCH 28, 1975
U.L.	ULA401-78	PORTABLE SPRAY HOSE NOZZLES FOR FIRE PROTECTION SERVICE MAY 25, 1978

FIGURE 2.15. Examples of standards. (Courtesy University of Michigan Libraries)

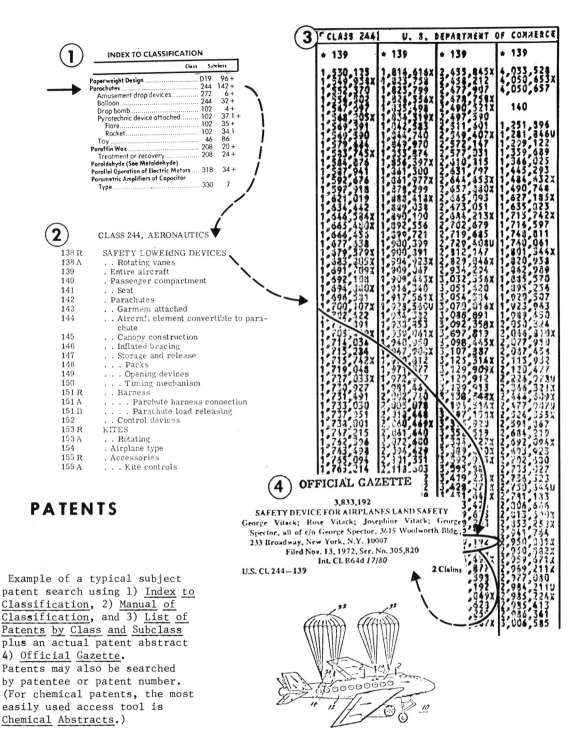

PATENTS

Example of a typical subject patent search using 1) <u>Index to Classification</u>, 2) <u>Manual of Classification</u>, and 3) <u>List of Patents by Class and Subclass</u> plus an actual patent abstract 4) <u>Official Gazette</u>. Patents may also be searched by patentee or patent number. (For chemical patents, the most easily used access tool is <u>Chemical Abstracts</u>.)

An airplane is provided with folded parachutes and means when actuated for causing the parachutes to be opened to lower the airplane safely to the ground in the event of engine failure or the like.

FIGURE 2.16. An example of the complexity of a patent search. (Courtesy University of Michigan Libraries)

COMPUTER SEARCHING

Using Boolean operators, many different bases can be searched by
subject, author, organizational affiliation, report number, journal
title or patent number, date, language, etc. Most bases are created
from printed sources and cover the period 1970 to date.

In engineering some examples are Metadex (<u>Metals Abstracts</u>), NTIS
(<u>Government Reports Announcements and Index</u>), INSPEC (<u>Electrical &
Electronics Abstracts</u> and <u>Computer and Control Abstracts</u>), and CAB
(<u>Comprehensive Dissertation Abstracts</u>).

```
        FILE8:COMPENDEX 70-79/SEP
        (COPR. ENGINEERING INDEX INC.)
            SET ITEMS DESCRIPTION (+=OR;*=AND;-=NOT)
            --- ----- ------------------------------

? S GAS?HOL

        1    3 GAS?HOL
? T 1/2/1

1/2/1
ID NO.- EI790753372    953372
  BRAZIL'S GASOHOL PROGRAM.
  YANO, V.
  CENT DE TECHNOL PROMON, RIO DE JANEIRO, BRAZ
  CHEM ENG PROG  V 75 N 4 APR 1979 P 11-19   CODEN: CEPRA8
  ISSN 0009-2495
  DESCRIPTORS: (*LIQUID FUELS, *BRAZIL), (ALCOHOLS, MANUFACTURE),
GASOLINE, MIXING), (SUGAR CANE, FERMENTATION), AGRICULTURAL
ENGINEERING, ECONOMICS,
  IDENTIFIERS: ETHANOL, GASOHOL
  CARD ALERT: 523, 802, 804, 821, 901, 911

? S ETHANOL AND (FUEL OR AUTOMOBILE?)

        631 ETHANOL
      19371 FUEL
       9630 AUTOMOBILE?
    2    52 ETHANOL AND (FUEL OR AUTOMOBILE?)
? T 2/5/1

2/5/1
ID NO.- EI790969407    969407
  CRITICAL ANALYSIS OF THE TECHNOLOGY AND ECONOMICS FOR THE PRODUCTION
OF LIQUID AND GASEOUS FUELS FROM WASTE.
  CHIANG, S. H.; COBB, J. T.; KLINZING, G. E.
  UNIV OF PITTSBURGH, PA
  ENERGY COMMUN  V 5 N 1 1979 P 31-73   CODEN: ENCODM
  ISSN 0097-8159
  IN ORDER TO MEET THE GROWING DEMAND FOR GASEOUS AND LIQUID FUELS
WASTE SHOWS A SIZABLE POTENTIAL FOR SERVING AS A SOURCE.  THE CURRENT
AND PROJECTED TECHNOLOGIES HAVE BEEN CONSIDERED AND THEIR ECONOMIC
ASPECTS TREATED EXTENSIVELY.  ECONOMICS IN THE LITERATURE HAS BEEN
UPDATED TO 1977 VALUES FOR A UNIFORM COMPARISON OF THE VARIOUS
PROCESSES.  BOTH UTILITY AND PRIVATE FINANCING COSTS HAVE BEEN
CALCULATED SHOWING FUEL GAS TO RANGE IN COST FROM $3/10**6 BTUS TO
$5/10**6 BTUS.  THESE COSTS OF GAS FROM WASTE ARE COMPARABLE TO THE.....
```
sample of Lockheed DIALOG search: COMPENDEX

FIGURE 2.17. An example of a search on the word "gasohol" on a computer-based
reference system. (Courtesy University of Michigan Libraries)

Many libraries also maintain or subscribe to a computer-based bibliographic search service (Figure 2.17). The titles and abstracts of most technical papers and reports published during the last two decades are stored in several computer data bases throughout the country. For a small fee, libraries can remotely access these computers by using their own computer terminals. The searcher must identify the key words relating to the subject of interest. The computer then provides a listing of all the references whose title or abstract contain these key words. Computer-based bibliographic indexes of interest in various fields are listed in Figure 2.18.

Sciences

AGRICOLA—agriculture

ASFA (Aquatic Sciences and Fisheries Abstracts)—aquatic and marine biology

AVLINE—audiovisuals in health science

BIOSIS—biological sciences

BIOETHICS—ethical issues

CAB—crop and animal sciences

CANCERLIT—cancer medicine

CONFERENCE PAPERS—scientific and technical meetings

CA SEARCH (Chemical Abstracts)
   —CA PATENT CONCORDANCE—correlates patents issued by different countries
   —CHEMDEX—CA registry nomenclature
   —CHEMNAME—CA *Chem*ical *Name* Dictionary

CRIS—USDA sponsored research

EXCERPTA MEDICA—health sciences

GEOARCHIVE—geosciences

GEO—REF—geosciences

HEALTH PLANNING—health sciences

INSPEC-PHYSICS—physics

IPA—pharmacy

MEDLINE—health sciences

SCISEARCH—biological and applied sciences

SPIN—physics

SSIE—research in progress

TOXLINE—toxicology

Technology/Engineering

APTIC (Air Pollution Abstracts)—air pollution

CIN—chemical industry production and sales

COLD REGIONS—freezing temperature areas

COMPENDEX.—engineering

DMMS—U.S. Defense Department contracts

ENERGYLINE—energy and environment

ENVIROLINE—environment

FSTA—food science and technology

INSPEC-ELEC/COMP—electrical engineering, computer science and control engineering

ISMEC—mechanical engineering and engineering management

METADEX—metallurgical sciences

MGA (Meteorological Abstracts)—meteorology and geoastrophysics

MRIS—naval architecture, maritime research

MTIS—government research

OA (Oceanic Abstracts)—oceanography and marine-related literature

PIRA—paper: packaging, printing, management

POLLUTION (Abstracts)—pollution and the environment

SAE ABSTRACTS—automotive engineering

WAA (World Aluminum Abstracts)—aluminum sciences

Social Sciences

ASI—U.S. government statistical publications

CHILD ABUSE & NEGLECT—research projects, service programs

CIS—publications of the U.S. Congress

ERIC—education

CEC (Exceptional Child Education Abstracts)—education of handicapped and gifted children

GPO MONTHLY CATALOG—U.S. government documents

LISA—library and information science

LLBA (Language and Language Behavior Abstracts)—speech and language pathology.

NICEM—nonprint educational material

PA (Psychological Abstracts)—psychology and other behavioral sciences

PAIS—social sciences

SOCIAL SCISEARCH—social sciences

SOCABS (Sociological Abstracts)—sociology

USPSD—political science

Business/Economics

ABI/INFORM—business, finance, and related fields

ACCOUNTANTS—accounting and related areas

ECONOMIC ABSTRACTS INTERNATIONAL—economics

INTERNATIONAL TIME SERIES—foreign business and economic data (production, consumption, etc.)

LABOR DOC—international labor documentation

MGMT—management and business

PTS EIS PLANTS—U.S. industrial plants

PTS F&S INDEXES (F&S Index of Corporation and Industries, R&S International)—business research

U.S. ANNUAL TIME SERIES—historical time series forecasts, annual data

U.S. STATISTICAL ABSTRACTS—forecast abstracts, census of manufacturers

Humanities

AMERICA: HISTORY AND LIFE—American history

ARTBIBLIOGRAPHIES MODERN—fine arts

HISTORICAL ABSTRACTS—world history

MLA BIBLIOGRAPHY—literature

PHILOSOPHERS INDEX—philosophy

Multidisciplinary Coverage

COMPREHENSIVE DISSERTATION ABSTRACTS—doctoral dissertations

FOUNDATION GRANTS INDEX—grants records of U.S. philanthropic foundations

MAGAZINE INDEX—current events, popular interest

NATIONAL FOUNDATIONS—descriptions of foundations in the U.S.

NEWSPAPER INDEX—major newspapers

FIGURE 2.18.   The computer data bases of most use for engineering applications. (Courtesy University of Michigan Libraries)

SUMMARY

The technical library is an important tool in engineering problem solving. Engineers should learn to make effective use of this resource early in their education. These include card catalogs, technical journals, abstract indexes, standards, patents, and computer-based bibliographic search services.

Example An engineer wishes to rapidly scan the technical literature concerned with the design of a novel energy source in which high-powered ion beams are used to trigger thermonuclear fusion explosions in tiny fuel pellets. It is decided to access the *Energy Abstracts* data base, maintained by the United States Department of Energy on a computer at Oak Ridge, Tennessee. The engineer dials the long distance number of the Oak Ridge computer from her remote computer terminal and then types in the appropriate identification and commands to interrogate the data base. First a search on the words "ion beam" and "fusion" is requested. The computer responds that each word appears in the title or abstract of 4517 and 7312 technical references, respectively. The engineer then queries the intersection of the words "ion beam," "fusion," and "reactor." The computer responds that there are 215 references that contain all three words simultaneously. The engineer then asks for the abstracts of only those references published since 1980, and the computer proceeds to print out 43 such abstracts on the engineer's high-speed printer.

2.2.3. Modeling and Simulation

The complexity of analyzing or designing engineering systems can be overwhelming. It is often futile to attempt a head-on attack at solving problems involving complex systems. Imagine the task of writing down the force balance equations for every beam or truss in a modern skyscraper, not to mention attempting to solve these equations. It would clearly be unthinkable to design full-scale experiments to test the integrity of a nuclear power plant under earthquake conditions. Therefore a critical aspect of engineering analysis involves stripping away the complexity of such problems by developing simple models of the system that represent only the essential elements of interest. Indeed, one of the most distinguishing traits of an engineer is the tendency to utilize models in the analysis of complex problems.

In general a *model* can be defined as any simplified description of an engineering system or process that can be used to aid in analysis or design. A model might be a physical realization of the actual system, such as a scale model of an aircraft to be used in wind tunnel tests. It might be an abstract mathematical model consisting of a set of equations describing the behavior of the system. It might also be a complex computer program or code that can simulate the essential features of the system. Models can assume more familiar forms such as the diagrams or graphs used by engineers in design and analysis. The form of a model is called its *representation*.

The use of models allows the engineer to avoid the complexities of studying realistic engineering systems in their entirety and focus instead only on the essential elements. Models also enable the engineer to predict the behavior of systems under various circumstances. They can be used to communicate the results of an engineering design or to train personnel involved in system operation. The application of models to acquire an artificial experience with the behavior of a system is known as *simulation*. Models can also be used to evaluate the effectiveness of control or safety systems. They allow the engineer to evaluate many possible design alternatives in an effort to optimize the final design or solution. The use of modeling accomplishes all of these tasks at only a fraction of the cost and effort that would be required if one were forced to work with full-scale engineering systems. Imagine the staggering costs that would be incurred if a large dam or electrical power plant had to be built and tested before its design could be optimized or verified.

SUMMARY

A model is any simplified description of an engineering system or process that can be used to aid in analysis or design. The form of a model is known as its representation. The use of models allows the engineer to avoid the complexities of studying a realistic engineering system in its entirety and enables him to focus instead on only its essential elements. Models can be used to predict the behavior of systems, communicate the results of engineering analysis, acquire an artificial experience using the behavior of the system, and evaluate possible design alternatives.

Example The use of scale models has become a common practice in large construction projects such as electric power plants (Figure 2.19). These models can be used to visualize complicated details, such as piping or equipment layouts, far more easily than engineering drawings. In fact, these scale models are frequently taken to the construction site and used to document changes in the design as the project proceeds. With the rapid development of computer-aided design and computer graphics methods, these scale models will eventually be replaced with two- and three-dimensional images of various components of the project stored in a computer memory and available for display or analysis on a monitor screen. (See Section 7.5.)

Types of Models To some people the word "model" conjures up images of plastic airplane kits or sailing ships in bottles. However, the engineer uses the term "modeling" to refer to the general process of abstracting and simplifying a complex engineering system by representing this system using only those elements that are essential to an understanding of the problem at hand. Models can assume a variety of forms or representations.

FIGURE 2.19.   Craftsmen building plastic models of power plant components. (Courtesy Bechtel Power Corporation)

FIGURE 2.20.   Windtunnel testing of a bus model. (Courtesy University of Michigan College of Engineering)

A model may bear an actual physical resemblance to the system of interest. Scale models of ships in towing tanks or aircraft in wind tunnels have been used for many years to study the flow characteristics of a given design. Models that physically resemble the system under study are called *iconic* representations. This term is also used to describe two-dimensional "models" such as sketches, photographs, or blueprints. It can also be applied to the use of computer graphics.

On somewhat more abstract level are *diagrammatic* representations of the system for example, electrical circuit diagrams, piping diagrams, and process flow diagrams. Although the diagrams and the elements that comprise them do not physically resemble the system of interest, they do accurately describe the function of the components of the system and prove invaluable in its analysis and design.

On a somewhat higher level of abstraction are *graphical* models, which represent system characteristics by plotting system behavior in graphical form. For example, the input-output characteristics of a circuit element can be a very faithful model of the performance of the device. Graphical representations are examples of *analog* models, which attempt to simulate behavior while not physically resembling the actual system.

FIGURE 2.21. A simulator for a nuclear power plant control room. (Courtesy Combustion Engineering)

In the most common form of model used in engineering analysis the system of interest is represented by a set of mathematical equations. *Mathematical* models represent the highest level of abstraction in modeling. They are also probably the most useful form of model since powerful methods of mathematical analysis can then be applied.

A more recent form of modeling or simulation involves the use of computers. In its most primitive form, one can think of the *computer* model as simply one method for solving and analyzing mathematical models. In practice computer models have become far more sophisticated. These models may be based on the mathematical equations developed to describe the system. But they may also be based on alternative approaches such as probabilistic simulations of the system behavior. Most commonly such models are implemented on digital computers in the form of large computer programs. But engineers also use analog computer models in which electrical circuit elements are used to simulate the components of the system.

SUMMARY

Models come in many forms or representations. Iconic models bear a physical resemblance to the system of interest. Diagrammatic models are used to represent the function of the system. Graphical models represent system characteristics by plotting system behavior in graphical form. Mathematical models represent the system by the set of mathematical equations describing its behavior. Computer models are based either on mathematical equations or probabilistic simulations of system behavior.

Example   Nature cooperates with certain forms of modeling in the sense that many physical laws exhibit the property of *similarity*. Such laws can be scaled in size or dimension. For example, the laws of fluid dynamics are scaled in such a way that one can accurately model the flow of air across the wing of an airplane by using a scale model in a wind tunnel.

Model Development   How does the engineer develop a model to represent a complex engineering system? The first step is to isolate the essential components or elements of the physical system under study and then represent them as abstract, idealized elements possessing all of the important characteristics of the real system. At this stage the engineer brings into play an understanding of the basic physical laws characterizing the system. The next stage involves the selection of suitable analogs to these idealized components. One can select from many possible representations by drawing from a knowledge of similar physical systems and past experience. For example, the engineer might choose to represent system components by a set of equations that adequately describe the basic physical principles governing this behavior. Or a graphical representation might be chosen that reflects empirical data obtained on similar components.

Another useful aid in the development of idealized components is to draw upon the analogies that arise between electrical, mechanical, fluid, and thermal systems. For example, it is frequently possible to construct an electrical circuit that simulates the behavior of a mechanical system comprised of masses and springs (e.g., the shock absorber of an automobile). In this way one can construct electrical models of the actual system components.

After the particular model has been developed, it must then be verified by comparison with the known behavior of comparable physical systems. Before the model can be used to predict aspects of system behavior used in actual design, the engineer must verify that the model will faithfully yield those aspects of system behavior that are already known. Certainly one would hesitate in extrapolating data obtained by shaking a small-scale model of a skyscraper on a vibrating table to the design of the full-scale structure without some confidence that the model was valid. Model verification might be accomplished, for example, by applying the model to analyze earlier designs or by comparing the predictions of the model with data obtained from the testing of subsystems or components.

Frequently the model will contain unknown or free parameters that can be adjusted in such a way that the model will predict known results. That is, the model can be calibrated to predict known behavior—it can be "fine-tuned" to yield the best possible agreement with previous experience.

SUMMARY

The major steps of model development are abstraction, synthesis, and verification. The engineer first isolates the essential components of the system of interest. A particular realization of these idealized elements is synthesized as a model. Finally this model is verified by comparing its predictions with the known behavior of actual systems. Many revisions of the model may be necessary before an acceptable predictive capability is achieved.

2.2.4.   Optimization and Iteration

We have introduced the concept of solution criteria to distinguish among the various solutions to an engineering problem. These criteria provide a quantitative measure of the attractiveness of a given solution—how well it meets the goals of the engineer. We have noted that the task of solution evaluation, that is, of selecting the most favorable solution, involves choosing the solution that optimizes the criteria. For example, this might involve maximizing performance or minimizing cost. The existence of such a solution suggests that the procedure of *optimization* plays an important role in engineering analysis and design. That is, the engineer attempts to evaluate or adjust the various solutions to an engineering problem to establish an optimum solution that will yield the best solution criteria.

To illustrate this idea, let us consider again the automobile fuel efficiency problem. Suppose we consider the criterion to be fuel efficiency in miles per gallon and focus only on one solution variable, the mixture of air and fuel used in the combustion process. Then our optimization problem would take the form of attempting to determine the mixture of air and fuel that would yield the maximum fuel efficiency. For very rich fuel mixtures, there is incomplete combustion and therefore poor fuel efficiency. As we decrease the fuel-to-air ratio, the fuel efficiency increases until it achieves a maximum value (Figure 2.22). Beyond this point, going to progressively leaner mixtures again decreases fuel efficiency because of decreased engine torque.

Unfortunately most optimization problems are not nearly so simple. The choice of solution variable that optimizes one criterion may yield an unfavorable value of other solution criteria. For example, the choice of fuel–air mixtures that yield the maximum value of fuel efficiency may lead to excessive exhaust emissions or engine wear. Therefore the task of optimization is much more complex since one will generally be faced with optimizing many different criteria, several of which may be of a conflicting nature. This situation of conflicting criteria and the need to find a compromise among them is common in engineering applications. It is usually referred to as a *trade-off* process.

An example of a trade-off process on a somewhat larger scale involves the development and implementation of new energy production facilities that will have a minimal environmental impact. This solution criterion is usually overlooked in the debate swirling about energy issues. For example, the demand for

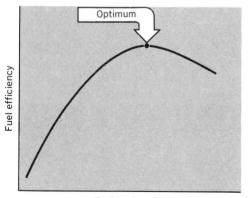

FIGURE 2.22. Optimization of the fuel efficiency of an automobile as a function of fuel-to-air ratio.

zero air pollution is certainly in conflict with the demand for an increase in coal-based energy production.

Faced with the complexity of the various criteria that are used to evaluate solutions and the conflicting nature of these criteria, one must examine formal methods for choosing among various solutions, that is, formal methods of optimization. Merely evaluating the various criteria for a number of randomly chosen solutions is extremely inefficient. Generally one attempts to formulate a mathematical model in order to obtain an explicit representation of solution criteria. Then optimization methods based upon calculus can be used to determine the choice of solution parameters that yields maximum values of the criteria.

However, many problems may be so complex that it is impossible to obtain an explicit functional representation of the criteria. Therefore one must adopt some kind of search procedure that samples various solutions, evaluates the corresponding criteria, and then uses this evaluation to choose a new solution for evaluation. In this sense, the process of optimization may acquire an *iterative* character. That is, one iterates between choosing possible solutions and calculating and comparing solution criteria by using prescribed rules until an optimum solution is obtained. Such an iterative process of optimization is basically an accelerated learning process since through a series of successive approximations, the engineer gradually closes in on the optimum solution.

While an optimum solution is always an objective for engineering analysis, it is also a goal that is seldom achieved. A serious constraint on engineering problem solving is frequently time. Although some effort should always be made to achieve an optimum solution, rare is the instance when engineers have the time necessary to determine the true optimum. Rather they seek to come as close to an optimum as possible in the time available. Eventually they either run out of time or reach a point of diminishing returns. Then it becomes necessary (or more profitable) to adopt the solution as it stands and direct further effort toward other problems requiring attention.

SUMMARY

The engineer attempts to evaluate and adjust various solutions to an engineering problem in order to determine the solution that optimizes the solution criteria. Frequently there are conflicting solution criteria that demand a compromise, a trade-off. Usually time or diminishing returns prevent the engineer from achieving a truly optimum solution.

Exercises

The Search

1. Identify examples of problems in which the solution was hindered by each of the following difficulties. (Use either your own experience or your knowledge of other engineering problems.)

    (a) Overly narrow initial problem definitions
    (b) Mind set
    (c) Premature evaluation of solutions
    (d) Isolation from the opinion of others
    (e) Limited nature of human memory
    (f) Insufficient incubation period

2. One interesting approach to problem solving is "brainstorming" in which a group discusses a given problem in a freewheeling dialogue, considering all ideas and solutions, no matter how far-fetched they may appear at first. This approach has been adopted by "think tanks" used in industry and government, such as the Rand Corporation and MITRE. Together with several other students, try a 10-minute brainstorming session and list the solutions you come up with in this period on any of the following problems:

    (a) Energy storage systems for solar power
    (b) Reducing inflation while maintaining full employment
    (c) Mining ore from the moon's surface
    (d) Designing an acceptable mass transit system for your community

3. A useful technique in problem solving is the use of a checklist in which the engineer identifies and lists the advantages and disadvantages of possible solutions. Try this approach on possible solutions for any of the problems listed in Exercise 2.

Technical Information

Most of the following exercises will make use of your university's technical library.

4. Find and list at least three references that would be of assistance in the following tasks:

    (a) Building a ham radio
    (b) Repairing the carburetor of your automobile
    (c) Evaluating a set of mathematical integrals
    (d) Obtaining information about the career of your local congressman

5. Find 10 articles in your technical library relevant to the Three Mile Island nuclear power plant accident that occurred in spring of 1979.

6. Trace the origin of the gyrocompass.

7. Identify the titles of three journals in which articles on fire prevention are published.

8. Using information in your technical library, find the different methods proposed in the past 10 years to remove oil slicks from the ocean surface.

9. Locate the earliest reference to Ohm's law.

10. Find out from your librarian how you could obtain a copy of a dissertation written at another university.

11. Find references concerning the safety standards of pressure vessels.

12. Where are government regulations on the use of paint sprayers published?

13. What is the thermal conductivity of brick and concrete?

14. Find the government regulations concerning the fuel economy standards of automobiles manufactured in the United States.

15. Find the procedures required to file for patents in the United States.

16. Find the titles and citations for one technical paper published in the period 1978–1980 by each of the authors of this text.

17. Learn from your librarian how to conduct a computer-based literature search.

Models

18. Classify each of the following models as iconic, diagrammatic, graphical, analog, mathematical, or computer and give the reasons for your classification:
    (a) A strip chart recorder
    (b) Newton's laws of motion
    (c) The use of billiard balls to model neutron-nucleus collisions in a nuclear reactor
    (d) A free body diagram
    (e) A roadmap of the United States

19. How would you model the dynamics of a long column of cars at an intersection as they respond to the change in a traffic light from red to green?

20. How might you model the brakes of an automobile?

21. How would you model the people standing in line, say, at a bank teller window or supermarket checkout counter?

22. Develop a predictive model to determine how much study time you should allocate per week to each of your courses.

23. Choose a common model used in your physics or chemistry courses. List all of the assumptions involved in the model, and identify those assumptions most likely to break down in real-world situations.

Optimization

24. Identify conflicting criteria that might lead to a trade-off or optimization problem in each of the following design problems:
    (a) An automobile
    (b) A text-reading machine for blind persons
    (c) The development of a new insecticide
    (d) The development of a major gasohol industry in the midwest grain belt

25. Identify five "optimization" problems you encounter in your everyday activities.

26. Identify the principal criteria involved in the following trade-off situations:
    (a) Energy supply versus environmental quality
    (b) Social welfare versus inflation
    (c) Lowering the speed limit for interstate travel by automobile
    (d) Oil supertankers

27. Ten years ago holography was widely touted as a great discovery with numerous applications. However it has not lived up to its expectations, and there are few uses of it today. Give the reasons that you expect have hindered the wider application of holography.

28. Within five years after they first appeared on the market, pocket calculators decreased in size and price significantly while vastly increasing their capability. Suggest several reasons for the outstanding technical and economic success of this product.

2.3. THE ENGINEERING DESIGN PROCESS

We have described a very general approach to engineering problem solving. This procedure can be applied to problems arising in any phase of engineering such as research, development, design, or even sales and management. In this section we shall consider how the methodology can be applied to the process of *engineering design*, that is, to the development of a physical system or process to perform a required function.

Design is the principal intellectual activity of the engineer. It is important to contrast this engineering activity with the primary activity of the scientist, *research*. The principal objective of research is the development of models, theories, or hypotheses to describe scientific phenomena. For example, suppose that a scientist observes that the incidence of serious crime tends to increase whenever a long period of hot, dry weather sets in. He might hypothesize that temperature and humidity affect human irritability. The scientist would then set out to verify or disprove this hypothesis by conducting psychological tests of individuals exposed to a controlled temperature–humidity environment.

As we have described it, scientific research is an *inductive* process because one attempts to draw general conclusions from specific experiences. In sharp contrast, engineering design is a highly specialized process of problem solving. It involves the development of a physical system or process that will perform a required function. Hence it is a *deductive* procedure that attempts to develop a specific solution to a given problem from a general set of principles. That is, the engineer first studies the fundamental scientific principles that govern the process of interest (e.g., the chemistry and thermodynamics of fuel combustion) and then uses these concepts to synthesize a particular design (an automobile engine with improved fuel efficiency). Whereas research proceeds from specific experience to general or abstract principles, design proceeds from general principles and abstract models to specific solutions (Figure 2.23).

Despite these differences the general process of problem solving used in design is similar to that used in research and development. However engineering design tends to make use of somewhat sharper, more analytical tools. It carries the problem-solving activity not only through the stages of problem definition and abstraction, solution synthesis and verification, but beyond to evaluation and decision and eventually to optimization, revision, and implementation. These latter activities distinguish the design process from other types of problem solving.

The design engineer's primary function is to create a structure, device, or process that will meet a practical requirement. In this sense design represents the culmination of the engineer process, the ultimate application of science and technology to the creation of new items for society. As an intellectual endeavor, engineering design is very similar to design activities encountered in the creative arts. As with all areas of design, one proceeds from general principles to develop a specific concept suitable for performing a given function or fulfilling a specific

FIGURE 2.23. The inductive logic of research compared to the deductive logic of design.

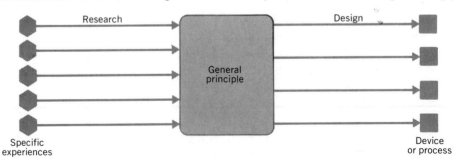

need. Engineering design is distinguished from other design activities primarily by the degree to which technological factors contribute to its implementation.

There are other important differences as well. Engineers usually do not produce the particular system they design. Rather they produce only a model, a replica, to be used as a template for reproducing the design as many times as needed. Of course this places an added responsibility on their shoulders, since any error in the design will be propagated to all reproductions of that design at the production stage. Furthermore the rapidly accelerating pace of technological development and innovation places strong pressures on design engineers to continually update their fundamental knowledge. In addition, the constraints on engineering design usually extend well beyond factors of a technological nature. Engineering design almost always requires a synthesis of technical, economic, and social factors.

Engineering design is a far more general process than invention, that is, than tinkering away in a basement laboratory until one stumbles across a new discovery. Certainly innovation, ingenuity, and creativity play vital roles in both design and invention. But inventors sometimes do not understand the physical principles behind their invention; they are usually unable to optimize its design or develop it

FIGURE 2.24.   There is a sharp contrast between the inductive logic of scientists and the deductive logic of engineers. (© 1978 by N. Kliban. Reprinted from *Tiny Footprints* by permission of Workman Publishing Company, New York)

into a form suitable for mass production. In this sense, engineering design is a far more general, powerful, and disciplined approach than mere invention. It requires great skill and training. It is not an activity left to happenstance, to accidental discovery. Rather engineering design is approached with the disciplined methodology of engineering problem solving. We proceed through the stages of problem definition, solution synthesis, verification, and evaluation. However for our discussion of the engineering design process, we group these activities in a somewhat different fashion, as shown in Table 2.1.

SUMMARY

Engineering design is the development of a physical system or process to perform a required function. Research is an inductive procedure in which one draws general conclusions from specific experiences to develop theories that describe scientific phenomena. Engineering design is a deductive procedure to develop a specific solution to a given problem from a general set of principles. Engineering design requires a synthesis of technical, economic, and social factors. It is far different from the process of invention, since it cannot be left to happenstance or accidental discovery, but rather requires a highly disciplined approach. We can approach engineering design as we approach other problem solving activities by employing the steps of problem definition, solution synthesis, and solution verification and evaluation.

TABLE 2.1   Comparison of Steps in Problem Solving and Engineering Design

STEPS IN PROBLEM SOLVING	STEPS IN ENGINEERING DESIGN
1. Problem definition Data collection Model development	1. Identification of need Information search
2. Solution synthesis	2. Synthesis of possible designs
3. Solution verification Solution evaluation	3. Feasibility study Preliminary design Final design
4. Presentation of results	4. Design documentation (technical specifications)

Example   During the 1930s European scientists investigated the behavior of heavy metals by bombarding them with a newly discovered particle known as the neutron. They found that a great many materials became radioactive when subjected to neutron bombardment. But two German chemists, Otto Hahn and Fritz Strassmann, discovered something else quite unexpected: when the heavy metal uranium was bombarded with neutrons, new elements appeared with roughly half the atomic weight of uranium. They postulated that a new process had been discovered, one in which a neutron actually split or fissioned a uranium nucleus. This hypothesis was borne out by later experiments and calculations. The discovery of nuclear fission is an excellent example of the inductive process of scientific research.

In contrast, engineers during the 1940s began to design systems to exploit the energy released in the nuclear fission process. This process was used in nuclear reactors to produce energy and also in nuclear bombs to produce explosions. The history of the Manhattan Project provides many examples of the deductive process of engineering design, of proceeding from general principles of nuclear physics to a specific device or process. Whereas the discovery of nuclear fission occurred essentially by accident, the design of systems to make use of this new energy source required a far more carefully planned and disciplined approach.

2.3.1.   Problem Definition: Identification of the Need

The first step of a design study is to determine whether a need exists, that is, to identify and define the design problem. Sometimes this is straightforward. It is obvious, for example, that astronauts cannot leave their spacecraft and walk on the surface of the moon without adequate protection. The design solution is easily identified: a pressurized suit.

Many times it is more difficult to identify a need. Progress in medicine is often hampered by the lack of suitable equipment. Engineers who could design the necessary equipment are usually unfamiliar with modern treatment and surgical methods and cannot identify the areas where an engineering solution would be helpful. Sometimes even the need for highly successful products is not evident. While the principles of electrophotography were developed in 1934, it was not until some 20 years later that a device was developed using this process to copy materials on ordinary paper. As late as 1960 the need for such photocopying devices was greatly underestimated and only became apparent with the remarkable success of the Xerox machine.

It sometimes requires great vision to recognize a need. It also requires foresight to anticipate future needs for a design. Some needs may be of a transient nature and may change appreciably over the lifetime of a design project. For example, in the early 1960s considerable attention was directed toward the development of supersonic passenger transport aircraft (SST). The staggering increases in fuel costs during the 1970s, coupled with the voracious fuel consump-

FIGURE 2.25. The need of lunar astronauts for protective clothing is obvious. (Courtesy National Aeronautics and Space Administration)

tion of these aircraft, may have doomed them to only a very limited role in the future. The energy crisis and the soaring cost of petroleum changed the conditions on which the original design was predicated.

After identifying or confirming the existence of a need for the design of an engineering system or process, engineers next formulate the goals of the design (the design criteria) and the restrictions (constraints) it must meet. This activity corresponds to the broad formulation of the design problem, and it serves to point the way toward more specific design criteria and may suggest possible approaches toward the synthesis of the design. At this stage the engineer studies the design problem in detail, identifying and acquiring the necessary information to support the design activity, and converges on more and more detailed statements of the specific design task including its constraints and design criteria.

SUMMARY

The first step of a design study is to determine whether a need exists for the particular project of interest. One must also assess the significance of this need and determine whether or not it is worth pursuing. Then the engineer identifies the general goals of the design (the design criteria) and the restrictions (constraints) it must meet.

Example   Occasionally a creative design innovation will trigger a new and unforeseen need. One of the most outstanding advances in the 1960s was the introduction of the pop-top aluminum beer can. No longer were engineering students required to carry can openers hanging from their belts along with their pocket calculators. They could simply pull open the tab opener on the top of the can.

But, as the environmental movement took hold, it became apparent that the pop-top can was ecologically unsound. Beer-loving engineers would discard the pop-tops (and later the empty beer can). The pop-tops scattered across the landscape began to present a serious hazard to our civilization. A very real need became apparent.

This need was met by two solutions, one political and one technological. First, many states passed returnable bottle and can laws that demanded a deposit on all beverage containers, refundable upon their return. Then product engineers

FIGURE 2.26.   Engineers and physicians collaborate to achieve advances in medicine. (Courtesy University of Michigan College of Engineering)

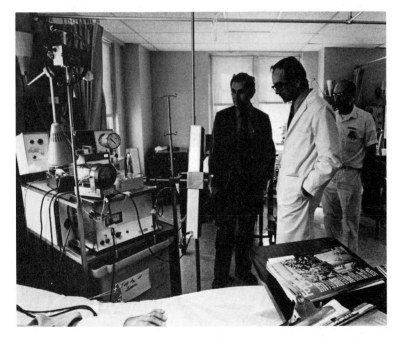

designed new cans that could be opened by merely depressing a spot on the top of the can, thereby eliminating the pull-tab openers altogether.

2.3.2. Solution Synthesis, Verification, and Evaluation: Feasibility Study, Preliminary Design, and Final Design

With the need identified and the design problem defined, the engineer now develops solutions. This, more than any other step in the design process, requires inventive and creative effort. By "creative" we do not necessarily mean that the individual elements comprising the design will be novel. Rather it is the design synthesis itself that is novel. Sometimes the design engineer will blend together a new scientific discovery with several more established ideas to arrive at a practical design.

As in any form of problem solving, engineers synthesize and evaluate several possible solutions. Usually the first design solutions are of a rather general and vague nature, addressing only the broader aspects of the design problem. As these solutions are studied and evaluated in greater detail, the engineer focuses on a more detailed statement of the specific design task. This activity usually takes the form of a sequence of design studies: a feasibility study, a preliminary design, and a final design. These design studies correspond to the steps of solution synthesis, verification, and evaluation in the more general problem-solving method.

The *feasibility study* involves formulating a variety of general solutions to the design problem and then evaluating the technical and economic feasibility of these solutions. A given solution may not be technically feasible because it violates a fundamental scientific principle, or because the technology necessary for its implementation is unavailable. An example of a design that violates a scientific law is the perpetual motion machine. This device produces a net surplus of energy. The jet engine is an example of a device whose feasibility was delayed by the development of suitable high temperature materials.

In addition to technical feasibility, the economic feasibility of a project must be evaluated. For a single item such as a bridge, building, or a structure, a reasonable estimate can usually be made of the cost. It is more difficult to make a cost estimate for mass-produced items. The manufacturing cost and the selling price for such products depend on the quantities sold. It is extremely difficult to anticipate the demand for a new product. Market surveys provide useful information, but they are not foolproof, especially when they are for as yet nonexisting products. It is often difficult to know if a competitive product will appear before the design can be marketed. The economic planner must make a forecast on the basis of the available, although frequently limited, information.

Finally, during the feasibility study the social, environmental, and safety impact of the design must be evaluated. Many engineering designs are found to cause more problems than they solve. A highway that requires the razing of buildings and the uprooting of many families may not be an acceptable solution.

Automobiles or plants that discharge pollutants may be unacceptable and may have to redesigned.

SUMMARY

The engineer begins the task of formulating a detailed statement of the specific design task from a broad formulation of the design problem and its possible solutions. The first aspect is a feasibility study to evaluate the scientific and economic feasibility of the design solution. Social, environmental, and safety factors must also be considered.

Example   One of the more interesting energy sources proposed for the future consists of large arrays of solar collectors placed in earth orbit (Figure 2.27). The energy collected by these solar satellites would then be beamed down to earth-based receivers as microwave radiation. A feasibility study of this proposal must examine several important questions. For example, how much material is needed to construct such a satellite; how can one beam down energy without significantly perturbing the atmosphere; how difficult will it be to maintain the solar collectors during their continual bombardment by cosmic debris? Of course the key question

TABLE 2.2   The Time to Develop a New Technology

INNOVATION	YEAR OF FIRST CONCEPTION	YEAR OF FIRST REALIZATION	DEVELOPMENT TIME (years)
Heart pacemaker	1928	1960	32
Hybrid corn	1908	1933	25
Hybrid small grains	1937	1956	19
Green revolution wheat	1950	1966	16
Electrophotography	1937	1959	22
Organophosphorus insecticides	1934	1947	13
Oral contraceptive	1951	1960	9
Magnetic ferrites	1933	1955	22
Video tape recorder	1950	1956	6
Average duration			19.2

SOURCE   Robert C. Dean, Jr., "Technical Innovation USA," *Mechanical Engineering* (November, 1978), p. 29.

involves economics. Can one construct such orbiting solar power stations at a cost comparable to earth-based facilities? Perhaps an even more fundamental question addressed in the feasibility study is whether the energy produced during the lifetime of the station will pay back the energy invested in manufacturing and placing into orbit its components. The public risk and environmental impact presented by such a massive undertaking must also be assessed.

Once the alternative solutions to a problem have been explored, the engineer selects the approach that appears most promising and performs a *preliminary design*. The purpose of the preliminary design is to evaluate the usefulness of the concepts and the ideas incorporated into the design and to determine whether or not the design works.

The preliminary design study usually makes extensive use of models of the design. Such models may take the form of drawings, diagrams, or mathematical calculations. Sometimes these models reveal serious flaws in the design, which require substantial revision. They may also support the initial feasibility, providing more confidence that the design will actually function as intended.

SUMMARY

The engineer performs a preliminary design to evaluate the usefulness of the concepts and ideas that are incorporated into the design and to determine whether or not the design will work. This stage usually involves building a working model or prototype.

Example   As a specific example, let us consider the design of an electric toaster. The basic idea behind the toaster is that bread will be toasted when placed near an electric heater. At the beginning of the design study we are confronted with several unknowns: (1) the type of wire to use in the electric heater, (2) the length and diameter of the wire, (3) the arrangement of the wires, (4) the amount of current provided to the wire, (5) the distance between the bread and the wires, (6) the time the bread is exposed to the heat, (7) the method used to remove the bread from the toaster. Answers to these questions must be found before the final product can be designed.

The first model of the toaster is usually in the form of a diagram. The heating coils and power supply are represented in Figure 2.28. From the known resistance of the wire and the current input, we can calculate the power input to the wires

FIGURE 2.27.   The solar power satellite is still in a very early stage of the design process. (Courtesy National Aeronautics and Space Administration)

and the heat they produce. We can also estimate the temperature of a slice of bread placed at various distances from the wire. Our calculations tell us how much power is needed to heat the bread to a given temperature with a given type of wire. We may find that the power requirements are too high, or that the maximum temperature achieved with the wire is too low for toasting a slice of bread. Changing the diameter or length of wire or using a wire with different electrical characteristics might alleviate the problem.

FIGURE 2.28.   Sketches representing the various stages in the design of an electric toaster.

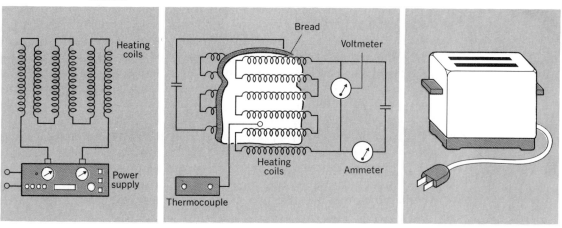

A working prototype of the toaster might consist of coiled wires supported in a frame, a power supply, and a slice of bread suspended from a positioning arm. Various measuring instruments, voltmeters, ammeters, and thermocouples are added to monitor the performance of the device. Tests are then conducted to determine the correct conditions for toasting the bread. The results will determine the appropriate material, length, diameter, and arrangement of the heating wires, the proper distance between the bread and the wires, and the time needed for toasting the bread. Once these questions are answered, we can turn to the design of a timer and a mechanisms for removing the bread.

Once design engineers are satisfied that all foreseeable problems have been eliminated, they then proceed to design the final form of the device or product. The steps involved in the *final design* are similar to those of the preliminary design. First the final version is specified by drawings. A prototype is then built according to these drawings. In contrast to earlier models, this prototype is identical in all respects to the final product. It performs the same functions. It even has the same size, shape, and color.

The prototype is a particularly critical aspect of the design since it represents the last stage at which problems can be identified and eliminated. Any problem not identified here will show up in the final product. Since it is difficult and expensive to make corrections after production has begun, it is of paramount importance that problems be detected and corrected during the final design stage. Some faults can be found only through comprehensive testing of the prototype.

The final stage of the engineering design process places a premium on economic feasibility. The product must be of sufficient value to repay the effort involved in its production. Design engineers should place themselves in the position of the producer, distributor, and the consumer. They should continually ask themselves whether the projected solution or design is the simplest that could accomplish the desired result.

Careful thought must also be given at this stage to questions of maintenance and service. Small design changes can sometimes greatly simplify a service procedure that may be required later. For example, one recent automobile model required partial lifting of the engine from its mounts in order to provide enough clearance to replace the spark plugs. One growing trend is to make design components as modular as possible, with a number of interchangeable subunits. Service in the field can then consist of simple replacement of a subunit, greatly reducing downtime and servicing costs.

Many design activities must meet the additional constraints imposed by government regulation. For example, the exhaust emission of any automobile must be tested and approved by the United States Environmental Protection Agency. New medical instruments must be approved by the Food and Drug Administration. The growing mountain of such regulations, which are continually changing, has come to represent a formidable challenge to the design engineer.

SUMMARY

When design engineers are satisfied that all foreseeable problems have been eliminated, they proceed to design the final form of the product and to document this design. A prototype of the design is usually built at this stage. This represents the last step where problems can be identified and eliminated before the design enters the production phase.

Example Consider how the test program for the prototype of a washing machine might be conducted. First the machine would be tested to determine whether its mechanical and electrical components operate properly. Then the machine would be loaded with clothing, and its washing performance evaluated; this might be done through tests that show the amount of water used, the amount of dirt removed, and any damage to clothes. Third, the machine would be thoroughly tested for possible electrical or mechanical safety hazards. Finally, the durability of the machine would be evaluated. To do this, the machine must be operated through many cycles. To simulate the long term use, durability tests may take several months or even years. Suppose the average household use of the washing machine is projected to be three times per week. If each cycle is 30 minutes long, then during 15 years of service life, the machine is in operation for 390 hours. Therefore this machine would have to be operated repeatedly and continuously for about 16 days to simulate usage equivalent to 15 years in a household.

A design flaw that goes unnoticed in the prototype and must be corrected after manufacture is very costly. For example, suppose that 10% of our washing machines prove defective after manufacture and must be repaired at a cost of $10 each. The total cost when this product is mass-produced at a level of 100,000 units a year is $1,000,000. This cost would have to be absorbed by the manufacturer or passed on to the consumer.

2.3.3. Design Iteration and Optimization

Engineering design is very much an iterative process. One must compare the analysis and evaluation of prospective solutions with the design requirements (the solution constraints). If these requirements are not adequately satisfied, a new design is attempted and evaluated.

In practice design is seldom performed in the sequence of distinct steps we have presented in this chapter. The boundaries between the different steps are not sharp; the work during different design steps overlap. Furthermore, the different phases of design are not accomplished independently. On the contrary, throughout the entire design process there is strong interaction between different stages. The early phases of design are performed with the ultimate objective in mind. As one progresses to later stages, new problems become evident or new solutions appear. Hence there is information flowing continuously from earlier to later design phases, and vice versa. Redesign does not stop even after the manufacture of the product. Products undergo changes with time and, in fact, most products go through many changes in almost a continuous, evolutionary process.

When the optimum design is finally chosen, the engineer must carry through a detailed analysis to arrive at a set of technical specifications for the design. These specifications consist of a detailed description, sketches, drawings, photographs, flow diagrams, in fact, any information that might prove of use to the person who is to assemble, use, or regulate the device or process.

We cannot emphasize strongly enough that an engineering design is only as good as the engineer's ability to describe the design to others. We shall return in

FIGURE 2.29. Design is very much an iterative process. Consider, for example, the evolution in the design of the bird. (© 1976 by B. Kliban. Reprinted from *Never Eat Anything Bigger Than Your Head* by permission of Workman Publishing Company, New York)

Chapter 6 to consider the very important topic of the documentation and presentation of engineering design.

SUMMARY

Engineering design is an iterative process since new design synthesis may be attempted and evaluated according to design goals and constraints until an optimum design is achieved. There is a strong interaction between the different stages of design. The final stage of design is the preparation of detailed technical specifications.

2.3.4. Some Concluding Remarks

Engineering design can be an extremely challenging and exciting activity. It requires imagination, creativity, and discipline. It also requires considerable knowledge, experience, and maturity. Therefore we have discussed only general aspects of engineering design to convey the flavor of its challenge and excitement. The development of the skills necessary for a design engineer is a gradual process, requiring continual exposure to design problems and the experience of seasoned engineers. It also requires the mastery of new tools such as computer-aided design. Engineers should learn to pull together the diverse components of their formal education and practical experience and focus them on the particular design problem of interest.

Exercise

Design, Research, and Invention

1. Thomas A. Edison had many patents to his credit. Two of his best-known inventions were the light bulb and the phonograph. Were these the results of research, invention, or both? Why?

2. How would you classify the inventors of the following devices, as scientist, engineer, or inventor:

Zipper	Laser
Gas turbine	Transistor
Sewing machine	Xerox machine
Typewriter	Jet engine

3. You are to develop a device helping patients breathe (i.e., a respirator). What type of background knowledge would you need to develop such a device?

4. Discuss the steps you would follow to invent a method or device that would do the same job as a safety pin, without the hazard posed by the pin (i.e., the pin can stick you).

Need Definition

Establish the need for a product in each of the following areas.

5. Medicine
6. Home heating
7. Temperature measurement
8. Electrical power measurement
9. Chemical pH measurement
10. Footwear
11. Sports

Steps in Design

Outline the design steps (feasibility study, preliminary design, final design) involved in the design and development of each of the following products.

12. A slide projector with coordinated (synchronized) sound.
13. A pocket portable telephone.
14. A backpack propulsion unit for flying people on short trips.
15. An electric car.
16. A pulse rate indicator for use during exercise.
17. A multicolored fountain pen.
18. An umbrella that can be used in strong winds.

General Design

Sketch how you would approach each of the following design tasks.

19. Develop a system that can be used to replace conventional traffic lights. Your system should be simpler and less expensive than the present system.
20. Develop a method for measuring pulse rate during exercise. Your device should be small, lightweight, and suitable for mass production.
21. Devise a system that would eliminate the need for seat belts in automobiles.

22. Design a device that will indicate when it is necessary to change the oil in an automobile.

23. Devise a method that will prevent pigeons from defacing the facade of buildings.

24. Trains lose time and waste energy when they stop at stations. Devise a scheme that would allow passengers to get on and off trains without the train having to slow down and stop at stations.

25. Design a system that could be used to desalinate water on life rafts.

26. Design a mechanical device for removing weeds.

27. Design a scheme for timing automatically the finishers in a marathon road race with 6500 entries. The scheme should provide the time of each finisher and display the order of finishers immediately after the end of the race.

28. In tournament tennis matches, eight persons are required to judge whether or not the ball was "in" or "out." Even then, calls are frequently disputed. Devise a system that would reduce the number of linesmen and, at the same time, improve the accuracy of the calls.

29. Design a device for locking up bicycles securely. The device should be lightweight and inexpensive.

30. It is desired to propel a boat quietly with human power. Oars, propellers, and paddlewheels are too noisy and cannot be used. Develop a method enabling one to propel a boat without these mechanisms.

31. Design a system to cool warm beer and soft drinks on a picnic when no electricity is available. The system should be small, lightweight, and inexpensive.

32. Design a system to better handle traffic flow into and out of Manhattan during rush hour.

33. Design a noise suppression barrier for a residential neighborhood surrounded by freeways.

REFERENCES

Problem Solving

1. Edward Krick, *An Introduction to Engineering: Concepts Methods, and Issues* (Wiley, New York, 1976).
2. Moshe F. Rubinstein and Kenneth Pfeiffer, *Concepts in Problem Solving* (Prentice-Hall, Englewood Cliffs, N.J., 1980).

Engineering Design

1. Morris Asimow, *Introduction to Design* (Prentice-Hall, Englewood Cliffs, N.J., 1962).
2. G. C. Beakley and E. G. Chilton, *Design: Serving the Needs of Man* (Macmillan, New York, 1974).
3. Percy H. Hill, *The Science of Engineering Design* (Holt, Rinehart and Winston, New York, 1970).

Modeling

1. Edward Krick, *An Introduction to Engineering: Concepts, Methods, and Issues* (Wiley, New York, 1976).
2. Russel L. Ackoff, *Scientific Method* (Wiley, New York, 1962).
3. D. W. VerPlanck and B. R. Teare, *Engineering Analysis* (Wiley, New York, 1954).
4. Marshal Walker, *The Nature of Scientific Thought* (Prentice-Hall, Englewood Cliffs, N.J., 1963).

Other References on Problem Solving and Design

1. Moshe F. Rubinstein, *Patterns of Problem Solving* (Prentice-Hall, Englewood Cliffs, N.J. 1975).
2. James L. Adams, *Conceptual Blockbusting: A Guide to Better Ideas* (Freeman, San Francisco, 1974).
3. Gary A. Davis, *Psychology of Problem Solving: Theory and Practice* (Basic Books, New York, 1973).
4. G. Polya, *How to Solve It* (Doubleday, New York, 1957).
5. Wayne Wikelgren, *How to Solve Problems: Elements of a Theory of Problems and Problem Solving* (Freeman, San Francisco, 1974).
6. J. Alger and C. V. Hays, *Creative Synthesis in Design* (Prentice-Hall, Englewood Cliffs, N.J., 1964).

7. A. D. Moore, *Invention, Discovery, and Creativity* (Doubleday Anchor Books, Garden City, N.J., 1969).

8. I. G. Wilson and M. E. Wilson, *From Idea to Working Model* (Wiley, New York, 1970).

9. John R. Dixon, *Design Engineering: Inventiveness, Analysis, and Decision Making* (McGraw-Hill, New York, 1966).

Other References of Interest

The Beńard fluid instability mentioned in the example on page 72 is discussed in detail in:

Manual G. Velarde and Christiane Normand, "Convection," *Scientific American* 243 (July, 1980), pp. 92–108.

PART III
The Tools

The engineer must approach the complex problems of modern society with skill and determination. The challenges of modern engineering require a careful, disciplined approach to problem solving. They also require a thorough mastery of the basic principles of mathematics and science, which comprise the *tools* of engineering.

Mathematics is the language of science and engineering. It provides the compactness and accuracy of expression so essential in the definition, abstraction, and analysis of complex problems. The rules of mathematics also help to structure a logical approach to problem solving.

The tools of engineering analysis are the scientific principles that describe the particular phenomena of interest. The engineer should acquire a solid grounding in basic sciences such as physics, chemistry, and biology, as well as in applied sciences such as thermodynamics, solid and fluid mechanics, and electrical processes.

Various methods can be employed to analyze engineering problems. Modern digital computing methods have come to play a particularly important role, both in analysis and design. Experimental and testing procedures are of considerable use in bridging the gap between concept and application. The final instruments of importance to the engineer are skills in written, spoken, and graphical communication. Even the most thorough and elegant solution or design is useless unless clearly presented to those who must implement it.

These are the principal tools of engineers, their unique set of skills. This knowledge is also the focus of the engineering curriculum. Most of these subjects are addressed in detail by the many courses taken during an engineer's formal education.

It is appropriate to survey these topics at a very early stage so that the engineering student can better appreciate the importance and interrelationship of later course work. This discussion is also intended to convey a taste of engineering by providing some experience in applying scientific principles to the solution of real and complex problems.

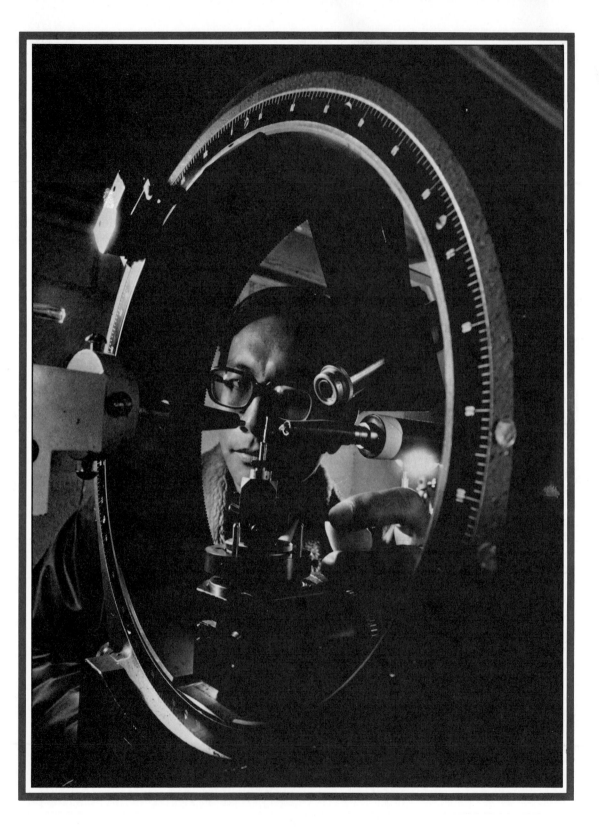

CHAPTER 3

The Tools of Mathematics and Science

The most powerful tools of the engineer are based on fundamental concepts from mathematics and science. Courses in these disciplines form the cornerstone of an engineering education.

The language of engineering analysis is mathematics. This discipline provides a compactness and accuracy of expression that proves essential in the definition and analysis of complex problems. The rules of mathematics also help to structure a logical approach to solving problems.

Of equal importance in engineering problem solving are the tools provided by the basic sciences. In many cases, concepts from the fundamental sciences such as physics, chemistry, and biology can be applied directly in engineering practice. To them engineers add the tools provided by applied scientific disciplines such as solid and fluid mechanics, thermodynamics, and electrical and materials sciences. These subjects have strong roots in basic science and mathematics. They provide the engineer with the powerful analytical methods necessary for engineering practice. But the concepts of mathematics and science are not easily mastered, and the formal curriculum requirements of an engineering program frequently resemble a complex maze of interrelated courses, many of which build upon one another as prerequisites. All too often engineering students become so bewildered and confused by this array of subjects in mathematics and science, this educational obstacle course, that they lose sight of the objectives of their education; they miss the forest for the trees.

Hence the purpose of this present chapter is to survey those subjects in mathematics and science that form the core of an engineering education. It is certainly not our intent to provide sufficient detail on any particular topic to duplicate other courses in these subjects. Rather we shall attempt to provide students with a general perspective of the role these courses will play in their engineering education, to show how the pieces of this puzzle fit together and, hopefully, how these courses will prove relevant to their future professional objectives as engineers.

3.1.   MATHEMATICAL ANALYSIS

Mathematics has become the language of modern science and engineering. Our traditional languages are simply not adequate to convey the precise concepts of science nor to implement the disciplined and rational approach to problem solving so characteristic of the scientific method. The symbolism of mathematics provides a compactness and accuracy of expression far beyond that of conversational language. The rules of mathematics also aid tremendously in structuring a logical approach to problem solving.

Instruction in mathematics plays an important role in the education of the engineer. Not only does mathematics provide a quantitative basis for engineering analysis and computation, but it also provides the foundation for a clear understanding of engineering phenomena.

Of course most engineers apply in professional practice only a small fraction of the mathematical concepts they so laboriously master during their formal education. In fact, complaints about the "excessive" stress placed on instruction in rigorous mathematics are common among engineering students and practicing engineers alike. Yet one should never underestimate the importance to the engineer of a thorough grounding in mathematics. Those who shy away from mathematics run the very real risk of becoming functionally illiterate. They would neither be able to communicate ideas in a coherent fashion nor to understand the activities of colleagues. Indeed, it is this mastery of mathematics and science that distinguishes the engineer from the engineering technologist or technician. It pro-

FIGURE 3.1.   Engineers make frequent use of clever mathematical methods in problem solving. (© 1977 by Sidney Harris/*American Scientist* Magazine)

vides the depth of understanding and versatility required to tackle the very complex problems that face modern society.

Mathematics is used in engineering analysis first as a concise language in which to express the physical laws governing the engineering process or system of interest. We first write the mathematical equations that represent the relevant physical laws. Then the standard tools of mathematical analysis are used to manipulate, simplify, and eventually solve these equations for the quantities of interest (Table 3.1). Provided that the mathematics has been performed correctly, the solution will be a direct consequence of the physical laws that we assume to govern the problem. In this sense, mathematics not only streamlines the logic of engineering problem solving, but it also allows us to solve problems far too complicated to be attempted with mental gymnastics alone.

Example   Suppose we wished to use mathematical analysis to study the trajectory of an unmanned deep space probe to be launched to the outer planets. We would first write the appropriate equations characterizing Newton's laws of motion as they apply to the probe, including the gravitational pull of the sun and the planets. In a simplified analysis we might obtain an approximate solution to these equations using analytical (i.e., pencil and paper) methods that would predict gross aspects of the probe trajectory. More precise predictions would require an accurate numerical solution of the equations of motion on a large digital computer. Provided that we had adequately accounted for all of the forces on the spacecraft, the resulting orbit prediction would be of acceptable accuracy for mission planning purposes.

Once a solution to the mathematical representation of the problem has been obtained, the engineer must then translate this solution back into physical terms. The solution should be checked very carefully for errors in mathematics or logic. Even the most trivial error in arithmetic, a misplaced decimal point or sign error, can have disastrous effects in engineering design. A particularly common source

TABLE 3.1   Methods of Mathematical Analysis

Analytical: Using "hand" methods such as pencil, paper, and hand calculator to write and solve mathematical equations representing problem.

Numerical: Rearranging mathematical equations into a form suitable for solution on a digital computer.

Statistical: Simulating the problem of interest using random sampling techniques (so-called Monte Carlo methods) rather than attempting to solve the relevant equations directly.

of error arises from flaws in the logic used to formulate the mathematical description of the problem.

The selection of the relevant physical laws and the assumptions necessary to simplify the mathematics describing them is a critical aspect of engineering problem solving. All too frequently engineers hurry to write down the equations characterizing the problem without giving adequate thought to its original formulation. And once the analysis of these equations is set into motion, there is a natural tendency to become so obsessed with mathematical manipulations that further consideration of the original problem statement ceases.

Although mathematics presents the engineer with a powerful tool for making logical deductions from a given set of facts, it cannot provide the basic scientific principles governing a problem. Nor can mathematics distinguish between correct and incorrect formulations of the problem. For the mathematics to yield useful results, careful attention must be given to the definition of the problem and the assumptions that are necessary to arrive at a suitable mathematical representation. For this reason engineers are frequently encouraged to carry the verbal analysis of a problem as far as possible before resorting to mathematics. They should avoid jumping directly from a general problem statement to a mathematical formulation. Rather they should attempt to first arrive at a concise and accurate word statement of the problem of interest.

Engineering "computation," that is, mathematical analysis, involves more than the mere manipulation of numbers. It also includes the proper use of abstract models and the formulation of equations, the precision of the data and numbers used and the degree of accuracy of the answers, and the units of measurement of the physical quantities appearing in the analysis.

SUMMARY

Mathematics serves as both the language of engineering and the logical framework for engineering problem solving. The mathematical analysis of an engineering problem can be outlined as follows: (1) Determine a concise word statement of the problem of interest. (2) Translate this statement into mathematical form by writing the mathematical equations that represent the physical laws embodied in the statement. (3) Introduce simplifying assumptions and approximations to reduce the complexity of this mathematical description. (4) Apply the tools of mathematical analysis to solve the equations. (5) Check both the solution and the simplifying approximations. (6) Interpret the physical significance of the solution.

Example   A key parameter in spacecraft missions is the thrust or force that must be generated by the craft's engines. Recalling that force is proportional to acceleration (Newton's second law), we can calculate the minimum acceleration (and therefore thrust) required to propel a spacecraft from Earth to Saturn over a trajectory 2 billion kilometers in length in a time of 3 years. Let us follow the procedure outlined above:

1. The problem statement would be: "Determine the minimum acceleration necessary to propel a spacecraft a distance of 2 billion kilometers in 3 years."

2. The relevant physical law relates the acceleration to the distance traveled and the time by the equation

$$\text{distance} = \tfrac{1}{2}\,(\text{acceleration}) \times (\text{time})^2$$

or in an obvious symbolic notation

$$d = \tfrac{1}{2}\,at^2$$

3. We assume a constant acceleration a for the first half of the journey, followed by a constant deceleration of the same magnitude over the second half.

4. We must solve the equation above for the acceleration

$$a = 2\frac{d}{t^2}$$

where we take $d = \tfrac{1}{2}\,(2\text{ billion km})$ and $t = \tfrac{1}{2}\,(3\text{ years})$. If we substitute these values into our result and modify our units, we find

$$a = \frac{2[\tfrac{1}{2}\,(2{,}000{,}000{,}000\text{ km})(1{,}000\text{ m/km})]}{[\tfrac{1}{2}\,(3\text{ y})(31{,}536{,}000\text{ s/y})]^2}$$
$$= 0.0089\text{ m/s}^2$$

5. It is relatively easy to verify the algebra and the conversion of units in this simple calculation.

6. The implications of this result are rather interesting. The acceleration required is extremely low, for example, compared to the acceleration of gravity (9.8 m/s²) or the acceleration typical of automobiles (1 to 3 m/s²). This implies that such deep space missions can be accomplished with long duration, low thrust engines (in contrast to the short duration, high thrust engines required for near-planet operations).

This is a primary motivation behind the development of deep space propulsion systems with low thrust such as the ion drive.

3.1.1. Mathematical Symbols

Two stages can be identified in the mathematical description of an engineering problem. We first must assign mathematical symbols to each of the physical quantities involved in the process. Then the laws of science are used to express the proper relationship between these symbols in the form of equations.

There is nothing mysterious about mathematical symbolism. It is just the shorthand notation of the engineer. We can assign any symbols to characterize physical quantities, as long as we define them and use them in a consistent fashion. Fortunately, most mathematicians, scientists, and engineers attempt to employ standard notational conventions that can be easily recognized. For example, the common symbols for algebraic operations, $+$, $-$, $\times$, and $\div$, are familiar to all of us. There are also well-established conventions to represent most of the quantities met in engineering practice, such as the use of the letter t for time, m for mass, T for temperature, and so on. Since there are usually not enough Roman letters for all the variables needed, Greek letters, such as those listed in Figure 3.2., are also used. The student may already be familiar with Greek letters such as π, θ, and ϕ from their use in geometry. If even more symbols are needed, we can introduce subscripts or superscripts to distinguish among different quantities such as temperatures, T_1, T_2, and T_3, or times $t^{(1)}$, $t^{(2)}$, and $t^{(3)}$.

FIGURE 3.2. The Greek alphabet is commonly used for mathematical symbols. (Reprinted with permission from *A Guide for Wiley-Interscience Authors*, Wiley, New York, 1974)

Alpha	A	α α	Nu	N	ν
Beta	B	β	Xi	Ξ	ξ
Gamma	Γ	γ	Omicron	O	o
Delta	Δ	δ ∂	Pi	Π	π
Epsilon	E	ϵ ε ε	Rho	P	ρ
Zeta	Z	ζ	Sigma	Σ	σ s
Eta	H	η	Tau	T	τ
Theta	Θ	θ ϑ	Upsilon	Υ	υ
Iota	I	ι	Phi	Φ	ϕ φ
Kappa	K	κ	Chi	X	χ
Lambda	Λ	λ	Psi	Ψ	ψ
Mu	M	μ	Omega	Ω	ω

Example   Although assigning mathematical symbols to represent physical quantities is to some extent arbitrary, each field of engineering has developed a mathematical "dialect." For example, mechanical engineers denote force by F, mass by m, velocity by v, acceleration by a, temperature by T or θ, and heat flow rate by q. Electrical engineers write voltage as V, current as I, resistance as R, capacitance as C, and inductance by L.

A variety of mathematical symbols can be used to represent operations performed upon physical quantities. For example, the addition of two temperatures to yield a third temperature might be written symbolically as

$$T_1 + T_2 = T_3$$

This, of course, is an example of an *equation* in which the symbol "=" has been used to indicate equality. However there are other mathematical symbols that do not express equality but can also be used to construct equations. Examples include the symbols for "inequality" ($T_1 \neq T_2$), "greater than" and "less than" ($T_1 > T_2$ and $T_1 < T_2$), "much greater than" and "much less than" ($T_1 >> T_2$ and $T_1 << T_2$), and "approximately equal to" ($T_1 \cong T_2$). In addition, mathematical symbols represent common terms such as "implies" ($\Rightarrow$), "therefore" ($\therefore$), "and so on" (. . .). In Table 3.2 we have listed several of the more useful mathematical symbols encountered in engineering applications.

SUMMARY

In mathematical analysis engineers adopt a shorthand notation in which quantities of interest are represented by letters (Roman or Greek), occasionally augmented by subscripts or superscripts. A variety of specific symbols represent mathematical relationships among quantities such as inequalities, mathematical operations, or other shorthand notation.

Example   Engineers occasionally lapse into mathematical symbolism to represent a prose statement. Although this tendency is not as flagrant as it is in pure mathematics, the engineering student should become adept at deciphering this symbolism. A couple of examples illustrate this practice:

TABLE 3.2   Some common mathematical symbols

SYMBOL	MEANING	EXAMPLE
$\pm$	plus or minus	2 ± 3
$\mp$	minus or plus	2 ∓ 3
$=$	equals	$2 + 3 = 5$
$\neq$	is not equal to	$2 + 3 \neq 4$
$\cong$	equals approximately	$100 \cong 101$
$\equiv$	identical with, defined as	$a \equiv 50$
$>$	is greater than	$4 > 3$
$<$	is less than	$4 < 5$
$\geq$	is greater than or equal	$a \geq b$
$\leq$	is less than or equal	$b \leq c$
$\sim$	equivalent, similar	$a \sim b$
$\Rightarrow$	implies	
$\therefore$	therefore	
$\because$	because	
∞	infinity	
$n!$	factorial	$3! = 1 \cdot 2 \cdot 3 = 6$
$\sum_n f_n$	summation	$f_1 + f_2 + \ldots + f_n + \ldots$
$\lvert a \rvert$	absolute value or magnitude	$\lvert a \rvert = a$ if $a > 0$ $= -a$ if $a < 0$
$\ldots$	and so on	$1 + 2 + 3 + \ldots$
ε	is a member of	$a \, \varepsilon \, A$

1. $T_1 > T_2 > T_3 > \cdots > T_n \Rightarrow T_i > T_j$ if $i > j$
(If a sequence of temperatures is progressively decreasing, then any earlier temperature will be greater than a later temperature in the sequence.)

2. $\alpha_1 \gg \alpha_2$, $\therefore \dfrac{\alpha_1}{\alpha_1 + \alpha_2} \cong 1$
(The quantity α_1 is much greater than α_2. Therefore the expression is approximately equal to one.)

3.1.2.   Basic Mathematical Concepts

The engineering student is expected to master a large arsenal of mathematical concepts useful for the analysis of engineering problems. At the beginning of their studies engineering students typically have a background in algebra, geometry, and trigonometry. They should be relatively adept at simple algebraic manipulations such as solving algebraic equations or determining the roots of quadratic polynomials. Some exposure to both plane and solid geometry is useful. Trigonometry also plays an important role in engineering, although some students postpone the study of this subject until their first year of college. Some entering students will have acquired a familiarity with more abstract topics such as sets, the real number system, and logic. Others will have already had some exposure to calculus. Still other beginning students will have mastered a computer programming language such as BASIC, FORTRAN, or Pascal (see Table. 3.3).

The mathematical backgrounds of entering students are often quite diverse. Indeed, one of the purposes of the early years of the engineering curriculum is to strengthen the student's skills in both basic and advanced mathematics and to fill in any gaps present in their mathematics background before they study more advanced subjects. Below we briefly survey topics that form the basis for the mathematics curriculum of an engineering program.

Analytic Geometry

Equations or numbers can sometimes be a rather awkward way to present engineering information. For example, it would be rather confusing to represent the route of a ship as it moves from one port to another only in terms of the numbers representing its longitude and latitude at any instant of time. Instead we would draw a map. Then we could trace out the ship's route as a path on this map.

Maps are just one example of the use of *analytic geometry* to develop a relationship between numbers, equations, and geometry. Analytic geometry couples algebra and geometry by representing each point in a plane by a pair of numbers. These numbers correspond to the respective positions on the scales of perpendicular axes. The horizontal axis is known as the *abscissa*, and the vertical axis is the *ordinate*. This construction is called a *Cartesian* or *rectangular coordinate system*. For example, we could use analytic geometry to represent the relationship between the Celsius and Fahrenheit temperature scales, by assigning the

TABLE 3.3 The Mathematics Background of the Student Entering Engineering

Usually required:	Algebra (1.5 years)
	Plane geometry (1 year)
	Trigonometry (0.5 year)
Desirable:	Advanced algebra (0.5 year)
	Analytical geometry (0.5 year)
Optional:	Set theory
	Calculus
	Computer programming (BASIC, FORTRAN, Pascal)

Fahrenheit temperatures to the horizontal or x-axis (abscissa) and the corresponding Celsius temperature to the vertical or y-axis (ordinate) as shown in Figure 3.3.

Functions

Many quantities in engineering cannot be specified by a single number but will instead assume a range of values. For example, the temperature in a room could be 23°C, or 20°C, or 30°C, and so on. We refer to such quantities as *variables* and usually represent them by a letter symbol such as x or y or T. Frequently the values assumed by certain physical quantities or variables are related. That is, the values of one quantity depend on those of another. The pressure of the atmosphere depends on the altitude above sea level. The rate at which a chemical reaction proceeds depends on its temperature. We can characterize such relationships, such dependence between variables, in terms of the concept of a *function*. When we say that one quantity or variable is a function of another, we mean that a knowledge of the second quantity (say, time) is sufficient to imply the value of the first (temperature). If we denote the first variable by x (or t) and the second by y (or T), we could denote this functional dependence symbolically by writing $y = f(x)$. For our earlier example of temperature as a function of time, we could write $T(t)$.

Analytic geometry can be used to plot a *graph* of a function by assigning to each pair of numbers, x and $y = f(x)$, a point in the xy plane (Figure 3.4). For example, let us represent the functional relationship between the Celsius and Fahrenheit temperature scales by

$$C = \tfrac{3}{5} (F - 32)$$

As in Figure 3.3 we assign F to the horizontal axis (the x-axis) and C to the vertical axis (the y-axis).

One particularly important class of functions in engineering is the trigometric functions, such as $\sin x$, $\cos x$, and $\tan x$. These functions prove very useful in geometric applications such as surveying or structural analysis. They can also be used to describe periodic phenomena, that is, processes that repeat at regular intervals such as a rotating wheel or a vibrating string. Another useful class of functions in engineering is the exponential functions e^x (where e is an irrational number $e = 2.71828 \ldots$) and the logarithmic functions $\log x = \log_{10} x$ and $\ln x = \log_e x$. These functions arise quite frequently in the mathematical description of dynamic processes.

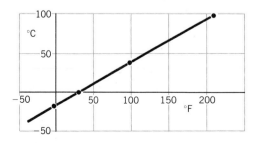

FIGURE 3.3. Analytic geometry can be used to represent the relationship between the Fahrenheit and Centrigrade temperature scales.

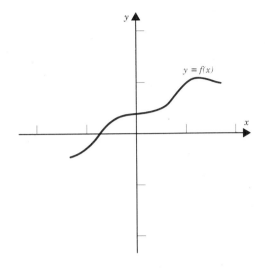

FIGURE 3.4. The graph of a function $f(x)$ versus x in which the abscissa is the variable x and the ordinate is the function $y = f(x)$.

Calculus

Much of engineering analysis involves dynamic processes in which physical quantities are continually changing. The position of a moving automobile or the strength of a signal received from a receding space probe are examples of time-varying quantities. The mathematics appropriate for describing changing quantities was developed some three centuries ago by Newton and Leibnitz and is known as *calculus*.

The study of calculus dominates the formal mathematical training of engineering students during their college education. Two aspects of this subject receive particular attention: differential calculus, which is concerned with determining the rate at which variable quantities change, and integral calculus, which is used to determine a function when only its rate of change is given. Closely associated with these aspects are the concepts of the derivative and the integral of a function. For example, if we know the velocity of an object as a function of time, $v(t)$, then we can calculate its acceleration, $a(t)$, as the time rate of change or *derivative* of the velocity function with respect to time:

$$a(t) = \frac{dv}{dt}$$

We could also calculate the displacement of the object as a function of time, $x(t)$, by calculating the *integral* of the velocity function

$$x(t) = \int v(t)\, dt$$

The derivative and integral of a function also have a geometric interpretation (Figure 3.5). The derivative can be identified as the slope of the graph of a function. The integral of the function can be interpreted as the area under this graph (Figure 3.6).

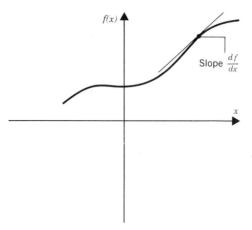

FIGURE 3.5. The slope of a function at a point is given by its derivative.

Since many engineering phenomena involve dynamic processes, it is common to encounter equations in which derivatives occur. These are known as *differential equations*. The study and solution of differential equations is one of the most important mathematical topics to be mastered by the engineering student.

Vectors

Thus far we have concerned ourselves with quantities or variables that can be characterized by magnitude alone. For example, the temperature of a room can be given in terms of the number of degrees; the volume of a box can be given in terms of the number of cubic meters it contains. Such quantities that can be completely specified by magnitude only are called *scalar* quantities, or more simply, *scalars*.

In engineering, we frequently encounter quantities that require the specification of both magnitude and direction. For example, the velocity of an airplane is fully specified not only by its magnitude or speed (say, 100 m/s), but also by its direction of flight (horizontal and north by northwest). Similarly, a force is characterized both by its strength and its direction of application.

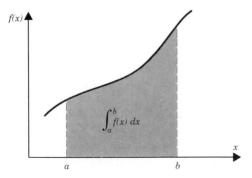

FIGURE 3.6. The integral of a function is equal to the area under its graph.

The mathematical study of processes that occur in physical space, such as the flight of an airplane or the application of a force, is greatly facilitated by the introduction of the concept of a *vector*. Vectors are simply mathematical quantities incorporating both a magnitude and a direction. In fact, vectors are most conveniently represented by arrows pointing in the direction of interest whose lengths are proportional to the magnitude of the vector. For example, a vector representing a force is drawn as an arrow in the direction of the applied force with a length equal to the strength of the force (Figure 3.7).

Many of the usual algebraic operations that are familiar from working with scalar variables can be extended to vectors. The concept of vector addition can be used to calculate the total or resultant force exerted on an object due to several applied forces. There are also operations between vectors analogous to multiplication (the dot product and cross product) that play an important role in many engineering applications. Vector analysis is an important topic in the undergraduate engineering curriculum for all fields.

Numerical Analysis

Most of the equations that are required to accurately describe an engineering system are far too complicated to yield to the "analytical" (i.e., pencil, paper, and sweat) methods in the mathematician's bag of tricks. Engineers are usually forced to introduce approximations to simplify the mathematics. Unfortunately, many problems are so complex that if we were to simplify them sufficiently to allow an analytical solution we would mutilate the formulation of the original problem so badly that our analysis would lose all touch with reality. In these problems our only recourse is the digital computer. Often it is only the enormous computational resources offered by modern digital computers that enables engineers to construct and analyze mathematical models that adequately describe complex engineering systems.

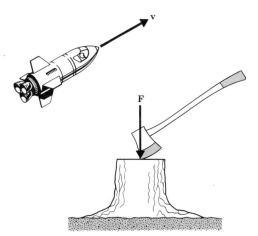

FIGURE 3.7. The velocity of a rocket and the force of an axe are examples of vector quantities.

FIGURE 3.8. The personal computer has become an important tool in mathematical analysis. (Courtesy Hewlett-Packard Corporation)

As we have noted, most engineering problems deal with changing quantities and therefore with calculus. Unfortunately, digital computers are terrible at calculus, that is, in handling derivatives or integrals. Their real talent lies rather in solving very large systems of algebraic equations. In fact, they must usually solve even these algebraic equations by using a sequence of numerical operations (additions or subtractions) rather than performing the straightforward manipulation of algebraic symbols that engineers might attempt. Therefore engineers must first recast the equations describing the system of interest into a form suitable for solution on a digital computer. They must then determine a sequence of numerical operations the computer can use to solve these equations. The sequence of mathematical operations needed to solve the equations is an example of an *algorithm*, a rule or sequence of steps for solving a problem. The particular approach taken by the computer consists of manipulations of numbers (as opposed to algebraic symbols), hence the name *numerical analysis*.

The task of casting equations into a form suitable for solution on a digital computer is typically accomplished by "discretizing" each of the independent and dependent variables in the mathematical description of the physical process of interest. We replace functions of continuous variables by a discrete set of values defined at a discrete set of points. The derivatives and integrals appearing in these equations must also be replaced by corresponding discrete representations (usually approximate formulas). In this way we arrive at a set of algebraic equations for the discrete representation suitable for solution on a digital computer.

Probability and Statistics

The mathematical concepts discussed thus far are appropriate for deterministic systems. By this we mean the physical system with which we are dealing can be described, at least in principle, with complete certainty. The results derived from a mathematical analysis are exact quantities that can be obtained to as great a degree of accuracy as desired. For example, the path of an ideal billiard ball can be predicted exactly if we know sufficient detail about the trajectory of the cueball, its spin, the friction of the table, and other relevant physical properties.

Not all systems in engineering are subject to such an exact analysis. In *stochastic systems* there is an element of randomness or chance that leads to results that cannot always be precisely predicted in advance. For example, it is not possible to say exactly when a machine part will fail, even though we may have excellent data on the average failure rate of such parts from previous experience.

Therefore a vital aspect of engineering analysis is the study of random phenomena; this is done by applying concepts from the mathematical subjects of probability and statistics. In such analyses we do not attempt to determine the precise behavior of a system but rather the expected or average behavior in a

FIGURE 3.9.   Methods of probability and statistics play important roles in engineering. (© 1979 by Sidney Harris/*American Scientist* Magazine)

statistical sense. It is important that the engineering student develop some familiarity with and appreciation for the methods of probability and statistics.

The concept of probability or likelihood occurs frequently in everyday life. We often make decisions among several alternative actions based on our estimate of the probability that each will lead to the desired result. In games of chance, for example, any serious gambler will carefully weigh the risks and probabilities of success in choosing a strategy. In engineering some of the same concepts can apply whether planning the proper marketing strategy for a product or deciding among different approaches to a problem in research or development. The concepts of probability also extend into the field of statistics, where the behavior of large groups or populations (of objects or people) can be analyzed using the probabilities governing the behavior of each individual.

Simple concepts from probability theory are introduced quite early in an engineering education. The likelihood of occurrence of any random event can be characterized by a *probability p* ranging between values of zero (impossibility of occurrence) and one (certainty of occurrence) (Figure 3.10). Such probabilities characterizing events can be measured by recording the fraction of the time the event occurs in a great many observations.

Since the overall process of interest may be made up of a combination of possible events, it is important to know how to combine the probabilities for individual events. For example, the probability that all of a number of independent events (independent in the sense that the occurrence of one event will not influence the occurrence of another event) is obtained by multiplying together their individual probabilities. The probability that any of a number of mutually exclusive events occurs is obtained by adding the individual probabilities. These simple rules can be applied to analyze many of the common probability problems in engineering.

The elementary concepts of probability are useful in the general field of data analysis known as *statistics*. This term refers to techniques and analytical methods useful in the characterizing and interpreting data in which there is some degree of chance or random variability.

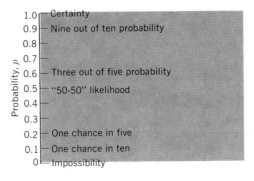

FIGURE 3.10.   The probability p characterizing the occurrence of an event ranges from a value of one (certainty) to zero (impossibility).

Statistical analysis can be used to catagorize data or observed behavior to facilitate its interpretation. It can also be used to predict the behavior of a group of elements (e.g., people, machines, playing cards), provided we know enough about the expected behavior of a typical member of that group. In Chapter 5 we shall consider the application of statistical methods in experiments and testing.

A particularly important application of probability and statistics is the use of random sampling methods to simulate the behavior of complex systems. These techniques, referred to collectively as *Monte Carlo methods*, are based on using a computer to simulate an actual physical process or event using random sampling techniques. Each event is assumed to be characterized by known probabilities. To simulate the occurrence of an event, the computer takes random samples from these probabilities to create a representation of an actual sample that might be observed in a real experiment. By repeating this process many times, a large number of random samples can be simulated, and the overall behavior of the process or system can be predicted from the statistical analysis of all of the samples. Such Monte Carlo simulations or computer "experiments" are commonly used to simulate complex processes ranging from the penetration of radiation to the results of political elections.

SUMMARY

Engineering students typically begin their studies with a background in algebra, geometry, and trigonometry. The mathematics curriculum of an engineering program adds to this background a variety of additional mathematical concepts: (1) Analytic geometry: the coupling of algebra and geometry. (2) Functions: mathematical relationships between variables. (3) Calculus: the mathematics used to describe changing quantities. (4) Vectors: quantities characterized by both magnitude and direction. (5) Numerical methods: methods used to cast a set of equations into a form suitable for solution by digital computer. (6) Probability and statistics: mathematical methods for studying phenomena in which there is an element of randomness or chance.

3.1.3. Equations

A key step in the mathematical analysis of an engineering problem is the expression of the physical laws describing a system in terms of equations. The engineer then attempts to solve these equations to obtain information about the system.

Mathematical *equations* are essentially just a symbolic representation, a model, of the physical laws known to govern the process we are studying. However the use of such mathematical symbolism or modeling is extremely powerful since it allows the straightforward application of mathematical logic to determine the quantities of interest and information about the system.

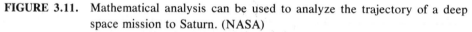

Example   Let us return to our design of a space probe to orbit one of the outer planets such as Saturn (Figure 3.11). We might first write the equations governing the gravitational forces on the probe as it travels through space. By solving these equations, we find the necessary earth escape velocity and therefore the thrust required to boost the probe out of earth orbit and on its journey. In this case the variables under our control might include the probe characteristics (particularly its weight), the intended time of launch, and rough characteristics of the trajectory (e.g., whether we wish a ''grand tour'' of intermediate planets along the way). The unknown variables might then include the required thrust, the allowable scientific payload, and the maneuvering capability (or requirements).

There are very few general rules to help us derive the equations characteriz-

FIGURE 3.11.   Mathematical analysis can be used to analyze the trajectory of a deep space mission to Saturn. (NASA)

ing an engineering problem. The general approach is to assign symbols to represent the variables of interest and then set up the proper logical relationships, the equations, between the known and unknown quantities.

Derivation of Equations

The derivation of the relevant equations is usually the key step in the mathematical analysis of an engineering problem. Indeed, the engineer will sometimes find it more valuable to be able to derive the basic equations characterizing the problem than to be able to solve them. Even without the solution, the equations may reveal quite a bit of information about the problem. Fortunately, modern computational techniques and digital computers have made possible the solution of even very complex equations.

It is best to begin by making as precise a verbal statement as possible of the various scientific principles that apply to the particular situation of interest. A sketch or diagram showing the relationship between the variables that characterize the system is frequently useful. In our earlier example of a space probe, we would sketch the positions of the sun and the planets to guide us in setting up equations representing the gravitational forces on the probe.

The mathematical representation usually takes the form of a set of equations. Therefore we must determine the quantities of interest in the problem and choose symbols to represent each of these variables. Frequently, we find it necessary to introduce additional variables to facilitate the mathematical analysis, even though these variables are not of direct interest and will not appear in the answer. Engineers should be wary of attempting to employ shortcuts (i.e., eliminating variables in their heads) since these can lead to mistakes that are quite difficult to detect. Throughout the analysis the safest approach is always to write out the equations and manipulations in detail.

Simplifying Assumptions

In most cases of practical interest the complete set of equations describing a system are too complicated to be solved, even on the largest computer. Therefore a premium is placed on simplifying the problem by introducing various assumptions and approximations to facilitate mathematical analysis. These assumptions also eliminate unnecessary effort expended on calculating quantities that are not of interest to the problem at hand.

The introduction of simplifying assumptions is one of the most important aspects of engineering analysis. If the assumptions are overly restrictive, they may invalidate the solution. For example, if we ignore the perturbations of the gravitational forces of the planets relative to that of the sun, we would drastically simplify the equations characterizing the trajectory of the space probe (Figure 3.12). (Indeed, these would be the equations solved by Kepler and Newton over three centuries ago.) Unfortunately, these equations would also fail quite badly in predicting the detailed trajectory of the space probe journey to the outer planets.

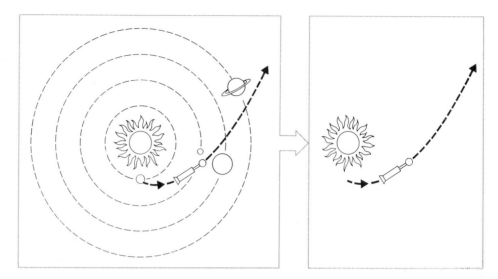

FIGURE 3.12.   We might ignore the forces exerted by the planets in describing the space-craft trajectory.

Engineers should always state carefully all assumptions introduced into the analysis. After the mathematical solution they should return to check whether the assumptions remain valid in light of these results. Perhaps the influence of several of the inner planets on the probe trajectory is negligible, or perhaps nonuniformities in the gravitational fields of large planets such as Jupiter cannot be ignored.

It is frequently of use to first solve a very simplified (perhaps intentionally oversimplified) problem in the hope of learning enough from the simplified case so that this solution may be extended to the more complicated situation of real interest. We might initially include only the effects of the Earth and the larger outer planets to determine the rough features of our spacecraft's trajectory. Then other forces on the probe could be introduced as small perturbations on this solution.

We frequently have considerable latitude in choosing the physical principles that describe the process of interest and serve as the basis for the mathematical representation of the problem. Experience is an invaluable aid in the appropriate choice of both the fundamental principles involved and suitable assumptions or approximations.

Methods of Solution

Although the most important aspects of engineering problem solving involve the problem definition and the derivation of the mathematical equations representing the problem, frequently the most laborsome task involves the solution of the equations themselves. Here one brings into play the enormous variety of analytical and numerical methods developed by the mathematician (Table 3.4).

**TABLE 3.4   Methods for
                     Solving Equations**

---

1. Try to solve by inspection.
2. Look up the answer.
3. Approximate equations or solutions.
4. Detailed analytical solution.
5. Computer solution.

Usually the first approach will be to attempt to solve the equations "by inspection." Sometimes by staring at the equations for awhile, you can spot their solution or recall an easy method to solve them. For simple problems, you might even be able to write down this solution directly.

If a simple solution is not obvious, the next approach might be to look up the answer in a book. Most of the common equations that arise in engineering problems have already been solved by mathematicians or other engineers. The problem is knowing where to look for these solutions and recognizing that your equations are of a form similar to other equations that already have been solved. Here the engineer can take advantage of the fact that the equations describing many quite different physical processes are identical. For example, the same type of equations characterizes the vibration of a spring, the electric current in a circuit, or the flow of fluid around objects. Hence the engineer may seek out these analogies to find solved equations of a form very similar to that of a particular application of interest.

A third approach is to reduce the complexity of the equations of interest by suitable approximations or assumptions until they are amenable to mathematical analysis and solution. Large classes of equations can be solved by the straightforward methods treated in undergraduate level mathematics courses. Unfortunately, the solution of the equations sometimes requires such brutal approximations that they no longer adequately describe the situation of interest. Then the engineer has only one recourse: to turn to the power of the digital computer and solve the equations directly by means of numerical analysis. Even if it is necessary to resort to a digital computer, an entirely new computer program is rarely required. A number of general computer programs exist that are capable of solving a wide variety of complex equations. Furthermore, it is becoming more common for engineers and scientists to develop specialized computer programs or "codes" that are capable of solving the specific types of equations relevant in particular fields. Examples include the stress analysis codes used in mechanical engineering, chemical reaction rate analysis and process control codes used in chemical engineering, and fluid dynamics codes used in the study of hydraulic systems. There are computer code "libraries" (both public and commercial) throughout the world that will supply computer programs (so-called computer "software") for a wide variety of applications. Because of their major importance to modern engineering analysis and design, computer-based methods are the topics of Chapters 4 and 7.

Checking

Once a solution to the equations characterizing a problem has been obtained, this solution must be verified by checking it carefully. For any mistake, whether it is due to a simple arithmetic error or a flaw in the logic used to derive or solve the equations, can have a disastrous effect on the results of the analysis. Since everyone is likely to make mistakes from time to time, procedures for checking mathematical calculations have special importance.

The most direct approach is to compare the results of these calculations with actual experimental data or the results obtained by independent calculations. However this step is not always practical. Therefore it is important to be familiar with less direct methods for checking the analysis, for example, checking the consistency of units appearing in equations or solutions, studying the plausibility of various limiting cases, or utilizing various symmetries that appear in the problem.

It is essential in problem solving to examine the validity of all assumptions and approximations that are introduced to simplify the mathematical analysis. Frequently one can work back through the intermediate stages of the mathematical analysis to verify each approximation introduced along the way.

We cannot stress strongly enough the importance of careful checking of mathematical calculations. Frequently students acquire a mistaken impression of the importance of speed in performing calculations (due, no doubt, to the time limits placed on most examinations or homework assignments). However in engineering analysis, what really counts is reliability, not speed.

Students should also be wary of the "pitfalls of prejudice" in checking calculations. After they have slaved long hours on the mathematical analysis of a problem, they may not be disposed to believe that anything could be wrong with this analysis. This attitude has become particularly prevalent in the blind acceptance of the results from computer calculations. The fact that computers usually churn out numbers to many decimal places gives a false impression of accuracy. A computer will present both correct and totally erroneous calculations in the same degree of detail. The old computer proverb, "Garbage in, garbage out," is still true. (Actually, one might do well to adopt an opposing attitude and question whether any number resulting from millions upon millions of successive calculations can possibly be correct.)

The tendency to feel that a particular analysis is so thoroughly worked out that checking is unnecessary must be strongly resisted. Such an attitude can prove disastrous in engineering analysis, as numerous engineering misadventures in the past have illustrated. Indeed, one of the most valuable personality traits in engineers is a "professional paranoia," a cautious skepticism, not only of the results obtained by others, but also of their own analysis.

Interpretation of Results

Once the equations have been solved for the quantities of interest, the engineer must translate these results into physical terms. The very conciseness of mathematical notation and analysis tends to disguise the implications of the solutions. Frequently a considerable amount of work is required to extract the desired information from these solutions. The physical interpretation of mathematical results is a very important facet of engineering analysis.

Rough sketches are among the most useful tools for determining trends of the solution for various limiting cases. It is usually not necessary to go to all of the trouble to plot accurately a numerical evaluation of the mathematical formulas. Rather only the rough behavior of the quantities of interest need be determined. Limiting cases can also play a useful role in interpreting the physical implications of mathematical results.

Usually several numerical cases are studied using general mathematical solutions. However these numerical calculations should not be performed until the end of the mathematical analysis. Leaving work in symbolic form for as long as possible facilitates checking by dimensional analysis and examining limiting cases. Finally, if numbers are not inserted until the end of the analysis, less numerical work is required, and there is less chance for error. This also allows the solution to be arranged into a form most convenient for numerical calculation.

SUMMARY

The physical laws describing an engineering problem can be represented by mathematical equations. To translate physical laws into mathematics, one begins by making a precise verbal statement of the relevant scientific principles. Sketches or diagrams are useful to indicate the relationship between variables characterizing the problem. Symbols are introduced to represent each variable. The engineer uses physical insight to introduce simplifying assumptions and approximations into the equations. Such assumptions should always be very carefully stated and reexamined after the solution is arrived at to assess their validity. Sometimes the equations can be solved by inspection, by looking up the answer, or by well-known analytical techniques developed in mathematics courses. In more complex problems, the equations must be solved on a digital computer. The checking and verification of both methods of solution and simplifying assumptions is a vital aspect of engineering problem solving. Usually reliability is more important than speed in engineering analysis. The final stage of analysis involves the physical interpretation of mathematical results (frequently assisted by the use of sketches, limiting cases, or numerical studies).

Exercises

Mathematics

1. Search through your college catalog and identify the advanced mathematics courses of possible importance to engineering. Suggest one engineering application that would make use of the mathematical topics developed in each of the courses you have listed.

2. Suggest an engineering application for each of the following mathematical topics: plane geometry, solid geometry, analytical geometry, trigonometry, calculus, set theory, topology, matrix algebra.

3. Many mathematical concepts were first developed to describe physical phenomena. Using your textbooks or an encyclopedia, list the physical phenomena that stimulated the following mathematical developments: (a) Newton's development of calculus, (b) Fermat's principle, (c) Lagrange's equations, (d) Stokes' theorem.

4. The Green Monster dragster is capable of covering a quarter-mile from a standing start in 5.9 seconds. Set up the mathematical procedure to determine the acceleration of this dragster over the quarter-mile course, following closely the analysis of the space probe example in this section.

5. A rocket launch vehicle is capable of achieving an acceleration of 5 m/s^2 for a thrust duration of 30 s. If the rocket was launched on a vertical trajectory, what is its altitude in meters when its engines finally shut off? (Ignore the action of gravity and set up the analysis of this problem following the space probe example.)

6. The present rate of growth of the world's population corresponds roughly to a 2% annual increase. If the world population was 4.5 billion in 1980, in what year would you predict this population to have doubled to 9 billion?

Symbols

7. Suggest mathematical symbols to characterize each of the following quantities: volume, weight, stress, intensity of light, automobile fuel efficiency, angle of spacecraft reentry, the velocity of a tractor measured at six different times, the height of beer remaining in a can after one sip, two sips, three sips, and so on.

8. "Translate" the following mathematical expressions into plain English:
   (a) $1000 \sim 1001$
   (b) $1000 > 990$
   (c) $1000 \gg 10$

(d) $m_1 >> m_2$

(e) $T_1 > T_2$ and $T_2 > T_3 \Rightarrow T_1 > T_3$

(f) If $a \in A$ and $b \in A$, then $(a + b) \in A$.

9. Write symbolic mathematical expressions for each of the following prose statements:

  (a) The maximum speed of a Ferrari is much greater than that of a Volkswagen Rabbit.

  (b) The resource base represented by the United State uranium reserves is roughly comparable to domestic reserves of petroleum.

  (c) The fact that coal-generated electricity is cheaper than oil-generated electricity, and that oil-generated electricity is cheaper than solar electricity, implies that coal-generated electricity is cheaper than solar electricity.

  (d) The sum of the number of college course credit hours required for the typical baccalaureate degree in engineering is roughly 120.

Analytic Geometry

10. Construct a Cartesian (rectangular) coordinate system and locate the following points (x,y): $(0,3)$, $(-2,12)$, $(-4,11)$, and $(2,4)$. By connecting these points with straight lines, demonstrate that they lie at the corners (vertices) of a rectangle.

11. Determine the distance between each of the following pairs of points: (a) $(-1,-3)$, $(4,2)$ (b) $(1,2)$, $(6,7)$ (c) $(0,8)$, $(6,-1)$ (d) $(-4,6)$, $(6,6)$

12. An aircraft is on its final approach for a landing when it suffers engine failure at a distance of 3 km from the end of the runway. If the aircraft has a glide ratio of 12 to 1 (i.e., the plane will drop 1 m for every 12 m it travels horizontally), and the engine failure occurs at an altitude of 800 m, will the plane make the runway? Sketch the trajectory of the plane on a Cartesian coordinate system to demonstrate your conclusion.

Functions

13. If $f(x) = 2x^2 - x + 3$, determine

  (a) $f(0)$ (c) $f(3)$

  (b) $f(1)$ (d) $f(h)$

14. If $f(x) = (1 - x)^{-1}$, determine

  (a) $f(0)$ (c) $f(\pi)$

  (b) $f(-1)$ (d) $f(+1)$

15. Sketch a graph of each of the following functions:
    (a) $f(x) = 3x + 2$ (c) $f(x) = e^x + e^{-x}$
    (b) $f(x) = \tan x$ (d) $f(x) = x \ln x$

Calculus

16. Using a table of derivatives, compute the derivatives of each of the following functions:
    (a) $f(x) = x^5$ (c) $f(x) = e^{-2x}$
    (b) $f(x) = x^{-5}$ (d) $f(x) = x^{-1/2}$

17. Using a table of integrals, compute the integrals of each of the following functions:
    (a) $\int x^5 \, dx$ (c) $\int (2x^2 + 3) \, dx$

    (b) $\int e^{-3x} \, dx$ (d) $\int \frac{1}{x} \, dx$

18. An automobile initially at rest begins to move along a straight road. The driver accelerates the car at a uniform rate until he suddenly spots a concrete barrier across the road ahead. He immediately applies the brakes, and the car decelerates rapidly—but not rapidly enough. It crashes into the barrier. Sketch the position, velocity, and acceleration of the car as functions of time.

Vectors

19. Determine whether each of the following quantities is a vector or a scalar:
    (a) Mass   (d) Temperature
    (b) Torque   (e) Pressure
    (c) Current   (f) Momentum

20. Draw a two-dimensional coordinate system and sketch the vectors with initial points taken at the origin and end points taken at the coordinates (x,y):
    (a) (2,0)   (c) (−3,1)
    (b) (1,−5)   (d) (0,0)

21. Repeat the exercise in Problem 20 for a three-dimensional coordinate system with vectors at coordinates (x,y,z):
    (a) (1,4,−2)   (c) (6,3,1)
    (b) (0,1.0)   (d) (−5,−2,−5)

22. A particle moves in a plane such that its x and y coordinates are given by: $x = \cos t$, $y = \sin t$. Sketch the trajectory traced by the particle.

23. Determine the velocity and acceleration for the particle motion described in Problem 22.

Numerical Analysis

24. Graph the discretized representation at 11 equally spaced points on the interval $0 < x < 1$ (i.e., $x = 0, 0.1, 0.2, \ldots, 1.0$) for each of the following functions:
(a) $f(x) = x^2$
(b) $f(x) = e^{-x}$
(c) $f(x) = \sin \pi x$
(d) $f(x) = x^2 e^{-x^2}$

25. Conjecture as to how you might use a digital computer to differentiate or integrate a function.

Probability and Statistics

26. What is the probability of being dealt a face card from a 52-card deck?

27. A full moon occurs about once every 28 days. If the probability is 1 in 1000 that your neighbor is a werewolf, what is the probability that he is out stalking tonight?

28. Most computer systems have built in random number generators such as the RND(I) function available in the BASIC language on many microcomputers. Using this function, write a computer program that will simulate the toss of two 6-sided dice. More ambitious students might wish to include sufficient interactive dialogue in this program to simulate a craps game.

Equations—General

Write mathematical equations to represent each of the following physical laws, taking care to define the symbols used to represent variables and constants in the equations:

29. Newton's law of motion: The force exerted on a body is equal to its mass times its acceleration.

30. Static equilibrium: The sum of the forces on a stationary body must be zero.

31. Velocity: The velocity of a body is equal to the time rate of change of its position.

Derivation of Equations

Follow the general procedure below in deriving equations for each of the following situations: (a) Make a precise verbal statement of the relevant physical principle(s). (b) Use a sketch to identify known and unknown variables. (c) Introduce mathematical symbols for relevant quantities. (d) Write mathematical equations representing the physical principle(s).

32. Determine the acceleration of an automobile subject to a given force. (Use Newton's law.)

33. Determine the mass of water remaining in a leaky bucket when the rate of leakage is given. (Use conservation of mass.)

34. If we are given the velocity of a body at any time, we are to determine its position as a function of time. (Use the definition of velocity.)

Assumptions

Carefully state the assumptions necessary to answer the following problems:

35. Estimate the amount of gasoline purchased by the student body of your university in one year's time.

36. Estimate the number of trees in North America.

37. Recall the famous quote of Archimedes: "If I had a lever long enough, I could lift the world." How much does the world weigh? How long a lever would Archimedes need to lift the world?

Interpretation

Venture an interpretation of the implications of each of the following solutions:

38. The temperature T at long distances x from a heat source is found to be:

$$T(x) = T_0 e^{-kx}$$

39. The height of a water wave as a function of position x and time t is given by:

$$H(x,t) = H_0 + H_1 \sin (kx - \omega t)$$

40. The real value of the dollar($) as a function of the year(s) after 1970 is given by:

$$\$(t) = \frac{1990 - y}{20} \qquad (y \text{ in years})$$

3.2. NUMBERS, DIMENSIONS, AND UNITS

The end result of an engineering calculation is usually a number. In fact, engineering is distinguished by the degree to which it introduces quantitative methods into the analysis of problems. Therefore let us briefly consider how numbers are used in solving engineering problems. We shall first look at the various ways in which the engineer writes and interprets numbers. Then we shall examine the manner in which characteristics of physical systems and processes are quantified in terms of dimensions and units.

3.2.1. Numbers

We are so used to writing the symbols for numbers, 1, 2, 3, . . . that we rarely recognize that this is but one system for "counting" quantities. There are not only alternative symbols we can use for numbers but entirely different systems of numbers. We should be thankful to the Arabic scholars who invented our present number symbols. Imagine the difficulty of performing engineering calculations using Roman numerals. For example, 123 + 1825 = 1948 would become CXXIII + MDCCCXXV = whatever—anyway, it would be very cumbersome.

Number Systems

When we think of number systems, we typically think of the decimal system (based on units of 10) used in everyday computation. But this is only one of an endless variety of possible number systems, although it is the system most convenient for a beast with 10 fingers. We could just as well use number systems based on other units, such as an octal system based on 8 units or a hexadecimal system based on 16 units. We shall find in Chapter 4 that the most convenient number system for computers is the binary system, based only on the two units 0 and 1. In this system we represent zero by 0, one by 1, two by 10, three by 11, four by 100, and so on. The binary number system is the most logical choice for digital systems that rely on devices with only two possible states—on or off.

Significant Figures

When dealing with numerical data, the number of accurate figures or digits, the way that data are rounded or truncated, and the way that numbers are written all must be considered. The number of accurate digits in a numerical result is called its *number of significant figures*. For example, if we measure the length of a pencil and find it to be 14.1 cm, then this result is said to contain three significant figures. Suppose we had represented this length as 0.000141 km. Then, despite the presence of 4 zeros, this number would still have only three significant figures. The zeros only serve to locate the decimal point, that is, to indicate the size of the unit used in making the measurement. They are not regarded as significant figures.

The number of significant figures in a result (excluding the zeros used to locate the decimal point) indicates the number of digits that can be used with confidence in an engineering calculation. There is no point in carrying digits beyond those that are significant since they give rise to unnecessary calculation effort. They can also convey a false sense of the true accuracy of the results.

For example, suppose we add the following column of numbers representing measurements of limited accuracy:

$$
\begin{array}{r}
3.51 \\
2.205 \\
0.0142 \\
\hline
5.7292
\end{array}
$$

The true accuracy of the resulting sum is only to three significant figures. Therefore we should limit the sum to this number of digits. We can either truncate the sum to three digits, 5.72, or round it off as 5.73 (using the prescription that if the digit following the last significant figure is 5 or greater, we increase the last figure by 1; if it is 0 to 4, we leave the last figure as it is).

A similar procedure can be used to handle multiplication or division. Once again the number of significant figures in the result should be no more than the fewest in any number involved. Therefore we should write

$$6.3 \times 3.471 \times 2.371 = 52$$

In problems that involving a series of calculations leading to the final answer, one more significant figure is carried through in the intermediate answers than will be used in the final answer.

Example   A mechanical engineer must choose among three different brands of electrical motors based on a comparison of their efficiencies of (a) 85%, (b) 85.2%, and (c) 85.28%. Which motor should he choose?

Answer

The efficiency numbers strongly suggest that only the first two figures are significant. Then all three motors would have effectively the same efficiency (85%). Another criterion would have to be used to select among them.

Example   A civil engineer must decide upon the relative economic merit of three different highway bid estimates: (a) $3.2 million, (b) $3.25 million, and (c) $3.258 million. Which bid should she accept?

Answer

The estimates strongly suggest that there are only two significant figures in all three calculations. However since the contractor is required by law to perform the work for the amount indicated in the bid, the engineer should obviously choose the lowest bid: (a) $3.2 million.

Scientific Notation

The engineer frequently must perform calculations using very large or very small numbers. It would be quite cumbersome always to write such numbers in decimal form. For example, consider the multiplication of

$$1\ 340\ 000\ 000 \times 0.000\ 000\ 000\ 31 = 0.0415$$

It is more convenient to utilize *scientific notation* in which the significant digits (excluding zeros immediately preceding or following the decimal) are retained, and the numerical value is written as a number between 1 and 10 (the *mantissa*) multiplied by a power of 10. In scientific notation, our previous example would be written as

$$1.34 \times 10^9 \times 3.1 \times 10^{-11} = 4.1 \times 10^{-2}$$

An equivalent notation consists of representing the power of 10 by the notation E for exponent. Then our example would be written as

$$1.34\ E+09 \times 3.1\ E-11 = 4.1\ E-02$$

This latter notation is used in most computer applications and is called *floating point* format.

The use of scientific notation greatly simplifies computations since multiplications of numbers with powers of 10 are performed by adding the exponents. Division of numbers with powers of 10 is performed by subtracting the exponents. Scientific notation also stresses the importance of significant figures in the numbers.

Example Scientists who delve into natural phenomena on the microscopic level of atoms and molecules or on the cosmic scale of the universe sometimes invent special units to characterize the very small or very large numbers with which they deal. For example, the cross-sectional area presented by an atomic nucleus to a fast subatomic particle is roughly 10^{-24} cm². Hence nuclear physicists have introduced a unit known as the barn (as "big as a barn door") to characterize nuclear cross sections: 1 barn = 10^{-24} cm². In contrast, astrophysicists deal with distances comparable to the diameter of the galaxy and therefore have introduced a unit of length known as the parsec: 1 parsec = 3.0857×10^{16} m. It is interesting to think of a volume 1 barn in cross-sectional area and one million parsecs in length. Although just thinking of such a volume tends to strain the imagination, in actual magnitude it is just

$$1\ \text{barn} \times 1\ \text{million parsecs} = 10^{-24}\ \text{cm}^2 \times 3.0857 \times 10^{24}\ \text{cm}$$
$$= 3.0857\ \text{cm}^3$$

—only three cubic centimeters.

SUMMARY

The end result of an engineering calculation is usually a number. Engineers use a variety of number systems in addition to the familiar decimal system. Of most use in computer (digital) applications is the binary system. The number of accurate digits in a result is termed its number of significant figures. The number of significant figures in a result should be no more than the fewest in any number involved in intermediate stages of the calculation. To accommodate very large or very small numbers, engineers commonly employ scientific notation, in which the result is written in terms of powers of ten, such as 1.34×10^9 or $1.34 \ E+09$.

3.2.2.   Dimensions and Units

Dimensions

The laws of science are based on characteristics of physical systems and processes that can be measured. These characteristics are collectively called *dimensions*. The space measurements of an object such as its length, width, and height are dimensions that characterize the object's size. The time interval between two events is another dimension. The amount of electric current flowing through a wire can also be measured and, therefore, electric current could be identified as a dimension.

Certain types of dimensions assume a more fundamental nature in the sense that they can be used to describe all other physical relationships. One such set of *fundamental* dimensions is composed of length (L), mass (M), time (T), electric current (I), temperature (θ), the amount of a substance (mole), and the luminous intensity of light (l). Notice here that we have chosen to denote the type of fundamental dimension by a letter symbol.

All other dimensions characterizing physical quantities can be formed by combining such fundamental dimensions. These are referred to as *derived* dimensions. Examples would include volume (L^3), velocity (L/T), and force (ML/T^2). Table 3.5 shows a set of fundamental dimensions and examples of derived dimensions.

It is advantageous to use as few fundamental dimensions as possible. But the selection of what is to be a fundamental dimension and what is to be considered derived is not unique. A *dimension system* can be defined as the smallest number of fundamental dimensions that will form a consistent and complete set for the field of interest. For example, in mechanics three fundamental dimensions are necessary. But the choice of these dimensions is not unique. In an *absolute* system of dimensions, length (L), time (T), and mass (M) are regarded as fundamental dimensions. Force would be a derived dimension (ML/T^2). However in the alternative *gravitational* system, the fundamental dimensions are length (L), time (T), and force (F), and mass is given by (FT^2/L).

Some quantities in engineering calculations have no dimensions, for example, numerical constants such as π or e. Geometrical angles are also dimensionless since they represent only a change in direction rather than a distance or length; angles are measured by taking the ratio of two lengths. Another example is percentage, which can be expressed as the ratio of two quantities characterized by the same dimensions.

SUMMARY

Characteristics of physical systems and processes subject to measurement are known collectively as dimensions. All physical measurements can be described in terms of a set of fundamental dimensions: length, mass, time, electric current, temperature, amount of substance, and intensity of light. All other dimensions can be derived in terms of fundamental dimensions.

Units

The magnitude of dimensions of physical quantities can be quantified only when compared against reference amounts known as *units*. The result of any measurement of a dimension is to determine how many of these reference amounts or units are present. That is, when we measure a dimension we must specify not only the

TABLE 3.5 Fundamental and Derived Dimensions

FUNDAMENTAL DIMENSIONS		DERIVED DIMENSIONS	
Length	(L)	Velocity	(L/T)
Mass	(M)	Acceleration	(L/T^2)
Time	(T)	Area	(L^2)
Electric current	(I)	Mass density	(M/L^3)
Temperature	(θ)	Force	(ML/T^2)
Amount of substance	(mole)	Energy	(ML^2/T^2)
Luminous intensity	(l)	Power	(ML^2/T^3)

magnitude of the dimension, but also the units in which this magnitude is expressed. For example, we can choose to measure the length of an object in units of the meter, centimeter, or kilometer, whichever is the most convenient.

When engineers analyze or design physical systems, they must always be certain that the units used to characterize physical quantities are consistent. The proper units must be assigned to constants and variables, and valid mathematical operations must be performed on these quantities. If all of the dimensions or units of each term in an equation are consistent, the equation is said to be *dimensionally homogeneous*. Dimensional consistency is an extremely useful tool for checking the validity of equations and is important in engineering measurements. Furthermore, since measurements involve a comparison with various reference amounts, one must maintain an accurate set of "standards" of units.

Once a consistent dimension system has been selected, a corresponding system of units must be introduced to quantify the measurement of these dimensions. In this sense, units are relative quantities and are defined only by comparison with other measurements of like quantities. For example, the meter is defined with respect to the wavelength of light emitted by a krypton atom, while the kilogram is defined as the mass of a cylinder of platinum–iridium alloy kept by the National Bureau of Weights and Measures in Paris.

Although a variety of different systems of units have been used (indeed, different systems of fundamental dimensions have also been used) throughout the world, the increasing interdependence of nations brought about by modern technology (travel and communications, in particular) have made obvious the need for a common system of units with which to measure physical quantities. The accepted worldwide standard, a refinement of the familiar metric system, is known as the *Systeme International d'Unites* or *SI System*. Essentially every advanced nation in the world—with the exception of the United States—has now adopted the SI system for all scientific and engineering activity. For social, economic, and political reasons, the United States has been very slow in introducing the SI unit system. This country continues to rely heavily on the British unit system (foot-pound-second).

SUMMARY

The magnitude of physical dimensions is expressed relative to reference amounts of similar quantities known as units. Many different systems of units have been employed in engineering practice. However the accepted worldwide standard system is a refinement of the metric system known as the SI system.

The SI System of Units

The Systeme International d'Unites (International System of Units) or SI system was adopted in 1960 and is presently maintained by the Conference General des Poides et Measures (the CGPM or General Conference on Weights and Measures). This system assigns one and only one unit, known as a *base unit*, to each of the seven fundamental dimensions. For example, the base unit of length is the meter, that of mass is the kilogram, and that of time is the second. These base units and their definitions (as established by the CGPM) are given in Table 3.6.

All other physical quantities can be expressed as algebraic combinations of these base units known as *derived units*. For example, the derived unit for velocity is meter per second, while that for force is kilogram meter per square second. Many derived SI units have been given special *unit names* and *unit symbols* (or abbreviations) such as the newton (N) for force, joule (J) for energy, and pascal (Pa) for pressure. Table 3.7 lists the unit names and unit symbols for derived units approved by the CGPM.

The units of radian and steradian for plane and solid angles can be regarded either as base or derived units and are sometimes referred to as *supplementary units* (Table 3.8).

One important advantage of the SI system is the manner in which it assigns special prefixes to the unit symbol to form new units that are decimal multiples or submultiples of the original unit. The prefixes for decimal multiples are given in Table 3.9. The choice of the appropriate multiple of an SI unit is a matter of convenience. The multiple is usually chosen so that the numerical value of a dimension lies between 0.1 and 1 000. For example, 1.45×10^4 m would be written as 14.5 km, while 0.003 5 s would be 3.5 ms.

The use of the SI system is governed by a number of rules that should be

TABLE 3.6 Base Units and Definitions for the Systeme International d'Unites (SI)

BASE UNITS		
DIMENSION	**BASE UNIT**	**SYMBOL**
length	meter	m
mass	kilogram	kg
time	second	s
electric current	ampere	A
temperature	kelvin	K
amount of substance	mole	mol
luminous intensity	candela	cd

DEFINITIONS OF BASE UNITS

Meter: The meter is the length equal to 1 650 763.73 wavelengths in a vacuum of the radiation corresponding to the transition between the levels $2p_{10}$ and $5d_8$ of the krypton-86 atom.

Kilogram: The kilogram is equal in mass to the international standard, a cylinder of platinum-iridium alloy kept at the National Bureau of Weights and Measures in Paris.

Second: The second is the duration of 9 192 631 770 periods of the radiation corresponding to the transition between the two hyperfine levels of the ground state of the cesium-133 atom.

Ampere: The ampere is that constant electric current that, if maintained in two straight parallel conductors of infinite length, of negligible circular cross section, and placed 1 meter apart in a vacuum, would produce a force between these conductors of 2×10^{-7} newton per meter of length.

Kelvin: The kelvin is the fraction 1/273.15 of the thermodynamic temperature of the triple point of water.

Mole: The mole is the amount of substance of a system that contains as many elementary entities as there are atoms in 0.012 kilograms of carbon-12. When the mole is used, the elementary entities must be specified and may be atoms, molecules, ions, electrons, or other particles.

Candela: The candela is the luminous intensity, in the perpendicular direction of a surface of 1/600 000 square meters of a black body at the temperature of freezing platinum under a pressure of 101 325 newtons per square meter.

---

learned and followed very carefully in practice. We have summarized these rules below along with illustrative examples.

Unit Names and Unit Symbols

1. Unit names are always written in lowercase unless they appear at the beginning of a sentence: meter, joule, not Meter, Joule.

2. Plurals are used as required with unit names: meters, joules.

3. The proper symbol should always be used to represent the unit: s, g, not sec, gm.

4. Unit symbols should always be written in lowercase letters except when the unit is derived from a proper name, in which case the first letter is capitalized: s, g, W, N, not S, G, w, n.

5. Unit symbols should be separated from numerical values by one space: 1.01 m, not 1.01m or 1.01   m.

TABLE 3.7   Derived Units in the Systeme International d'Unites (SI)

QUANTITY	DERIVED UNIT	SYMBOL	BASE UNITS
Frequency	hertz	Hz	$1 \text{ HZ} = 1 \text{ s}^{-1}$
Force	newton	N	$1 \text{ N} = 1 \text{ kg m/s}^2$
Pressure, stress	pascal	Pa	$1 \text{ Pa} = 1 \text{ N/m}^2$
Energy, work, heat	joule	J	$1 \text{ J} = 1 \text{ N m}$
Power	watt	W	$1 \text{ W} = 1 \text{ J/s}$
Electric charge	coulomb	C	$1 \text{ C} = 1 \text{ A s}$
Electric potential	volt	V	$1 \text{ V} = 1 \text{ J/C}$
Electric capacitance	farad	F	$1 \text{ F} = 1 \text{ C/V}$
Electric resistance	ohm	Ω	$1 \Omega = 1 \text{ V/A}$
Electric conductance	siemens	S	$1 \text{ S} = 1 \Omega^{-1}$
Magnetic flux	weber	Wb	$1 \text{ Wb} = 1 \text{ V s}$
Magnetic flux density	tesla	T	$1 \text{ T} = 1 \text{ Wb/m}^2$
Inductance	henry	H	$1 \text{ H} = 1 \text{ Wb/A}$
Luminous flux	lumen	lm	$1 \text{ lm} = 1 \text{ cd sr}$
Illuminance	lux	lx	$1 \text{ lx} = 1 \text{ lm/m}^2$

6. Periods are never used after unit symbols unless they occur at the end of a sentence: s, m, g, not s., m., g.

7. Unit symbols are never written in plural form. The same symbol is used to represent both singular and plural forms: 25W, not 25 Ws.

8. Unit symbols are preferable to unit names.

Prefixes and Compound Units

9. Products of unit names can use either spaces or hyphens: newton-meter or newton meter.

TABLE 3.8   Supplementary Units in the Systeme International d'Unites (SI)

QUANTITY	NAME OF UNIT	SYMBOL
Plane angle	radian	rad
Solid angle	steradian	sr

TABLE 3.9   Decimal Multiple Prefixes for SI units

MULTIPLIER	PREFIX NAME	SYMBOL
10^{18}	exa	E
10^{15}	peta	P
10^{12}	tera	T
10^{9}	giga	G
10^{6}	mega	M
10^{3}	kilo	k
10^{2}	hecto	h
10^{1}	deka	da
10^{-1}	deci	d
10^{-2}	centi	c
10^{-3}	milli	m
10^{-6}	micro	μ
10^{-9}	nano	n
10^{-12}	pico	p
10^{-15}	femto	f
10^{-18}	alto	a

10. Products of unit symbols can use dots or spaces: N m or N.m or N·m.

11. Compound unit names using division are represented using "per": meters per second.

12. Compound unit symbols using division are represented using "/" (solidus), division, or inverse power: m/s or $\frac{m}{s}$ or ms^{-1}. One should never use more than one solidus in a compound unit: $m\ s^{-2}$, not m/s/s.

13. To avoid errors in calculations, prefixes may be replaced with powers of ten: 1 MJ becomes 10^6 J.

Decimals and Digits

14. A dot on the line is used as the decimal marker. For numbers less than one, a zero should be written before the decimal point: 0.125, not .125.

15. In some countries the comma is used to represent the decimal point. Therefore one should group digits into groups of three, separated by a space rather than a comma: 5,000,000 becomes 5 000 000; 0.124456 becomes 0.124 456.

Significant Figures

16. Either prefixes or scientific notation can be used to indicate the number of significant figures:

    10 000 m

    10 000 km   or 10.000×10^3 m

SUMMARY

> *The SI system is composed of seven base units characterizing the fundamental dimensions. Derived units can be expressed as algebraic combinations of base units. Special unit names and unit symbols are introduced to represent the more common derived units. Prefixes can be used to represent decimal multiples or submultiples of these units. Precise rules have been adopted to govern the use of SI units.*

Conversion of Other Unit Systems

While the SI system of units (meter, kilogram, second) has been accepted by the rest of the world, the engineer entering professional practice in the United States should be prepared to encounter the more awkward British unit system (foot, pound, second) on occasion. As we noted earlier, this is a gravitational (rather than an absolute) system of units in which force replaces mass as a fundamental dimension. The conversion factors between the British and SI system are given in Appendix C.

The conversion between unit systems is straightforward, although occasionally frustrating and tedious. One sometimes confusing aspect should be noted. In the British system, the terms "weight" and "mass" are sometimes confused. When someone speaks of a person's weight in pounds, they frequently mean mass. However when we speak of weight in scientific terms (or the SI system), we are referring to the force of gravity on an object. For example, in SI units, weight would be measured in newtons, not kilograms. It is important to keep these quantities distinct in engineering calculations.

FIGURE 3.13. Conversion from the English unit system to the SI unit system is straightforward. (© 1978 by B. Kliban. Reprinted from *Tiny Footprints* by permission of Workman Publishing Company, New York)

SUMMARY

Engineers should be prepared to perform calculations in either SI or British units. The conversion between unit systems is straightforward.

Exercises

Number Systems

1. Write the binary representation of the following decimal numbers:
   - (a) 12
   - (b) 23
   - (c) 54
   - (d) 5.34

2. Convert the following binary numbers into decimal representation:
   - (a) 1101
   - (b) 1000110
   - (c) 1111111
   - (d) 11.011

3. Develop a table comparing the first ten digits of number systems based on units of 3 and 6.

4. Although digital computers utilize circuit elements to process binary numbers, the computer engineer frequently uses hexadecimal num-

bers to analyze the design and operation of computers (e.g., for memory addresses and contents). Why?

Significant Figures and Scientific Notation

5. Determine the number of significant figures in each of the following numbers:

(a) 8.030

(b) 104.3

(c) 0.000 671

(d) 0.010 03

(e) 807.003

(f) 2.130 0

6. Perform the indicated calculation, rounding off each answer to the proper number of significant figures:

(a) $1.21 + 3.1 + 2.579$

(b) 6.35×2.103

(c) $8.210 + 1$

(d) $367 \div 0.021$

(e) $67.24 - 0.0002$

(f) $(45.21 + 67.1) \div 2.2$

7. Express each of the following numbers in scientific notation (in both regular and computer or "E" form):

(a) 134.2

(b) 0.003 2

(c) 62 000 000

(d) 31 416.001

(e) 0.000 000 32

(f) 0.097 2

8. Write the following numbers in decimal form:

(a) 3.26×10^{-3}

(b) 9.827×10^{1}

(c) 1.006×10^{5}

(d) 2.718E+02

(e) 1.417E−6

(f) 7.1E+10

9. Perform the indicated calculation, expressing each answer in both scientific and decimal form (rounded to the proper number of significant figures):

(a) $(3.14 \times 10^{-5}) \times (1.32 \times 10^{6})$

(b) $(2.16 \times 10^{1}) - (5.62 \times 10^{-3})$

(c) $(0.92 \times 10^{15}) \div (1.23 \times 10^{16})$

(d) $(130.5) - (1.29 \times 10^{2})$

(e) $(897.1) \times (2.65 \times 10^{-2})$

(f) $(0.000\ 7) \div (1.3 \times 10^{-3})$

10. An airplane travels a distance of 891 km in a time of 1.3 hours. Calculate the average speed of the plane.

11. An atomic nucleus has a radius of 1.3×10^{-14} m. Calculate the volume of the nucleus assuming that it is spherical in shape.

12. If the nucleus in Problem 11 has a mass of 3.21×10^{-27} kg, calculate its density (mass per unit volume).

13. The Nova laser system at the Lawrence Livermore Laboratory is capable of producing a light pulse of energy 3×10^{4} J in a time interval of $1 + 10^{-10}$ s. Calculate the power level of this laser, recalling that one joule per second is equal to 1 watt. (It might be noted for comparison that the present electrical generating capacity of the United States is roughly 5×10^{11} W.)

Dimensions and Units

14. Verify the following equations using the concept of dimensional homogeneity:

(a) $F = ma$ (force = mass × acceleration)

(b) $KE = \frac{1}{2} mv^2$ (kinetic energy = $\frac{1}{2}$ mass × velocity²)

(c) $E = mc^2$ (energy = mass × velocity of light²)

(d) $P = I^2R$ (power = electric current² × resistance)

15. Determine the dimensions of the following coefficients defined by the corresponding equation:

(a) universal gas constant R:

$$pV = nRT$$

(pressure × volume = number of moles × R × temperature)

(b) drag coefficient c_D:

$$F = c_D A\, \rho\, \frac{v^2}{2}$$

$\left(\text{drag force} = c_D \times \text{area} \times \text{density} \times \frac{1}{2}\, \text{velocity}^2\right)$

(c) heat transfer coefficient:

$$Q = hA\,(T_1 - T_2)$$

(heat transferred = h × area × temperature difference)

16. One commonly encounters dimensionless parameters in engineering applications, that is, parameters without dimensions. Verify that each of the following are dimensionless parameters:

(a) Reynolds number:

$$\text{Re} = \frac{\rho v d}{\mu} \qquad \frac{\text{density} \times \text{velocity} \times \text{diameter}}{\text{viscosity}}$$

(b) Prandtl Number:

$$\text{Pr} = \frac{\mu c_p}{k} \qquad \frac{\text{viscosity} \times \text{specific heat}}{\text{thermal conductivity}}$$

(c)   Nusselt Number:

$$\text{Nu} = \frac{hD}{k} \cdot \qquad \frac{\text{heat transfer coefficient} \times \text{length}}{\text{thermal conductivity}}$$

17. Correct each of the following dimensions given in SI units:

(a)   15,000 W
(b)   10 g/cm³
(c)   87 Joules
(d)   155kPa
(e)   1.5 μf

(f)   100 km/hr
(g)   60 cps
(h)   55 k m
(i)   30 meters/second
(j)   9.81 m/s/s

18. Write each of the following quantities in SI units:

(a)   The speed of light
(b)   The federal highway speed limit
(c)   Atmospheric pressure
(d)   Freezing point of water
(e)   Your weight
(f)   Your mass

Conversion among Units

19. Convert the following quantities into SI units:

(a)   15 inches
(b)   9 yards
(c)   25 acres
(d)   100 horsepower
(e) $\frac{4}{5}$ quart
(f)   60 calories
(g)   4500 angstroms
(h)   6 slugs

(i)   105 poundals
(j)   15 psi
(k)   12,000 BTU
(l)   15 lbs$_m$
(m)   15 lbs$_f$
(n)   98.6° F
(o)   2 teaspoons
(p)   6 ounces

20. You are driving down the highway at 65 miles per hour when you suddenly spot a sign indicating that the posted speed limit is 100 km/h. Should you slow down?

21. Determine the weight in SI units of the following objects (assuming that the local acceleration of gravity is $g = 9.81$ m/s²):

(a)   A 1000 kg automobile
(b)   A 2.2 ton automobile
(c)   A quarter-pounder (0.25 lb$_m$)

22. The force of gravity varies slightly from place to place. For example, while $g = 9.81$ m/s² at the equator, $g = 9.83$ m/s² in Greenland. Calculate the difference in weight of a 100 kg object weighed at each location.

23. An astronaut weighs 800 N on earth. Determine the weight of the astronaut on the moon where the acceleration of gravity is $g = 1.6$ m/s^2.

3.3 SCIENTIFIC CONCEPTS

Many of the tools of engineering spring directly from fundamental scientific principles. Frequently engineers are identified with a particular branch of science. For example, chemical engineers are associated with chemistry, mechanical engineers with mechanics or thermodynamics, and electrical engineers with electrical sciences such as electronics. However, all engineers need a broad foundation in fundamental scientific principles ranging across subjects such as physics, chemistry, materials science, electrical sciences, atomic and nuclear physics, and systems theory. This scientific knowledge is a primary focus of an engineering education.

It would be unrealistic to attempt any more than a cursory discussion of the many important scientific concepts of importance to engineering. Rather we shall only sketch a path through the scientific subjects covered in the engineering curriculum. We shall illustrate how these principles are applied in engineering practice.

Engineering is most frequently considered a *macroscopic* science, concerned with the bulk behavior of materials and processes, with the design and building of large structures, the dynamics of machines, or with industrial production. Yet most of the recent advances of pure science, of physics, chemistry, and biology, have been on a *microscopic* level, at the level of atoms or molecules or atomic nuclei. The remarkable progress achieved during the twentieth century in explaining phenomena on this level has revolutionized our understanding of the macroscopic world more familiar to human experience. Therefore we shall depart

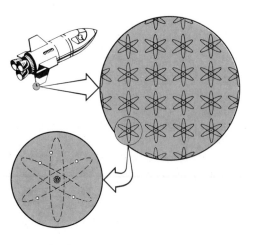

FIGURE 3.14. The macroscopic behavior of materials is determined by their microscopic structure.

somewhat from the usual pedagogy of engineering courses and begin by considering scientific phenomena at the microscopic level. We shall then use this knowledge as the basis for the study of macroscopic phenomena, such as machines, structures, and processes, that are of more direct concern to the engineer.

SUMMARY

Engineers should acquire a broad knowledge of fundamental scientific principles. Engineering is most frequently thought of as a macroscopic science, concerned with the gross behavior of materials and processes familiar to everyday human experience. The engineer should also have an understanding and appreciation of the advances in our knowledge of processes that occur on the microscopic level of atoms and nuclei.

3.3.1. The Structure of Matter

Fundamental to modern science is the concept that all matter is composed of *atoms*, microscopic particles typically about a billionth of a centimeter in diameter. These atoms also have a structure that consists of an even smaller atomic *nucleus* with a positive electric charge, surrounded by several *electrons* possessing negative electric charge. Atoms are frequently pictured (by both physicists and commercial artists alike) as a kind of a miniature solar system, with the electrons orbiting about the nucleus, much as the planets orbit about the sun. Although the atomic nucleus contains essentially all of the mass of the atom, being roughly a thousand times more massive than the electrons, it is only about one ten-thousandth as large as the atom (about 10^{-14} m in diameter as compared to 10^{-10} m). If we were to magnify a uranium atom to the size of the Rose Bowl, then the uranium nucleus would be about the size of a football (and, incidently, about the same shape as a football as well) (Figure 3.15).

The atomic nucleus is made up of two types of particles: *protons,* which carry a positive charge, and *neutrons*, which are electrically neutral. The number of protons in the atomic nucleus is denoted by the *atomic number Z*. There will be an identical number Z of negatively charged electrons orbiting the positive nucleus, attracted to it by the electric forces that arise between charged objects. Since the electron charge is equal in magnitude but opposite in sign from that of the proton, the atom itself is electrically neutral.

Atoms can interact with one another by means of electric forces that arise between their electrons and nuclei. They can bang into one another, knocking electrons loose, or acquiring and sharing electrons with one another. Such electric forces can cause the atoms to stick or bind together into groups of atoms known as *molecules*. The binding together of atoms into molecules is essentially the subject

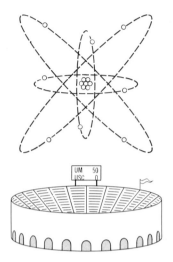

FIGURE 3.15. If an atom were magnified to the size of the Rose Bowl, then its nucleus would be about the size of a football.

of chemistry. Because the interaction among atoms depends on their electric charge, atoms with the same atomic number Z will display identical chemical behavior. Each different value of Z corresponds to a unique chemical *element*. The chemical properties of an element are determined by the number of electrons in the outermost orbit of the atom, since these are the electrons that participate in the atomic (or chemical) reactions. This behavior gives rise to a classification scheme of chemical properties of the elements displayed by the Periodic Table of the Elements (a familiar tool to any chemistry student).

We have noted that atoms may bind together to form more complicated structures known as molecules. In a similar manner either atoms or molecules can attach themselves together to form macroscopic-sized samples of matter. At low temperatures, the atoms will bind to one another in a more or less rigid pattern forming a *solid* material. Of course, even though solids appear to be quite rigid and inanimate forms of matter, on a microscopic level the atoms are in rapid motion, vibrating about a fixed position in the solid structure (in the language of physics, the crystal lattice). As the solid is heated to a higher temperature, these vibrations become more agitated until the atoms break away from their lattice position and begin to move about in a random fashion. At this point the solid loses its structure, its rigidity. We find that on a macroscopic level the solid has melted; it has become a *liquid* (Figure 3.16).

Suppose we continue to heat the liquid to still higher temperatures. The atoms move more and more rapidly until they reach speeds that are too high to allow them to be attracted to one another and clump together. The liquid vaporizes into the *gas* phase. Hence, the form taken by matter—the phase of the matter—depends sensitively on its temperature, that is, the average speed of motion of the atoms or molecules comprising its microscopic structure. In fact we are accus-

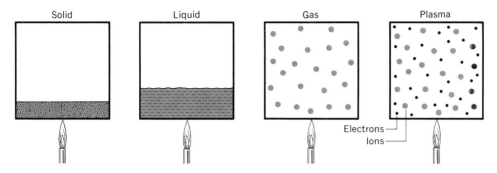

FIGURE 3.16.   Matter assumes several forms or states as it is heated.

tomed to thinking of matter as existing in one of these three states: solid, liquid, or gas.

There is yet another fundamental phase of matter. Suppose we continue to heat the gas to still higher temperatures. Eventually we shall supply sufficient energy to the molecules of the gas that they knock one another apart when they collide; they dissociate into individual atoms. If we continue to add energy, eventually the collisions will become so violent that one or more electrons may be stripped off from the atom. This process of removal of an electron from an atom is known as *ionization*. The net result is a free electron plus an atom missing an electron from its normal structure, an *ion*. If we add sufficient energy, all the available electrons may be stripped away, leaving a fully ionized gas consisting of free electrons and positively charged nuclei. Such an ionized gas is called a *plasma*. Plasmas are not ordinarily encountered (or at least recognized) in our daily experiences since the temperatures required to ionize a gas are very large, typically ranging upwards of 10 000 C. Nevertheless, plasmas are quite common in physics. For example, the glow in a neon light is a plasma. On a more grandiose scale, the sun and, indeed, all stars are giant blobs of plasma. In fact, on the scale of the universe the most common form of matter is plasma.

The explanation of the macroscopic character of matter in terms of its microscopic structure, the interactions between the very atoms that comprise it, is one of the stunning intellectual accomplishments of civilization. Since matter in its various forms is the clay of engineers, the substance they mold into devices and processes to meet the needs of society, this microscopic view of science has had a most dramatic impact on the engineering profession. It has led to the development of new materials such as plastics, high strength metals, and ceramics. It has also stimulated the development of entirely new technologies such as microelectronics, lasers, and nuclear energy. Although engineers today are still concerned primarily with the macroscopic world, they have become strongly dependent on their knowledge of microscopic processes to provide them with the tools necessary to ply their craft.

SUMMARY

Atoms interact with one another by means of electrical forces. Atoms and molecules bind together to form macroscopic-sized samples of matter. The form taken by matter depends on its temperature, ranging from solids at low temperature, to liquids and gases at higher temperatures, and eventually to a form of matter known as a plasma in which molecules become dissociated and atoms become ionized into a mixture of free electrons and ions. The explanation of the macroscopic properties of matter in terms of its microscopic structure has had a dramatic impact on engineering. It has provided the engineer with new concepts, materials, processes, and devices to apply as tools in engineering practice.

Example While developing a theory of light absorption and emission by atoms over a half-century ago, Albert Einstein arrived at an interesting prediction. When a unit of light energy (called a "photon") of exactly the right wavelength hits an excited atom, there is a possibility that this will stimulate the emission of a second light photon of exactly the same wavelength. In 1957 three physicists, Charles Townes, Nikolai Basov, and Alexander Prokhorov, proposed a device for exploiting Einstein's theory of *l*ight *a*mplification by *s*timulated *e*mission of *r*adiation. Two years later another scientist, Theodore Maiman, built the first such "laser" using a ruby crystal excited by flash lamps. Nevertheless for many years afterwards the laser was regarded as a scientific curiosity, a "solution in search of a problem," since its features were so unusual. However today engineers have applied the microscopic phenomenon of laser action in a wide variety of ways, ranging from precise measurements of distances in civil engineering to communication over fiber optics cables to the playback of video disks. What was once only a curiosity in atomic physics is now an important tool of the engineer.

3.3.2. Mechanic.

Most engineering students begin their study of science with the subject of *mechanics*, the physics of motion and dynamics. The fundamental principles governing the motion of physical objects subject to applied forces form the foundation for much of the later scientific education of the engineer. The primary concepts of interest include motion, forces, and energy.

Mechanical Systems

A mechanical system may consist of solid or fluid parts. It is useful to idealize the

FIGURE 3.17. Laser systems can be used to measure precisely angular and linear displacement. (Courtesy Hewlett-Packard Company)

description of such a system by introducing the concept of a *body*, a system of fixed identify whose composition cannot change. Frequently the detailed structure of a body is of no concern. In determining the effects of gravitational forces on the trajectory of a spacecraft, we can usually ignore the rotation of the craft about its own axis. In these cases it usually suffices to treat the body as a *particle* without extended size, located at a point in space.

In more complex problems the detailed structure of the body may be important. Certainly the orientation of a tennis racket is just as important as its trajectory in determining the success of a serve. If the particles constituting an object such as the racket do not change their position relative to one another, then we refer to the object as a *rigid body*. While a rigid body will always retain its size and shape, a *fluid* is a substance that can deform or flow. Here we use the term fluid in its most general sense to include liquids, gases, and plasmas. A liquid has a definite volume and is nearly incompressible. On the other hand, the volume of gases and plasmas will expand to fill a closed container, and they are easily compressed. (Figure 3.18)

Kinematics

The simplest aspect of mechanics involves the motion of a body, that is, the changes in its position or orientation. The study of motion is known as *kinematics*. Motion is a relative concept. To describe the motion of objects, we must first introduce a point of reference. The position of the body is represented by a

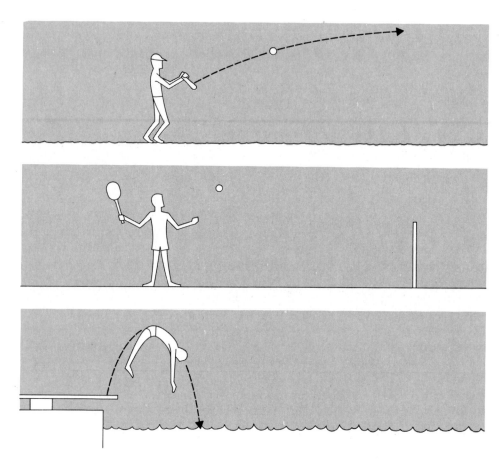

FIGURE 3.18. Three different mechanical bodies: (*a*) the motion of a baseball modeled as a particle, (*b*) a tennis racket modeled as a rigid body, and (*c*) the fluid in a swimming pool.

coordinate specifying its *displacement* from this position as a function of time, say, denoted by $x(t)$.

The time rate of change of the position of the body is given by its *velocity*. For uniform motion, the velocity can be calculated as the distance traveled divided by the time required to travel this distance. More generally the velocity must be calculated as the time derivative of the displacement function $x(t)$:

$$v(t) = \frac{dx}{dt}$$

In a similar manner one can define the *acceleration* of a body as the time rate of change of its velocity. Mathematically, we would calculate the acceleration as the time derivative of the velocity function $v(t)$:

$$a(t) = \frac{dv}{dt} = \frac{d^2x}{dt^2}$$

When the motion of the body is along a straight line, we refer to the motion as *linear* or one-dimensional. In this case we can always describe the motion in a coordinate system with one axis aligned along the direction of motion so that only one distance variable, say x, is involved. In more general terms the displacement of the body is a vector quantity, characterized by both magnitude and direction, that can be visualized as an arrow leading from the old to the new position. Velocity and acceleration are similarly vector quantities, possessing both magnitude and direction. For example, to characterize the motion of a ship, we have to give two coordinates specifying its position (e.g., the longitude and latitude on a map), and similarly two coordinates specifying its velocity (or direction of motion). This is a case of two-dimensional motion, since the ship movement is restricted to the surface of the ocean (provided it does not sink, of course). For the more general three-dimensional motion of a submarine, three displacement and three velocity coordinates are necessary.

Of particular importance are the kinematics of bodies moving along curved paths. The motion of a body along a circular path is sometimes called *angular motion* since it can be characterized by a single coordinate corresponding to the angle through which the body moves as it traces out a circular arc. A special case is the angular motion of a rigid body as it rotates about an axis. One can define the concepts of angular velocity and angular acceleration in a manner similar to that used in linear motion.

The traditional concepts of kinematics (i.e, motion) have been well established since Galileo and Newton. However one of the most fundamental discoveries of this century involved modifying these concepts to describe motion on either the microscopic scale of atoms and nuclei or the astronomical scale of stars and galaxies. Albert Einstein developed his *special theory of relativity* to describe motion on these scales, when velocities become comparable to the speed of light, $c = 3 \times 10^8$ m/s. This theory leads to several surprising predictions. It implies that no object can travel faster than the speed of light. Furthermore, as an object approaches the speed of light, it appears shorter in length while time in its moving frame of reference appears to slow down to a stationary observer. Finally, Einstein's theory implies that mass can be converted into energy according to the famous relationship, $E = mc^2$. These predictions have since been verified many times by experimental measurement. Many engineers (in particular, astronautical engineers and nuclear engineers) encounter such relativistic effects in their work.

Dynamics

Dynamics concerns the study of forces, the causes of motion. A force applied to the body in the direction of motion speeds up or accelerates the body; if forces are applied in a direction opposing the motion, the body decelerates. Forces applied transverse to the motion will change the direction of motion.

A force acting on a body is the result of the mutual interaction of this body with other bodies. If a force results when two bodies are in physical contact, it is called a *surface force* (Figure 3.19). For instance, when we push against a wall, the force between our palm and the wall is a surface force. Forces can also be exerted on a body from a distance, such as the forces exerted on an airplane by gravity or the force exerted on a solenoid by magnetic fields. These are known as *body forces,* since they act through a distance upon every part of the body, not just its surface.

Force is a vector quantity, characterized by both magnitude and direction just as are displacement, velocity, and acceleration. Hence to determine the sum or *resultant* of several forces acting on a body, we cannot add the magnitudes of the forces directly but rather must use vector addition. This addition process is simplest for parallel or perpendicular forces. The calculation of the total force on a body can also be simplified by drawing a diagram of the body of interest, detached

FIGURE 3.19.   Examples of surface and body forces.

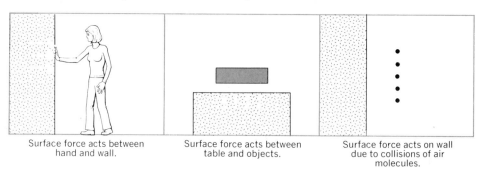

Surface force acts between hand and wall.

Surface force acts between table and objects.

Surface force acts on wall due to collisions of air molecules.

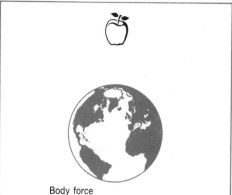

Body force

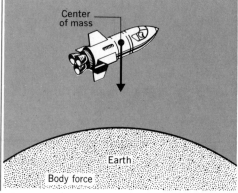

Center of mass

Earth

Body force

from its surroundings, with all of the forces acting upon it represented as vectors. Such figures are known as *free body diagrams* (Figure 3.20).

The laws governing forces and motion were postulated by Newton in the late seventeenth century:

First Law: A body remains either at rest or in a state of uniform motion unless acted upon by forces.

Second Law: A force acting on a body causes an acceleration of the body in the direction of the force and of magnitude proportional to the force and inversely proportional to the mass of the body.

Third Law: When any force acts upon a body, there is an equal force acting in the opposite direction.

In a sense, all three laws can be summarized in a simple equation

$$F = ma = m\frac{d^2x}{dt^2}$$

that represents a mathematical statement of the second law, although the first and third laws can be interpreted as special cases of this formula. Here F is the resultant (total) external force acting on the body, m is its mass, and a is its acceleration. As we have also noted, acceleration a can be expressed mathematically as the second time derivative of the displacement function $x(t)$.

FIGURE 3.20. A free body diagram showing the forces acting on an airplane in flight.

Newton's laws find many applications in engineering practice, from the dynamics of macroscopic objects such as cars or planes, to the behavior of large structures such as bridges or buildings, or to the microscopic effects of subnuclear particles as they cascade through materials. A special case of particular interest involves the forces exerted on a body immersed in a fluid due to the impact of the fluid molecules on its surface. The component of these forces parallel to the surface of the body are known as friction and appear only when the body is in motion through the fluid. The normal or perpendicular component of force on the surface is known as *pressure*. These forces are of importance in the motion of fluids, a subject known as fluid dynamics.

Yet another important case involves forces applied to objects in angular motion (rotation). A force applied to a body in a direction that does not intersect the axis of rotation results in a *torque* that produces an angular acceleration. The magnitude of the torque (and hence the angular acceleration) depends on the magnitude of the applied force and the distance between the force and the axis of rotation.

SUMMARY

Mechanics is the study of motion and dynamics. Mechanical systems can be idealized as bodies of fixed composition. The simplest concepts in mechanics involve kinematics, the study of motion. Here the mathematical relationships between the displacement, velocity, and acceleration of a moving body are analyzed. In general these are vector quantities, characterized by both magnitude and direction. Dynamics concerns the study of forces, the causes of motion. Force is also a vector quantity. Newton's laws can be used to determine the motion that results from forces applied to a body. Special types of forces include forces exerted by fluids (friction and pressure) and forces exerted on rotating objects (torque).

3.3.3. Work, Energy, and Heat

Work

Many of us are inclined to define work as the necessary and unpleasant activity that takes time away from play. However for scientists and engineers, who require a more precise concept, *work* is the product of the force acting on a body and the distance through which the body moves (Figure 3.21):

$$\text{Work} = \text{force} \times \text{distance}$$

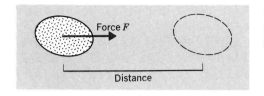

FIGURE 3.21. Work is defined as the product of the distance moved and the force in the direction of the motion.

Only the component of the force in the direction of motion enters into this definition. For example, if we were to tilt the surface of a table on which a book is lying so that it begins to slide down the sloped table surface under the action of its own weight, only the force component parallel to the table surface does work. No work is performed in the direction perpendicular to the table surface because there can be no displacement in this direction. Work is a scalar quantity characterized only by magnitude. If work is performed on a body, then it is taken as negative; if work is performed by the body, then it is positive.

The concept of work does not include time. The same amount of work is done when a body is moved a distance d by a force F, regardless of how long it takes to cover the distance. Therefore it is useful to introduce a new quantity, *power*, to characterize the time rate at which work is performed. Power can be defined mathematically as the time derivative of the work function. When the work is performed at a uniform rate for a period of time, the power is just the amount of work performed divided by the time.

A final concept of some importance is the *efficiency* of a device. Efficiency is the ratio of the power produced by the device to that provided by the device. For example, the efficiency of an electrical motor is given by the ratio of its shaft mechanical power output to the electrical power it consumes. Efficiency can also be defined as the work output to work input for a device.

Although we have introduced the concepts of work, power, and efficiency within the context of mechanical devices, these concepts also apply to other phenomena such as electricity, heat generation, and chemical processes.

Energy

Energy is a measure of the capacity for performing work. A body is said to possess energy if it is capable of doing work. The energy of the body may be defined as the work that could be obtained from the various forms of energy possessed by the body.

We can distinguish between two principal types of energy: stored energy and energy in transition. *Stored energy* is present in various forms such as potential energy, kinetic energy, energy associated with molecular or atomic motion, chemical energy, electrical energy, and nuclear energy. *Energy in transition* exists only at the instant when energy is crossing the surface of a body. Examples of energy in transition include work and heat. Once this transient energy enters the body, it becomes stored energy.

Energy can be produced, stored, transformed from one form to another, and of course, wasted. Indeed, since no machine is 100% efficient, there will always be some waste in the utilization of energy.

Stored energy is conveniently classified into three primary forms: potential energy, kinetic energy, and internal energy (Figure 3.22). The *potential energy* of a body is a measure of the body's capacity for performing work as a result of the body forces exerted upon it. For example, the potential energy of an elevated body due to the force of gravity is equal to the product of the weight w of the body and its elevation above the earth's surface:

$$PE = \text{weight} \times \text{elevation} = wh = mgh$$

(Here we have noted that the *weight* w of a body is equal to its mass m multiplied by the acceleration of gravity g.) If the body were lowered a distance h, it could perform work of magnitude wh.

Kinetic energy corresponds to the energy of motion. The kinetic energy of a body of mass m moving with velocity v is given by

$$KE = \tfrac{1}{2} mv^2$$

The potential and kinetic energies of a body are specified by the bulk or macroscopic properties of mass and velocity. The *internal energy* U represents the microscopic energy associated with the atoms and molecules comprising the body. This component includes all forms of energy other than bulk kinetic and potential energy. Examples of work produced by changes in internal energy include the flow of steam or other hot gases past the blades of a turbine and the combustion of gasoline in a cylinder that produces hot gases that expand against a piston. In both of these cases, work is performed at the expense of internal energy.

Potential, kinetic, and internal energy can be stored in a body. The amounts of each kind of energy can be changed by performing work on the body or by letting the body perform work on its surroundings. Work, however, cannot be stored. It can be identified only at the instant when it crosses the surface of the body. The amount of internal energy can be changed by another energy flow

FIGURE 3.22. Various forms of stored energy.

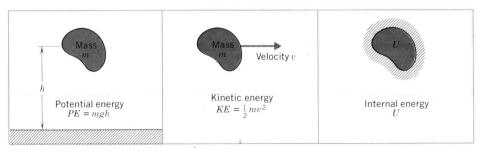

process: heat transfer. To discuss heat transfer, it is necessary to introduce the concept of temperature.

Temperature

Temperature is a property intuitively familiar to most people. It is difficult, however, to give a simple yet precise definition of this quantity. From a microscopic viewpoint, *temperature* is a measure of the average velocity of the atoms or molecules comprising a substance. The faster the molecules move about, the higher the temperature of the substance. But it is extremely difficult to measure molecular motions directly. Instead we must use indirect methods that measure the change in some property of the medium. For example, when a mercury thermometer is placed in contact with a hot medium such as air, the molecules in the glass interact with those of the air and begin to move more rapidly. They then transfer this energy of motion to the mercury atoms. As a result of this increased motion, the mercury expands. This expansion, indicated by a change in the position of the surface of the mercury column, enables us to measure the temperature of the air.

Four scales are generally used to quantify temperature measurements. The Celsius and Fahrenheit scales are based on the freezing and boiling temperatures of water at standard pressure (1.01 MPa). On the *Celsius* scale the freezing point of water is designated as 0°C and the boiling point as 100°C. The difference between these two temperatures is divided into 100 parts or degrees. On the *Fahrenheit* scale, the freezing point is taken as 32°F and the boiling point as 212°F. The difference between these two points is divided into 180 degrees (Figure 3.23).

In scientific work it is more convenient to use so-called *absolute temperature*

FIGURE 3.23.   A comparison of temperature scales.

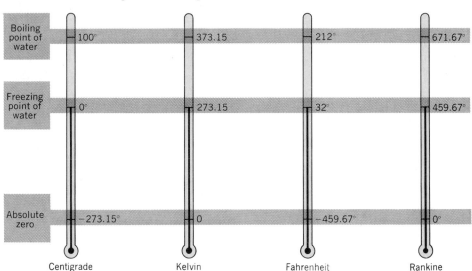

scales. The French physicist, J. A. C. Charles noted that if a fixed mass of gas contained in a fixed volume is cooled, both its temperature and pressure decrease. For each Celsius degree drop in temperature, the pressure decreases by about 1/273 of its value at 0°C. Thus, near −273°C (more precisely, −273.15°C) the pressure of the gas would be zero. This is called the absolute zero temperature. In Fahrenheit this absolute zero point would be −459.67°F. Absolute temperature scales are based on these zero points. The *Kelvin* scale utilizes degrees of the Celsius scale, starting from −273.15°C as its zero point. Hence the Kelvin and Celsius temperature scales are related by

$$K = C + 273.15$$

The corresponding absolute temperature scale based on the Fahrenheit degree is the *Rankine* scale, defined by

$$R = F + 459.67$$

The temperature scale adopted for all scientific and engineering work (the SI system) is the Kelvin scale.

The Ideal Gas Law

There is a definite relation between the temperature, pressure, and volume characterizing each substance. This is known as the *equation of state* for the substance. The relationships between these properties are usually very complex and cannot be described by a simple formula. One important exception is the *ideal gas*. In this model the gas molecules are assumed to move about freely without colliding with one another. Of course this model is only a crude idealization of a real gas in which molecular collisions can become quite important. Nevertheless, many gases approximate an ideal gas when their temperature is sufficiently high or the pressure is sufficiently low.

The equation of state for an ideal gas takes the simple form

$$pV = nRT$$

where p is the pressure of the gas, V is its volume, T is its temperature, n is the number of moles of the gas (the mass of the gas m divided by its molecular weight M), and R is a constant known as the universal gas constant, $R = 8.314$ joule/mole K.

Heat

Heat is another form of energy that is transferred from one body to another because of a temperature difference. A body does not contain heat. Rather heat is a form of energy that can be identified only as it modifies the nature of a substance (its temperature or state) or crosses the boundaries of the substance. When a hot block of copper is placed in a tank containing cold water, heat is transferred from the copper block to the water. Because of this heat transfer, the temperature (and therefore the internal energy) of the copper decreases and that of the water increases. If we wait long enough, the copper block and the water come into thermal equilibrium with one another at the same temperature. At this point the heat transfer ceases, since it can only occur because of a temperature difference, flowing from the higher to lower temperatures.

There are many similarities between heat and work. Both are transient phenomena. Both are identified only at the surface of a body, not within the body, since both represent energy crossing a surface. And both heat and work are measured in the same unit, the joule, as are all forms of energy.

Heat can be transferred by three different mechanisms: conduction, radiation, and convection (Figure 3.24). Thermal *conduction* is heat flow within a body or between two bodies in physical contact. This process involves the microscopic motion of atoms and electrons in a material. It is quite similar to the flow or conduction of electric current. In fact, materials that are good electrical conductors or insulators also happen to be good conductors or insulators of heat.

Radiation heat transfer is the transfer of energy through space in the form of electromagnetic radiation such as light or infrared radiation. *Convection* involves the transfer of heat by the mass motion of a fluid, in which the energy stored in the fluid is transported to a region of lower temperature by the fluid motion.

In practice, all three modes of heat transfer frequently occur simultaneously. A good example is provided by the transfer of heat through the roof of a building. First, heat is transferred to the inside surface of the roof by convection in the air

FIGURE 3.24.   The principal forms of heat transfer: conduction, radiation, and convection.

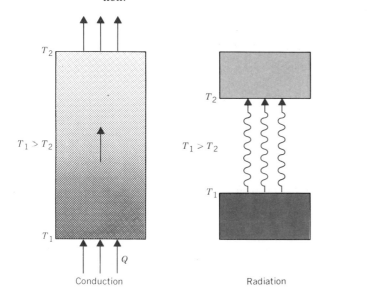
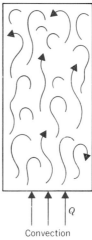

Conduction   Radiation   Convection

inside the building. Then the heat is conducted through the materials comprising the roof. Finally, heat is transported away from the outside surface of the roof by air convection and by radiation (Figure 3.25).

Thermodynamics

In any physical process, energy can neither be created nor destroyed; it can only be converted from one form to another. This is known as the *law of conservation of energy*. This law states that the total energy transferred through the surface of a body must equal the change in stored energy of the body. We have seen that stored energy may be in the form of potential, kinetic, or internal energy. Thus, according to the law of conservation of energy, the sum of the amount of heat transported to (or away from) and work performed on (or by) the body during a given time period must equal the change in the total energy of the body:

$$
\begin{matrix}
\text{Net heat} & \text{Net work} & \text{Change in potential,} \\
\text{input to} & - \text{performed} = & \text{kinetic, and internal} \\
\text{body} & \text{by body} & \text{energy of body}
\end{matrix}
$$

The law of conservation of energy, when stated in this form, is known as the *first law of thermodynamics* (Figure 3.26).

Notice that this law states that work can be produced by changing the potential, kinetic, or internal energy of a body. However this energy can only be entirely coverted into work if there is no heat transfer. As a practical matter, heat transfer can never be eliminated entirely in any physical process. Some transformation of energy into heat is always present. For example, friction accompanying motion will produce heat transfer. As a result, the useful work produced by a body is always less than the change in the total amount of energy. Consequently the efficiency of any work-producing device or machine is always less than 100%.

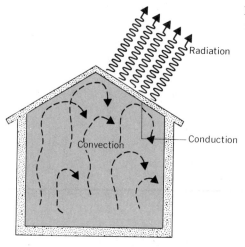

FIGURE 3.25. The heat transfer processes involved in a house.

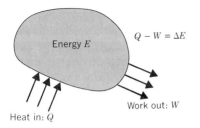

FIGURE 3.26. The work, energy, and heat balance described by the first law of thermodynamics.

Energy E

$Q - W = \Delta E$

Work out: W

Heat in: Q

The efficiency of any energy conversion process is intimately related to the *second law of thermodynamics*. According to this law, it is impossible to convert a given quantity of internal energy completely into work. In any energy conversion process, some energy will always be degraded in availability or quality so that the ability to perform work is reduced.

Devices that convert internal energy into work are known as *heat engines*. For example, consider the production of electricity in a steam-driven power plant. The heat produced by the combustion of a fuel, such as coal or perhaps the heat of a nuclear fission chain reaction in uranium, is used to turn water into steam. This steam is then allowed to expand against the blades of a turbine, spinning the turbine shaft and producing mechanical work (Figure 3.27). The spinning turbine shaft drives the rotor of an electric generator to produce electricity, thereby converting mechanical energy into electrical energy. The steam leaving the tur-

FIGURE 3.27. A steam-driven turbine generator. (Courtesy General Electric Company)

bine is condensed back into water and pumped back to the boiler to repeat the process.

In such a heat engine a "working fluid" (in this case, water or steam) is used to absorb the heat produced by the fuel as internal energy. The fluid then expands against the blades of the turbine, converting this internal energy into work. These processes are governed by the laws of thermodynamics. The second law of thermodynamics restricts the efficiency that can be achieved in this transformation of internal energy into mechanical work. As a result, any heat engine operating at temperatures characterizing most industrial processes will convert only a fraction of the fuel energy into work (typically from 20 to 40%). The remainder of this heat will be rejected to the environment.

Conservation of Mass and Material Balance

Another conservation law of considerable importance to engineers is the *law of conservation of mass*. This states that mass can neither be created nor destroyed (at least on a macroscopic scale). This law can be applied to situations that involve the movement, distribution, sorting, mixing, and separation of materials. For example, it might be applied to the flow of materials through a complex chemical processing facility. This conservation law is most commonly applied to a designated volume in space with well-defined boundaries known as a *control volume*. A balance between the material flow into and out of the surface of this volume is then established. The law of conservation of mass implies that the time rate of change of the mass inside the control volume must equal the difference in the rates with which mass flows into and out of the volume (Figure 3.28). In the case of a steady process with no mass accumulation in the volume, the rates of mass inflow and outflow must be equal.

The laws of conservation of energy and mass are two of the most important and useful scientific principles to the engineer. However on the microscopic level of atoms and nuclei, these laws must be somewhat modified. Einstein's special theory of relativity implies that on this scale, both laws are combined into a single law that allows mass to be converted into energy and vice versa. More precisely, the amount of energy produced by the conversion of an amount of mass m is

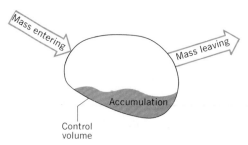

Control
volume

FIGURE 3.28. It is most convenient to apply a material balance to a region in space known as a control volume.

given by the famous formula, $E = mc^2$, where c is the speed of light, $c = 3 \times 10^8$ m/s. This result implies that the energy that can be obtained from a small conversion of mass is truly staggering. For example, the amount of mass converted into energy each day in a gigantic billion watt nuclear power plant is only 0.15 g.

As a practical matter, however, engineers are normally concerned with macroscopic phenomena. On this scale the laws of conservation of energy and mass can be applied in a separate and independent fashion. It is only for microscopic processes such as atomic or nuclear energy release that the more general conservation law becomes necessary.

Machines and Engines

Energy can be converted from one form into another. Devices designed to produce energy transformations are known as *machines*. For example, pulleys and levers are simple machines that redirect mechanical energy. The water wheel and hydraulic turbine are somewhat more complex machines that convert the potential energy of water at higher elevation into the kinetic energy of a turning shaft.

An *engine* is a device that transforms internal energy into mechanical work. The steam-driven turbine is one example we have already noted. Other familiar examples are internal combustion engines or gas turbines that convert the internal energy produced by the combustion of fuels into mechanical work by allowing the hot combustion gases to expand against the surface of pistons or turbine blades. The concept of a heat engine can be reversed in a refrigeration unit, which is designed to cool a substance. In this case, mechanical work must be performed first to cool a working fluid to a temperature below that of the substance to be cooled, and then to pump this fluid past the surface of the substance to cool it by convection.

SUMMARY

Work is defined as the product of the force acting on a body times the distance (in the direction of the force) through which the body moves. Power is defined to be the time rate at which work is performed. The efficiency of a device or process is the ratio of its work (or power) output to its work (or power) input. Energy is the capacity for performing work. Energy can be stored in one of three forms: potential, kinetic, or internal energy. It can also be transferred either as work or heat. Heat is a form of energy transferred from one body to another as a result of a temperature difference. Heat transfer can occur via conduction, radiation, or convection processes. Of key importance in mechanical processes is the first law of thermodynamics, which states that the difference between the heat transferred to a body and the work it performs must be equal to the change in its stored energy. This is simply a restatement of the law of conservation of energy. A related concept is the law of conservation of mass. The second law of thermodynamics restricts the efficiency with which internal energy can be converted into work. Devices that transform mechanical energy from one form to another are known as machines; devices that convert internal energy into work are known as engines.

3.3.4. Electrical Processes

The technology for producing and utilizing electrical energy has had a most dramatic impact on twentieth century society. In a like manner, the importance of electricity and electrical processes to the modern engineer cannot be overemphasized. Electricity is one of the most convenient forms of energy. It can be easily produced, transmitted, and converted into other forms of energy. Electrical signals have proven to be the most versatile and convenient medium for communication and control. Most instrumentation and measurement devices convert measured quantities such as temperature, displacement, and speed into electrical signals that can be processed into a form suitable for analysis (e.g., amplified and converted into analog or digital form for display). Because of their speed of operation, low power requirements, and flexibility, electric circuits have become the most common logic devices in computing systems, including even biological "computers" such as the human brain.

Electric and Magnetic Forces

Electrical processes make use of the forces that arise between electrically charged objects. Experimentally these forces are found to depend not only on the magnitude of the charges and their distance of separation, but also upon their velocity. For convenience, we distinguish between forces arising between stationary charges, known as *electric forces*, and those forces depending on the motion of the charges, known as *magnetic forces*.

The electric or *Coulomb force* between two stationary charges is found to be proportional to the magnitude of the charges and inversely proportional to the square of the distance of their separation. Particles of like charge repel one another; those of opposite charge attract one another. Since matter is made up of atoms in which the positive charge of the nucleus is exactly balanced by the negative charges of the electrons, matter on a macroscopic scale tends to be electrically neutral and therefore not subject to electric forces under ordinary circumstances. However it is possible to separate the charges of a few atoms (say, by rubbing a glass rod with a piece of silk) to build up a positive or negative charge on a sample of material so that it will experience electric forces.

Although we have described the electric forces arising between stationary charges as an "action at a distance" type of phenomenon, it is sometimes convenient instead to imagine that one of the charged particles establishes a region of influence about it known as an *electric field*. This field then acts to exert a force on the second charged particle.

This concept of a force "field" established by an electric charge is particularly useful in dealing with the more complicated form of the force law characterizing moving charged particles. These forces are quite complex, since they depend not only on the charge and distance of separation, but also on the velocities (and direction of motion) of the charges. Furthermore, in more complex situations the forces arising between moving charged particles will also depend on the acceleration of the charges. The forces arising between moving charges are known as *magnetic forces*. As with electric forces between stationary charges, one can imagine that a moving charged particle produces a region of influence known as a magnetic field. When both the electric and magnetic components of the forces (or fields) are treated simultaneously, they are called *electromagnetic forces* (or fields).

Electrical machinery takes advantage of such forces on charged particles as they move through a magnetic field. Since an electric current is nothing more than the flow of a large number of charged particles (electrons), current-carrying conductors can produce and interact with magnetic fields in such a way as to convert electrical energy into mechanical work. In a similar sense, time-varying magnetic fields, caused by the mechanical motion of a magnet, for example, will induce current flow in wires, thereby generating electricity.

Electric Currents and Voltages

On a macroscopic scale matter is electrically neutral with most electrons being tightly bound to atomic nuclei. However certain materials such as metals are characterized by relatively weak attachment of the outermost atomic electrons. In these materials some outer electrons can become temporarily separated from the atoms and drift randomly about in the material, occasionally colliding with other electrons, knocking them loose, or becoming reattached to atoms. We can influence the motion of these free or "conduction" electrons by applying an electric field to the material. This field will exert a force on the electrons, thereby influencing their motion in such a way as to cause a net drift or flow of electric charge. This charge flow is known on a macroscopic level as an *electric current*.

As the conduction electrons drift through the material, they will experience numerous collisions with other atomic electrons. This resistance to current flow must be overcome by the applied electric field. The work per unit charge necessary to bring like charges together or induce current flow is known as the *electric potential* or *voltage*. The magnitude of the current that will flow in a conductor is directly proportional to the potential difference or voltage across its ends. This can be stated in terms of *Ohm's law*:

$$V = IR$$

where V is the voltage, I is the current, and R is the *resistance* of the conductor.

The resistance of a material depends upon many factors such as the density of free or loosely bound electrons and the degree to which atomic collision processes inhibit electron flow. Materials with low values of resistance such as copper, silver, and aluminum are known as *conductors*. Other materials such as glass and carbon are characterized by tightly bound electrons and large resistances and are known as *insulators*. Of great interest in modern electronics applications is a type of material intermediate between a conductor and an insulator known as a *semiconductor*. Only small concentrations of free charge carriers are normally present in a semiconductor. Their relative number can be strongly influenced by adding a controlled amount of a selected impurity (doping) or absorbing a small amount of energy (such as incident light). Devices such as transistors and solar cells are based on semiconducting materials.

The application of a voltage to cause the flow of current in a wire corresponds to a certain rate of performing work, that is, to a certain *power*. More precisely, the power required to maintain the current flow is given by the product of the voltage V and the current I:

$$P = VI$$

This power investment appears as heat in the wire. This is the most common form of energy dissipation in electric systems. It is the basis for the operation of electric heaters. But it is also the source of inefficiency in the operation of electric motors and generators.

Electrical Circuits

An electrical *circuit* is a closed path about which current flow is confined. The simplest type of circuit involves steady current flow through wires or devices characterized only by resistance. Although every wire or electrical device will have some amount of resistance associated with it, it is convenient to represent an electrical system by an idealized circuit in which all resistance is lumped into separate resistors connected by perfectly conducting wires. The simplest electrical circuit would then consist of an electrical source such as a battery or a generator that can produce electrical energy as a voltage, connected by perfectly conducting wires to a resistor that dissipates this energy as heat (Figure 3.29).

A resistor is a circuit element that can only dissipate electrical energy. However other types of circuit elements can store electrical energy in the form of electric or magnetic fields. One such circuit element known as a *capacitor* consists of electrically isolated plates on which charge can be accumulated to establish electric fields. Another circuit element known as an *inductor* consists of coils of wire that can produce a magnetic field.

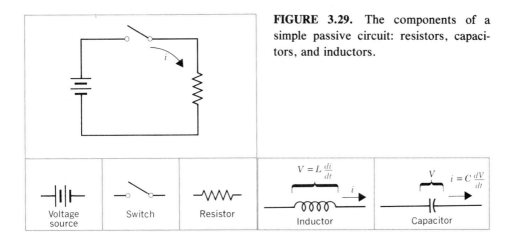

FIGURE 3.29. The components of a simple passive circuit: resistors, capacitors, and inductors.

Electrical circuits consisting of resistors, capacitors, and inductors are called *passive* circuits, while these devices are known as passive circuit elements. The term "passive" means that these devices only store or dissipate electrical energy. In contrast, "active" circuit elements such as transistors, amplifiers, and transducers can also serve as sources of energy.

Any material that carries an electrical current—even a straight wire—exhibits the characteristics of passive circuit elements: resistance, capacitance, and inductance. Fortunately the voltages and currents in even the most complex passive circuits can be analyzed using two simple laws that arise from the conservation of charge and energy in the circuit. These principles, known as *Kirchhoff's laws*, state that (1) the sum of the voltages around any closed path in the circuit must be zero; and (2) the sum of the currents into any node in the circuit is zero. These laws can be applied to the circuit to yield a set of equations to determine any voltage or current in the circuit.

Active Circuit Elements and Electronics

Passive circuit elements such as resistors, capacitors, and inductors can only store or dissipate electrical energy. *Active* circuit elements such as transistors and amplifiers can also serve as sources of energy. Furthermore, active circuit elements can sometimes change their state or characteristics in response to an applied signal.

Perhaps the most important example of an active circuit device is the *transistor*. These devices can amplify an electrical signal. They are characterized by a gain; that is, a signal entering the transistor can be amplified in magnitude by the device. This is in sharp contrast to a passive electrical circuit that modifies a signal in various ways but invariably reduces the amplitude of the signal.

Transistors as well as most other modern circuit elements are fabricated from semiconductors. These devices are occasionally referred to as *solid state* electronics to contrast them with older vacuum tube devices used in the early days of electronics. To make a semiconductor, one takes a material such as silicon, which is ordinarily an insulator, and injects into it or "dopes" it with minute quantities of impurity atoms that modify the charge carrier concentrations. For example, if we dope silicon with phosphorus, each phosphorus atom adds an extra electron to the silicon crystal, which act as the major carrier for conducting electric current. This is known as an *n-type* semiconductor, since the impurity has introduced a negative charge carrier. The addition of boron impurity atoms to the silicon creates electron deficiences or "holes" that act as positive charge carriers, thereby yielding a *p-type* semiconductor.

By joining together semiconductor materials of different types, one can build diode rectifiers or transistor amplifiers or switches. Such devices, known as *junction* or *bipolar transistors*, were originally fabricated by physically joining together different types of semiconductors. However they were replaced by a second type of transistor known as the *insulated gate field effect transistor* in which a thin insulating layer of silicon dioxide is placed between layers of doped silicon. Of most common use today are metal oxide semiconductor field effect transistors known as MOSFETs or simply as MOS. These devices can be created by diffusing or etching islands of *n*-type silicon on a substrate of *p*-type silicon. Rapid developments in surface science and materials have provided the capability to fabricate thousands of such transistors on a single tiny clip of silicon, yielding an *integrated circuit* or IC. These circuits can not only contain active elements such as transistors, but they can also simulate passive elements such as resistors and capacitors (Figure 3.30).

Integrated circuit or microelectronics technology has revolutionized electrical engineering. As we shall discuss in greater detail in the next chapter, the trend is toward using transistors in integrated circuits as switches (i.e., on/off devices) to process electrical signals in digital form. That is, rather than using a circuit to process continuously varying voltages and currents, modern integrated circuits process a series of discrete on/off signals that can be used to represent information in the form of binary numbers. The various transistors in the circuit are then arranged into digital logic devices known as *gates*, which process this digital information.

The development of integrated circuits and microelectronics has had a major impact on all areas of technology. The advances in this field have led to truly phenomenal developments, such as the *microprocessor*, a tiny computer on a single chip of silicon (Figure 3.31). Indeed, since the first introduction of integrated circuits in the late 1950s, the density of transistors that engineers have been able to achieve in microelectronic circuit fabrication has roughly doubled each year. The memory circuits in modern computers contain over 64,000 transistors in a silicon chip no larger than a fingernail. New capabilities in integrated circuit design and fabrication now promise to increase the density of integrated circuits over the next several years by as much as a factor of 1000, leading to *very large scale integrated circuits* (or VLSI).

FIGURE 3.30. Modern solid state electronics has made possible the development of integrated circuit chips containing hundreds of thousands of circuit elements on a single tiny chip of silicon. (Courtesy IBM Corporation)

FIGURE 3.31. Of major importance in microelectronics has been the development of the microprocessor, a computer on a tiny chip of silicon. (Courtesy Digital Equipment Corporation)

SUMMARY

Electricity is one of the most convenient forms of energy since it can be easily produced, transmitted, and converted into other forms of energy. Electrical processes make use of the forces that arise between charged particles. Forces between stationary charges are known as electric forces; those between moving charges are magnetic forces. An electric field or voltage can be applied to a conducting material to induce a flow of charged particles, that is, an electric current. An electrical circuit is a closed path about which current flow is confined. If the circuit contains only passive elements such as resistance, capacitors, and inductors, then it is capable only of storing or dissipating energy. Of more interest are circuits containing active elements such as transistors that can amplify or switch signals. Thousands of such elements can be fabricated on a tiny chip of semiconducting material such as silicon to make possible the integrated circuits that form the basis of modern microelectronics.

3.3.5. Chemical Processes

Chemistry is the study of the properties and transformations of matter. Chemical processes play a very important role in such diverse areas as the manufacturing of metals, alloys, and plastics; combustion; power generation; petroleum refining; photography; and integrated circuit technology. In this section we shall discuss briefly some of the more important principles relevant to chemical processes used in engineering.

Chemists classify the properties characterizing a substance into two general catagories. *Extensive* properties, such as mass and volume, depend on the size of the sample of material. *Intensive* properties such as melting point and density are independent of sample size. *Physical* properties such as density and melting point can be specified without reference to any other substance. *Chemical* properties characterize interactions between substances. It is this latter class of properties that is the concern of chemistry.

Compounds and Mixtures

On a microscopic level chemical processes involve the interaction of atoms as they bind together or rearrange themselves to form molecules. For example, the burning of fossil fuels such as coal involves the rearrangement of atoms from one molecular form (carbon and oxygen) to another (carbon dioxide).

Elements are the simplest form of matter, consisting of atoms of only a single type. The presently known 105 elements form the basic building blocks of all of

FIGURE 3.32. Chemistry involves the study of the structure, properties, and transformations of matter. (Courtesy Hewlett-Packard Company)

the more complex substances produced by chemists. Elements can be combined together to form *compounds*, substances having their constituent elements always present in the same proportions. When the compound water is formed, it will always be composed of the elements hydrogen and oxygen in the proportion by weight of one part hydrogen to eight parts oxygen. On a microscopic scale we know that this ratio arises from the composition of the water molecule, H_2O, and the relative weights of hydrogen and oxygen atoms.

In contrast, elements can also be combined into *mixtures*, which are characterized by a variable composition. Sea water is a mixture of salt and water, since we can vary the salinity at will by simply adding more salt. If the mixture is uniform or homogeneous, we refer to it as a *solution*. For example, sea water is a solution of salt in water. On the other hand, a heterogeneous mixture such as oil and water would not be classified as a solution since on a sufficiently fine scale we would still be able to distinguish between the individual properties of oil and water.

Atomic Weights and Moles

The mass of an individual atom or molecule is extremely small. For example, a single oxygen atom has a mass of 25.5590×10^{-27} kg. To avoid using such cumbersome quantities, chemists instead measure the mass of all atoms and molecules in terms of a convenient reference chosen as one-twelfth the mass of the carbon-12 atom: 1.67252×10^{-27} kg. More precisely, one defines the *atomic weight* of an element as the ratio of the mass of the atom to one-twelfth the mass of a carbon atom. The corresponding unit is known as the *atomic mass unit* or *amu*. The mass of an oxygen atom on this scale is 15.9994 amu. That is, the atomic weight of oxygen is 15.9994. The *molecular weight* of a compound is calculated in a similar manner by summing the atomic weights of the atoms comprising the compound.

Atomic units such as the amu are rather inconvenient to use in describing macroscopic chemical processes because in a typical chemical reaction, billions upon billions of atoms or molecules are combining or rearranging. Hence, to facilitate the transition to the study of chemical reactions among macroscopic-sized samples, the chemist defines a reference amount of substance, the *mole*, as the number of atoms contained in 12 g (0.012 kg) of carbon-12. This number has been determined experimentally to be 6.023×10^{23} atoms and is known as *Avogadro's number*.

The choice of carbon-12 as the basis for defining the mole is particularly convenient, since this atom also serves to define the atomic weight. Therefore one mole of any substance (that is, 6.023×10^{23} atoms or molecules) has a mass equal to its atomic or molecular weight in grams. A mole of atomic oxygen has a mass of 15.9994 g. Since the mass of the mole is usually expressed in grams, it is also common to refer to an amount of substance in gram moles. (Actually, a somewhat longer name, the "gram molecular weight," was the origin of the name "mole," rather than any humorous reference to a subterranean animal.)

Chemical Symbols and Formulas

Symbols and formulas play an important role in chemistry. In fact, the layperson usually associates the word "formula" with the symbol for some complex chemical compound (e.g., "Tell us where you hid the secret formula for the rocket fuel, Mr. Bond, or we will slice you into tiny pieces with this laser beam . . . "). Each element is assigned a chemical symbol associated with its name, such as H for hydrogen, O for oxygen, and Cl for chlorine. The only confusing departure from this prescription is the use of Latin names for some elements, for example, Na for sodium (natrium), Ag for silver (argentum), and Cu for copper (cuprum). A chemical compound could then be represented by a *chemical formula* that gives the actual number of each kind of atom found in the molecule, such as H_2O for water. For more complex chemical compounds such as organic materials, a chemical formula is used that represents not only the molecular constituents, but also provides information about the nature of the chemical bonds that hold the atoms together in the molecule. For example, the *structural formula* for acetic acid is:

$$
\begin{array}{ccc}
 & H & & O \\
 & | & & \| \\
H- & C & -C- & O-H \\
 & | & \\
 & H & \\
\end{array}
$$

Chemical Reactions

A chemical process involves the reaction of two or more constituents. Such a reaction can be represented by an equation that symbolizes the changes that occur to the reacting species. For example, the equation representing the reaction between carbon and oxygen to produce CO_2 is

$$C + O_2 \rightarrow CO_2$$

In this equation the substances on the left (C and O_2) are known as *reactants*, while those on the right (CO_2) are known as *products*. The arrow indicates that the reaction proceeds in the direction from C and O_2 as separate substances to CO_2 as product.

Such chemical reaction equations can also be interpreted as balance relations. For example, our equation above indicates that one molecule of carbon combines with one molecule of oxygen to form one molecule of carbon dioxide. Since a mole of any substance contains a fixed number of molecules, the reaction equation could also be read as stating that one mole of carbon reacts with one mole of oxygen to produce one mole of carbon dioxide. Sometimes it is necessary to insert coefficients in front of the reacting species to yield a balanced reaction equation. The determination of the proper coefficients to use, that is, the balancing of the reaction equation to yield the proper quantitative aspects of chemical composition and reaction, is known as *stoichiometry*.

Chemical Reaction Energetics and Kinetics

The study of chemical reactions not only involves the determination of the stoichiometry of the reactants and the products. It also involves determining why certain reactions take place and others do not. This particular topic is known as *chemical thermodynamics*, and it involves the energy changes that accompany chemical processes (just as physical thermodynamics involves the energy changes that accompany mechanical processes).

Of equal importance is the dynamics of the chemical reaction, the speed or rate at which the reaction occurs, the effect of outside agents such as temperature or catalysts, and the nature of the reactants and the products. This subject is known as *chemical reaction kinetics*. To determine the rate at which a chemical reaction proceeds, one must establish the mechanism of the reaction—the detailed sequence of steps that are followed along the path from reactants to products. Chemical reaction kinetics also involve consideration of the detailed interaction mechanisms among the chemical species at the atomic level, that is, atomic collision processes. These subjects are generally treated in courses on physical chemistry.

SUMMARY

Chemistry involves the study of the properties and transformations of matter. At the most fundamental level are the elements. These can be combined into compounds or mixtures. The chemist introduces the concept of atomic weight to characterize the mass of individual atoms or molecules. The mole is used to quantify the amount of a substance. Symbols represent various elements and chemical compounds. These symbols can be arranged into equations to describe chemical reactions. Although the chemical equations describe the relative amount of each substance participating in the reaction, the additional subjects of chemical reaction thermodynamics and kinetics are necessary to analyze the detailed energetics and rates of chemical processes.

3.3.6. Materials

If we identify engineers as the designers and builders of structures, machines, or processes, then we immediately understand why the subject of engineering materials plays such an important role in engineering activities. For engineering design depends heavily on available materials. Indeed, a design is only as good as the materials that comprise it.

The engineer is frequently faced with choosing the best material for a particular application. Some designs may require the use of high strength materials; others may stress ductility and flexibility; still others may require specialized properties such as electrical conductivity or resistance to radiation damage. Fortunately, engineers have a very wide range of materials from which to choose. Chemists and physicists have roamed through the periodic table, making countless combinations of elements into new materials with every conceivable property. The engineer can choose from metals, metal alloys, ceramics, plastics and rubber, electronic materials such as semiconductors, and even natural products such as wood and fiber.

It is important that the engineer acquire familiarity with the major types of engineering materials and the methods used to determine and characterize the properties of these materials. A variety of analytical skills should be mastered so that the materials requirements for a particular engineering application can be determined. We shall survey these topics, including a discussion of properties of materials (with an emphasis on mechanical properties), materials performance, and types of materials.

Mechanical Properties—Strength of Materials

The characteristics of materials that are usually associated with engineering design involve strength. Related mechanical properties, such as elasticity, ductility, hardness, and toughness determine the behavior of the materials under applied forces or loads. These properties not only vary greatly from material to material, but they are also very sensitive to the manufacturing processes used to prepare the material, the type of applied forces, and the environment of the material (e.g., temperature, chemical compatability with adjacent materials, and radiation exposure).

The forces or loads acting on a body can be classified as static or dynamic. *Dynamic forces* include the force of impact, alternating or cyclic forces, and vibration. These forces are of great importance in machine design. *Static forces* or loads are of more concern in structural design. There are three types of static forces: tensile, compressive, and shear. Under *tension* the material is pulled apart. *Compressive* forces push against the material, acting to compress it. *Shear* forces tend to slice or tear the material (Figure 3.33).

The effect of forces on materials is analyzed using the concept of *stress*. This is defined as the force applied to the material per unit area and is usually denoted by the Greek symbol sigma:

$$\text{Stress} = \frac{\text{Force}}{\text{Area}} = \sigma$$

The unit of stress is the same as that for pressure, N/m² or pascal.

Any stress applied to a material will result in some deformation. For example, the material might lengthen during tension or shorten during compression. This change in material dimensions under stress is called *strain*. More precisely, the strain resulting from an applied stress is its fractional change in dimension.

Every material undergoes strain when subjected to a stress. The extent of this strain is very important in engineering design. A small strain might be quite acceptable in a design, while a larger strain would exceed tolerances. To determine the magnitude of the strain expected for a material subjected to a given stress or load, we need to know the stress–strain relationship characterizing the material (Figure 3.34). This relationship can be determined by experimental testing in which a given stress (force) is applied to a specimen of the material, and the change in its dimension (length) is measured. If the strain increases linearly with stress, then we say the material is *elastic*. When the stress is removed, an elastic material will return to its original shape. However there is a limiting strain, called the *elastic limit*, beyond which a permanent change in the shape or plastic deformation of the material will occur. For engineering purposes, the elastic limit is frequently taken as the *yield point* of the material.

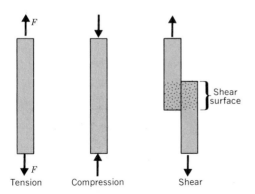

FIGURE 3.33. The three basic types of stress: tension, compression, and shear.

FIGURE 3.34. Strength testing of a specimen in civil engineering. (Courtesy University of Michigan College of Engineering)

There are other mechanical properties by which we may characterize materials. The amount of plastic deformation occurring at the failure point of the stress-strain curve (Figure 3.35) is known as the *ductility* of the material. To calculate the *tensile strength* of the material we divide the maximum load by the original cross-sectional area of the material at the point of fracture. The *toughness* of the material refers to the measure of energy necessary to fracture the material. The *hardness* is defined as the resistance of a material to the penetration of its surface. The *modulus of elasticity* of the material refers to the coefficient relating strain to stress in the regime of elastic behavior. All of these characteristics play an important role in the choice of materials for a given engineering design.

Other important characteristics must be considered when the material is to be subjected to dynamic or time-dependent loads. For example, when an airplane lands, there are large dynamic loads on the wheels and undercarriage of the plane. Repeated loadings may fatigue materials and cause eventual failure. Typically the fatigue life of a material is measured by subjecting it to repeated loadings until failure occurs.

Other Factors Involved in Materials Selection

A material is selected for an engineering application because of many factors in addition to mechanical properties such as strength. For example, the thermal

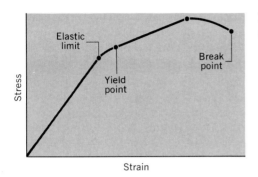

FIGURE 3.35. The stress-strain curve characterizing a material.

properties of a material (e.g., its melting point, thermal conductivity, expansion properties, and heat capacity) are frequently of importance. Chemical properties are also important, since most materials used by engineers are subject to chemical deterioration, oxidation, or corrosion. Many materials are chosen for their electrical properties. In nuclear systems materials must be able to maintain strength or ductility in high radiation environments. Other factors such as the fabricability, weight, durability, appearance, and cost of the the materials should be considered in engineering design.

Types of Engineering Materials

The most widely used materials in engineering are *metals*. Metals exhibit high strength, toughness, ductility, and good electrical and thermal conductivity. Furthermore, they can be easily processed into the desired shape. Rarely are primary metals such as iron, copper, aluminum, or titanium used alone. It has been found that the strength and toughness of metals can be increased markedly when they are combined or alloyed with other elements. The discipline of metallurgy involves the development of such metal *alloys*. Perhaps the most important example is steel which is composed of carbon and other elements added to iron. The mechanical properties of metals can also be strongly influenced by manufacturing and fabrication methods, such as by hammering, rolling, or heat treating.

High temperature applications in which strength is at a premium frequently require the use of *ceramics* or refractory (i.e., heat-resistant) materials such as oxides, nitrides, carbides, and borides of metals such as aluminum, beryllium, and titanium. Another useful class of materials include the *plastics*, which are made of long chains of hydrocarbon molecules known as polymers. The advances in developing plastics for various uses has been one of the impressive achievements of twentieth-century chemistry.

Glasses are noncrystalline or amorphous materials characterized by high strength and brittleness. Although most people know of silicon-based glasses such as window glass from everyday use, many materials including metals can be made into glasses by rapid cooling from molten form.

One can sometimes take advantage of the desirable properties of materials and compensate for their deficiencies by forming *composite* materials. For example, fiberglass is formed by embedding fibers of glass in plastic. One can use the strength of glass to reinforce the plastic and use the ductility of plastic to compensate for the brittleness of the glass. Various new fiber-reinforced plastic composite materials are now used in industry where low weight, high strength, and stiffness properties are desired.

More specialized materials find important roles in engineering. For example, many of the recent advances made in the area of electronics have depended on the development of new materials. Of particular importance have been the semiconducting materials such as silicon and gallium arsenide. Also important has been the development of prosthetic materials for use in biological systems; for example, many advances in medicine require the development of new materials for artificial limbs or organs that are compatible with the human body.

SUMMARY

Engineering design depends heavily on the availability of suitable materials. Of particular concern are the mechanical properties of materials. Forces applied to materials are characterized as tensile, compressive, or shear stresses. Associated with these applied stresses is some deformation of the material known as strain. If the strain is proportional to the stress, the material is said to be elastic. Eventually the body will reach its elastic limit or yield point, beyond which a permanent change in shape will occur. Materials are characterized mechanically by their modulus of elasticity, yield strength, and tensile strength. Other properties of concern include thermal, chemical, and electrical properties, as well as ease of fabrication, weight, durability, appearance, and cost. Engineers can choose from a wide variety of materials including metals, metal alloys, ceramics, plastics, glasses, and composite materials.

3.3.7.   Systems and Control

Dealing with problems of overwhelming complexity is a familiar engineering activity. Methods of abstraction are employed to represent the complicated behavior of real systems by simplified models that are more amenable to analysis. The methods used for developing and analyzing models of complicated systems have evolved over the past several decades into a distinct discipline known as *systems engineering*. The mathematical study of models of complex systems has led to the development of very general and powerful methods for the analysis of feedback control systems, stability, and optimization.

Although systems engineering was developed as a subdiscipline of electrical engineering for the design of electrical circuits, it has since been extended to other fields. Electrical systems ranging in complexity from microscopic circuit chips to massive computers or electrical power distribution networks have been designed using systems methods. Applications have also been made to mechanical systems ranging from automobiles to large manufacturing plants. Systems concepts have also been applied to systems outside of engineering. For example, there have been significant applications in biology, ranging from the study of simple cells to complex organisms (including pop culture applications such as "biofeedback"). Some of the most important (and controversial) applications of systems methods have occurred in the social sciences, involving systems as complex as the national economy and international politics.

We regard a *system* as any collection of interacting physical objects. Indeed, almost anything that serves a function can be regarded as a system. All systems are characterized by some form of operation. The description of the system can usually be represented as a mathematical model that includes provisions for various inputs and outputs. Frequently a *black-box* approach is adopted in which the detailed functions of the system are unknown (or not of direct concern), and only the input–output relationship is studied. In this sense the system of interest implies a cause–effect (or stimulus–response) relationship, and the engineer studies the relationships between system inputs and outputs. We have sketched the black-box diagrams of several systems in Figure 3.36.

Most studies of system behavior are concerned with dynamic processes, that is, processes that vary as functions of time. Indeed, steady-state or time-

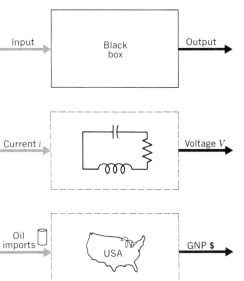

FIGURE 3.36. Examples of black box diagrams of systems.

independent studies of systems properties are rarely of interest since they imply equilibrium behavior. Dynamic models of systems usually are represented mathematically as a set of differential equations involving derivatives with respect to time, that is, time rates of changes of quantities. In this sense, then, systems analysis is closely related to the subject of differential equations. It has benefited significantly from mathematical developments in this field.

One of the most important applications of systems theory is to *automatic control*. Many complex modern systems cannot tolerate the imperfection that accompanies human intervention. Human capabilities are too primitive to effectively handle the rapid response required by modern aircraft or the intricate circuitry of a large computer. Therefore the engineer has tended more and more to remove the human element from systems and replace it by automatic controls that allow the system to adjust its behavior in response to changing conditions without human intervention.

Of course, an automatic control system is not new to nature. Consider a particularly primitive biological system, the weekend tennis player attempting to return a serve. The player first senses the position of the tennis ball by sight, then makes the necessary muscle movements required to bring the racket into position. During this movement the position of the racket is sensed both visually and through touch. As the ball moves closer and closer, the racket position is coordinated with the trajectory of the ball so that at the moment of impact the racket is in an optimum position to propel the ball deftly into the net.

The key features of the control system of the tennis player include a means for *detecting* the position of the tennis racket, *comparing* this with the desired position, and then making the necessary *adjustments* to bring the racket into the desired position. The inputs in this system are visual information concerning the trajectory of the ball and the position of the racket; the output is the position of the racket. The key feature that allows the tennis player to control the racket position is the ability to sense racket position and make adjustments. This is an example of *feedback*, the ability to sense a system's output characteristics, compare these against a desired output, and then act on the results of this comparison to modify the inputs of the system. A simple black-box diagram of a feedback system is shown in Figure 3.37.

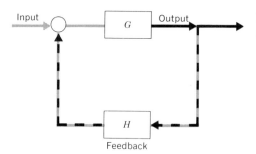

FIGURE 3.37. A black box diagram of a simple feedback system.

Nature has long used feedback control to allow biological systems to adapt to changing environments. Certainly one can identify many such control systems in nature, ranging from the simplest cell to complex organisms such as the human being. But only recently have engineers applied these same concepts to design self-regulating or feedback control systems into machines. Simple examples of automatic control systems surround us. The thermostats used to control temperature in a room, automatic pilots used in airplanes, automatic speed or combustion controls used in automobiles, and automatic color or frequency controls in television sets are familiar examples.

Closely related to systems control is the concept of the *stability* of the system. In any system there are inherent disturbances arising either from external perturbations or from internal properties of the system. Such random disturbances are referred to collectively as *noise*. It is important to examine the role played by noise in a feedback system. For example, if amplifier units are used to amplify the feedback signal, one must make certain that these feedback signals are not overwhelmed by inherent noise in the system.

The effect of noise on a system's behavior is important for another reason. It may happen that under certain conditions, disturbances applied to the system will result in a wildly increasing response. Such a system is said to be *unstable*. Such instability frequently arises because of a "positive feedback" situation. When the output signal from a system is returned as feedback, it is added in a positive sense to the input signal. As the output signal increases due to some input disturbance, it causes a corresponding increase in the input signal, leading to still further increase in the output signal, and so forth.

In order to achieve stability in a system, one requires a negative feedback in which the response to any disturbance of the system is gradually suppressed. Unfortunately, many systems with negative feedback may still be unstable because there is typically a time lag between sensing of the output signal and the feedback of an appropriate modification in the input signal. For example, in a thermostat unit there is usually an appreciable lag between the sensing of temperature and the time at which cooling or heating units respond to modify the temperature. Therefore one can imagine a situation in which the sensing unit determines that the temperature is low so that heating units are activated. This results in a gradual increase in temperature. But by the time that the sensing unit has determined that the temperature has reached the desired level and signaled the heating unit to shut off, the temperature may have continued to increase, overshooting the desired level. In this situation the feedback lag will cause the temperature to oscillate.

All control systems that have a lag in feedback are prone to oscillations. If the lag is too great, these oscillations may occur exactly out of phase so that the input signal is reinforced at exactly the wrong time, effectively creating a positive feedback situation and instability. Such phenomena are of considerable concern in the design of automatic control systems.

SUMMARY

A system is any collection of interacting physical objects. We can model any system by using a black-box approach that represents the system as a cause–effect or input–output relationship. The rapid response, reliability, or hostile environment required by many systems demand automatic control based on feedback systems. Such systems sense the output, compare this against a desired output reference, and then modify the input of the system accordingly. Automatic control systems feed back as input a portion of the output signal to control the system. Feedback systems must respond in a stable fashion when subjected to random disturbances or noise.

Exercises

Scientific Principles

1. Describe one engineering application of each of the following scientific principles:
   - (a) Metal expands when it is heated.
   - (b) A gas heated by chemical combustion will expand and exert a pressure on its surroundings.
   - (c) The force required to distort an elastic body is proportional to the amount of distortion.
   - (d) For every action there is an equal and opposite reaction.

2. Identify the scientific principle that plays the most significant role in each of the following engineering applications:
   - (a) The internal combustion engine
   - (b) Photocopying
   - (c) Solar photovoltaic cells
   - (d) Rocket propulsion

Structure of Matter

3. Protons and neutrons have masses of 1.67×10^{-27} kg and diameters of about 10^{-13} m. Calculate the density of these particles.

4. The earth has a mass of 6.59×10^{24} kg and a diameter of approximately 15 000 km. What would be the diameter of the earth if it had the same mass but was compressed to the density of nuclear particles?

5. Identify the various phases of matter that are present when a very hot ingot of steel is quenched in a tank of water.

6. List three examples of each phase of matter (solid, liquid, plasma) encountered in your everyday activities.

Mechanical Systems

7. Determine the simplest mechanical system that would each of the following situations.
   (a)   The motion of an airplane in flight.
   (b)   The rotation of a record on a turntable.
   (c)   The hydraulic brakes of an automobile.
   (d)   The fall of a parachutist.

8. For what situations would you represent each of the following objects as a particle, a rigid body, and a fluid.
   (a)   A drop of water
   (b)   A baseball
   (c)   An automobile
   (d)   The sun
   (e)   An atomic nucleus

Linear Motion

9. The Le Mans 24-hour endurance race was won in 1980 by a French racing team covering 2860 miles. What was the average speed of the winning team?

10. The average flight time for a round trip transcontinental flight (5000 km one way) is 10 hours. If there is a 400 km/h jet stream velocity flowing from west to east, what is the difference in flight time in each direction?

11. Graph acceleration, velocity, and distance versus time for uniformly accelerated motion.

Forces

12. Two tugboats are pulling against a large tanker with forces of 15 000 N each. The distance between the tugboats and the ship is 150 m, while that between the tugboats is 50 m. What is the total or resultant force exerted on the ship?

13. An object weighing 1 000 N is being lifted by two ropes attached to pulleys. If the angle between the ropes is 45°, then what is the force (tension) exerted on each rope?

14. Explain how one would set the sails of a boat to move northeast into a north headwind. In particular, determine the setting of the sails and the course of the boat that would lead to the most rapid motion into the wind. Explain the reasoning behind your answers.

Newton's Laws of Motion

15. A force gives a 50 kg mass an acceleration of 2 m/s². What would be the acceleration of a 125 kg mass if the same force was applied to it?

16. Explain why it is safest for a parachutist to land in a haystack than on the surface of a concrete road.

17. A person weighing 750 N is standing in an elevator moving upwards with an acceleration of 1.8 m/s². How much tension does the person contribute to the suspension cable of the elevator?

18. Could the elevator in Problem 17 accelerate downwards fast enough to remove all tension from its suspension cable? If so, then how large an acceleration would be required?

19. A car of mass 1500 kg covers 100 m from a standing start in 4 s. Under the assumption of uniform acceleration, determine the force acting on the car during this period.

Work and Power

20. How much work is performed in lifting a 20 kg sack of flour onto a table 1.2 m in height?

21. A coil spring is shortened by 4 cm when a 750 N weight is placed on it. How much work is done when this same spring is shortened 5 cm?

22. A mountain climber of mass 70 kg reaches the top of a 2000 m high peak in 10 hours. How much work did the climber perform? How much power was produced?

Energy

23. Does it require more work to increase the speed of a car from 20 km/h to 30 km/h or from 50 km/h to 60 km/h?

24. A tennis ball and a ping-pong ball are dropped inside a chamber where there is a vacuum. Which ball will have the higher kinetic energy just before the balls reach the bottom of the chamber?

25. Explain why high jumpers take a running start before attempting a leap.

Temperature

26. The average body temperature is 98.6° F. What is this temperature in °C? In K?

27. Find out the maximum and minimum temperatures on the previous day and express these in each of the four temperature scales (C, K, F, R).

Gas Laws

28. Air is contained in a spherical balloon of diameter 60 cm. The pressure in the balloon is 100 kPa. What will be the pressure in the balloon if it shrinks to a diameter of 35 cm?

29. A tire is inflated to a pressure of 30 psi while at the ambient temperature of 20° C. After several hours of driving, the tire temperature rises to 50° C. What is the new tire pressure at this temperature?

Heat

30. Why is a fan installed in front of the radiator of automobiles?

31. Why does it cost more to heat a house when it is windy outside than on a calm day, even when the temperature is the same?

32. Why will ice form of bridges before forming on the roadway pavement of a highway?

33. White clothing is supposed to be better on hot, sunny days. But Bedouins wear black clothing in the desert. Why?

Thermodynamics

34. From a height of 12 m a 5 kg stone falls into a container of water of dimensions 5 m × 5 m × 1 m. Calculate the change in internal energy of the water.

35. The doors of a refrigerator are left open while the refrigerator is operating. Will the average temperature of the kitchen rise, fall, or remain the same? What happens after the door is closed?

36. When a bullet is fired into a slab of wood, its temperature increases. Why?

Conservation of Mass

37. A tank contains 2.5 kg of sea water with a concentration of salt of 5.8%. Sunlight evaporates 600 g of water out of the tank. What is the salt concentration of the remaining water?

38. A tank of volume 8 m³ is filled with a pump at a rate of 100 liters per minute. There is a small leak at the bottom of the tank where water escapes at a rate of 1 liter per minute. How long does it take to fill the tank? How long will it take for the tank to empty due to the leak once the pump is halted?

Machines and Engines

39. List the work-producing machines present in a typical household and identify the principal energy conversion processes in each.

40. Discuss the major advantages and disadvantages of each of the following methods for generating electricity: hydroelectric power, steam engines, steam turbines, and solar photovoltaic power.

41. Diesel engines utilize fuel roughly 20% more efficiently than conventional spark-ignited (Otto cycle) internal combustion engines. Estimate the fuel that would be saved each year if one-half of the automobiles in the United States were converted over to diesel engines.

Electric and Magnetic Forces

42. How many electrons correspond to 1 coulomb of charge?

43. Two positively charged particles are moving at high velocity toward one another. If the effect of magnetic forces are ignored, will these particles be deflected away from each other? How would the consideration of magnetic forces effect the collision process?

44. Explain the behavior of a bar magnet in terms of the microscopic nature of the magnetic forces.

45. Provide a microscopic explanation of the generation of electromagnetic waves (radiation) from a television broadcast antenna. (This will require some additional reading in a physics or electrical engineering text.)

Electric Currents and Voltages

46. Determine the current in a wire carrying a charge of 100 C each second.

47. A car battery can supply 5 amperes of current for a period of 1 minute. How many coulombs of charge are produced?

48. Conduction electrons move very slowly in a material (about one meter per hour). How, then, do we explain the fact that lights in a room come on almost instantly when we turn on the light switch?

Circuits and Devices

49. Identify the circuit elements involved in a typical household appliance.

50. Give five examples of major changes in engineering technology caused by the introduction of integrated circuit technology.

Compounds and Mixtures

51. Categorize each of the following properties as extensive or intensive: mass, volume, melting point, boiling point, density, color, thermal conductivity.

52. Classify the following materials as compounds or mixtures: water (pure), sea water, beer, a penny, rubber cement, wool, a silver spoon.

Atomic Weight and Mole

53. Suppose the atomic mass unit were based on the mass of oxygen-16 rather than carbon-12. Calculate the molecular weight of water in the oxygen and carbon systems.

54. How many moles of carbon are required to produce 1 mole of propane, C_3H_8?

55. What is the mass of one mole of: O_2, H_2SO_4, NH_3, C_8O_{16}?

Chemical Symbols and Formulas

56. Write the chemical symbols for each of the following elements: gadolinium, mercury, ytterbrium, radon, radium, kurchatovium, hahnium.

57. How many atoms of each kind are in the molecules of each of the following compounds: H_2SO_4, Na_2CO_3, H_2, $(NH_4)_3PO_4$, $K_4Fe(CN)_6$?

Chemical Reactions

58. How many moles of nitrogen (N_2) and hydrogen (H_2) are needed to produce two moles of ammonia in the reaction:

$$N_2 + 3H_2 \quad \rightarrow \quad 2NH_3$$

59. Potassium chlorate, $KClO_3$, decomposes into potassium chloride and oxygen when heated according to the reaction

$$2KClO_3 \quad \rightarrow \quad 2KCl + 3O_2$$

How much oxygen is released from 40 g of potassium chloride?

Mechanical Properties of Materials

60. Analyze and classify the type of stress in each of the following situations:
    (a) Suspension cables on the Golden Gate Bridge
    (b) Roadbed girders on the Golden Gate Bridge
    (c) Landing gear struts on an airplane
    (d) Wings on an airplane
    (e) Chin of a boxer

61. A vertical metal rod of diameter of 1 cm supports a load of 90 000 N. What is the stress in the bar? What type of stress is this?

62. The metal rod of Exercise 61, has a length of 5 cm. When subjected to the load of 90 000 N, it is compressed by an amount of 0.003 cm. What is the strain of the rod?

63. A steel elevator cable of diameter 3 cm supports an elevator car containing 20 people. If the average weight of each person is 500 N, and the car has a weight of 1500 N, what is the stress on the cable? What type of stress is this?

Other Material Properties

64. Why is the Golden Gate Bridge in San Francisco orange?

65. Discuss the advantages and disadvantages of painting the surface of an airplane.

66. In designing a bumper for a car, would you select a material with a large or small modulus of elasticity?

67. What material considerations are important in each of the following: baseball bats, nuclear reactor fuel elements, contact lens, frying skillets?

Types of Materials

68. List as many different types of materials as you can think of that are used in the following items: clothing, bridges, home computers, dental work.

69. Some composite materials have long continuous filaments embedded inside a resin such as epoxy. Do you expect the modulus of elasticity to be higher in the direction parallel or normal to the fibers?

70. What are the possible advantages of using plastics and composite materials in airplanes instead of metals?

71. What material properties would be important in the design of an artificial heart?

Systems

72. Give three examples of systems in each of the following areas:
(a) Engineering   (e) Biology
(b) Society   (f) Medicine
(c) Business   (g) Economics
(d) Politics   (h) Universities

73. Consider a television set as a system. Draw a black-box representation and identify inputs and outputs.

74. Draw a black-box representation of each of the following "systems," taking care to identify the most important inputs and outputs:
    (a) The United State economy   (d) Your instructor
    (b) Your own personal economy   (e) An automobile
    (c) Your college education

Feedback and Control

75. Discuss the use of feedback in each of the following situations:
    (a) Automatic pilot of an aircraft
    (b) AFC on an FM radio
    (c) Plant leaves turning toward the sun
    (d) The pro football college draft

76. Give three examples of oscillation phenomena in common feedback control systems encounted in your everyday experience.

REFERENCES

Basic Mathematical Concepts

1. Howard Anton, *Calculus with Analytic Geometry* (Wiley, New York, 1980).

2. George B. Thomas, Jr. and Ross L. Finney, *Calculus and Analytic Geometry*, 5th Edition (Addison-Wesley, Reading, 1979).

3. L. Salas and Einar Hille, *Calculus: One and Several Variables*, 3rd Edition (Wiley, New York, 1979).

4. Earl W. Swokowski, *Calculus*, 2nd Edition (Prindle, Weber, and Schmidt, New York, 1979).

5. Louis Leithold, *Calculus Book: A First Course with Applications and Theory* (Harper and Row, New York, 1979).

Graphical Analysis

1. George C. Beakley and H. W. Leach, *Engineering: An Introduction to a Creative Profession*, 2nd Edition (Macmillan, New York, 1972).

2. Arvid R. Eide, Roland D. Jenison, Lane H. Marshaw, and Larry L. Northrup, *Engineering Fundamentals and Problem Solving* (McGraw-Hill, New York, 1979).

Dimensions and Units

1. "Standard for Metric Practice," American National Standards Institute E 388–76 268–1976.

2. "Metric Editorial Guide," 2nd Edition, American National Metric Council, 1975.

3. "SI Units and Recommendations for the Use of their Multiples and of Certain Other Units," International Organization for Standardization (ISO), 1973.

Probability and Statistics

1. William L. Quirin, *Probability and Statistics* (Harper and Row, New York, 1978).

2. Ronald E. Walpole and Raymond Meyers, *Probability and Statistics for Engineers and Scientists*, 2nd Edition (Macmillan, New York, 1978).

3. A. D. Little, *Probability and Statistics for Engineering* 2nd Edition (Prentice-Hall, Englewood Cliffs, N.J., 1977).

General Physics

1. David Halliday and Robert Resnick, *Fundamentals of Physics,* 2nd Edition (Wiley, New York, 1977).

2. David Halliday and Robert Resnick, *Physics*, 3rd Edition (Wiley, New York, 1977).

3. Richard P. Feynman, Robert B. Leighton, and Matthew Sands, *The Feynman Lectures on Physics*, Vol. 1, 2, 3 (Addison Wesley, Reading, Mass., 1965).

4. Jay Orear, *Physics* (Macmillan, New York, 1979).

5. Paul A. Tipler, *Physics* (Worth, New York, 1977).

6. John P. McKelvey and Howard Grotch, *Physics* (Harper and Row, New York, 1978).

Mechanics

1. David Halliday and Robert Resnick, *Fundamentals of Physics*, 2nd Edition (Wiley, New York, 1977).

2. Morton Mott-Smith, *Principles of Mechanics Simply Explained* (Dover, New York, 1963).

Work, Energy, and Heat (Thermodynamics)

1. Gordon J. Van Wylen and Richard E. Sonntag, *Fundamentals of Classical Thermodynamics*, 2nd Edition Revised (Wiley, New York, 1978).

2. Jack P. Holman, *Thermodynamics*, 3rd Edition (McGraw-Hill, New York, 1979).

3. William C. Reynolds and Henry C. Perkins, *Engineering Thermodynamics*, 2nd Edition (McGraw-Hill, New York, 1977).

4. H. C. VanNess, *Understand Thermodynamics* (McGraw-Hill, New York, 1969).

5. Kenneth Wark, *Thermodynamics*, 3rd Edition (McGraw-Hill, New York, 1977).

Electrical Processes

1. Ralph J. Smith, *Circuits, Devices, and Systems: A First Course in Electrical Engineering*, 3rd Edition (Wiley, New York, 1976).

2. Ralph J. Smith, *Electronics: Circuits and Devices* (Wiley, New York, 1981).

3. Arthur E. Fitzgerald, *Basic Electrical Engineering*, 4th Edition (McGraw-Hill, New York, 1975).

4. Special Issue on "Microelectronics", *Scientific American, 237* (September, 1977).

5. James D. Meindl, "Microelectronics Circuit Elements," *Scientific American 237* (September, 1977), p. 70

6. William C. Holton, "The Large Scale Integration of Microelectronic Circuits," *Scientific American 237* (September, 1977), p. 82

Chemical Processes

1. James E. Brady and Gerard E. Humiston, *General Chemistry: Principles and Structure*, 2nd Edition (Wiley, New York, 1978).

2. William L. Masterton and Ernest J. Slowinski, *Chemical Principles*, 5th Edition (Saunders, New York, 1979).

3. Charles E. Mortimer, *Chemistry: A Conceptual Approach* (Van Nostrand, New York, 1975).

4. William H. Nebergall, F. C. Schmidt, and H. F. Holtzclaw, 5th Edition, *General Chemistry* (Heath, New York, 1976).

5. T. L. Brown, H. Eugene Lemay, Jr., *Chemistry: The Central Science* (Prentice Hall, Englewood Cliffs, N.J., 1977).

Nuclear Processes

1. James J. Duderstadt, *Nuclear Power* (Marcel Dekker, New York, 1979).

2. James J. Duderstadt and Chihiro Kikuchi, *Nuclear Power: Technology on Trial* (University of Michigan Press, Ann Arbor, 1979).

3. T. M. Connolly, *Foundations of Nuclear Engineering* (Wiley, New York, 1978).

4. R. L. Murray, *Nuclear Energy*, 2nd Edition (Pergamon, New York, 1981).

Engineering Materials

1. Lawrence H. Van Vlack, *Materials Science for Engineers* (Addison Wesley, Reading, Mass., 1970).

2. Lawrence H. Van Vlack, *Elements of Materials Science*, 3rd Edition (Addison Wesley, Reading, Mass., 1975).

3. C. R. Barrett, W. D. Nix, and A. S. Tetelman, *Principles of Engineering Materials* (Prentice Hall, Englewood Cliffs, N.J., 1977).

4. A. G. Guy, *Essentials of Materials Science* (McGraw-Hill, New York, 1976).

5. K. Ralls, T. Courtney, and J. Wulff, *Introduction to Materials Science and Engineering* (Wiley, New York, 1976).

6. "Materials: Advanced Technology," *Science 208*, No. 4446 (May, 1980).

Systems and Control

1. Otto Mayr, The Origins of Feedback Control (MIT Press, Cambridge, 1970).

2. Stafford Beer, *Cybernetics and Management* (Wiley, New York, 1964).

3. A. B. Carlson, Communications Systems, 2nd Edition (McGraw-Hill, New York, 1974).

4. Simon Haykin, *Communication Systems* (Wiley, New York, 1978).

CHAPTER 4

Digital Computers

Digital computers have revolutionized engineering practice, just as they have other aspects of modern life. Computers have become an absolutely essential tool in all phases of engineering, from research and development to marketing and management. The engineer of a generation ago would probably not even recognize many of today's routine engineering activities that rely heavily on digital computers. Even after three decades the computer revolution shows no sign of slowing. New discoveries and improvements in computer technology continue at such a pace that many products and techniques in the computer field become obsolete within a few years time.

The power of computers in most applications can be traced to two fundamental features: their capacity to store and display large amounts of information and their ability to process this information at a very rapid rate. Information storage and display are important in many engineering activities. Today's engineer relies on computer-based library reference services, computer data bases, and computer scheduling, accounting, and inventory control services. The ability of the computer to tackle large and complex calculations has also had a major impact on engineering practice, particularly in research, development, and design activities. Today the engineer routinely uses the computer to solve problems that would have been unthinkable even as recently as a decade ago. Moreover the microelectronics revolution has stimulated the development of powerful hand-held calculators and personal microcomputers. These devices have become as familiar to the engineer of today as the slide rule was to engineers several years ago.

Example A dramatic example of the rapid growth in information storage capability is provided by the computer system at the Lawrence Livermore Laboratory in California (Figure 4.1). This system can store roughly one trillion (10^{12}) bits of information—the equivalent of 100,000 books—in its computer memory devices. All of this information is available for retrieval and display within a fraction of a

FIGURE 4.1. The computer system at the Lawrence Livermore Laboratory. (Courtesy Lawrence Livermore National Laboratory)

second. Other somewhat slower storage devices such as magnetic tapes extend this memory capacity to roughly 10^{14} bits (the equivalent of 10 million books).

The ready availability of such mammoth information storage capacity has led to the establishment of public "information utilities." These computer-based services provide information to subscribers in much the same way that electrical utilities provide electricity. One can access these services by simply dialing the appropriate telephone number from any computer terminal (or home computer). Such services typically provide a myriad of information including daily news wire services, encyclopedia and library references services, educational materials, personal finance and consumer data, accounting and finance services, and games, all available at the request of the user for only a small fee.

The evolution in the speed of computer calculation has also been quite dramatic. Even the relatively inexpensive personal computer can perform roughly 100,000 operations (e.g., additions) per second. Larger machines average about 10 million operations per second, while the fastest machines can perform more than

100 million operations per second. Indeed, the speed of some advanced computers is now limited primarily by the length of time it takes an electrical signal to travel from one part of the computer to another, at nearly the speed of light.

Exercises

Introduction

1. List the interactions (either direct or indirect) you have with computers on a typical day.

2. What role might computers play in the following situations:
   1. Assembly line operations
   2. Regional electric power distribution
   3. Traffic light sequencing
   4. Supermarkets

3. How could computers be used in the following engineering activities?
   1. Literature search
   3. Modeling of engineering systems
   3. Model refinement
   4. Optimization
   5. Technical report preparation

4. List and describe all of the computers that you own. (Don't forget to include devices such as multifunction watches.)

5. How do you regard the computer philosophically: as a benefit or a threat? Contrast your views with those of your friends (particularly those with nonscientific backgrounds).

4.1. TRENDS IN COMPUTER DEVELOPMENT

4.1.1. What Is a Computer?

A computer is essentially a machine that processes information or data. More precisely, a *computer* is a device that can receive and store a set of instructions and then carry out those instructions to process data. This definition implies that a computer not only receives and processes *data*, but it also receives, stores, and executes *instructions*. The set of instructions that can be fed into the computer is known as a computer *program*. Both the data fed into the computer, as well as the program (the set of instructions) that governs the processing of that data, can be changed at the discretion of the user.

This definition suggests that the most primitive "black box" diagram of a

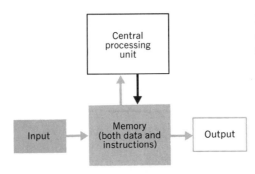

FIGURE 4.2. A black box diagram indicating the principal components of a digital computer.

computer would take the form shown in Figure 4.2. Here we have identified four components: an input device to receive the data and/or instructions (program), a memory device to temporarily store this data, a central processing unit that executes the instructions to process the data, and an output device to display the results of the data processing. Typically the input device might be a keyboard on a computer terminal, a card reader to read punched cards, a magnetic disk or tape device, or perhaps a speech identifier or light pen. Output devices include teletypes or printers, TV screens, magnetic disks or tapes, speech synthesizers, and so on. The computer memory most typically consists of tiny electric (integrated) circuits, but other devices such as magnetic disks and tapes are also used. The central processing unit (the "brains" of the computer) consists of electric circuits designed to carry out the data processing (e.g., arithmetic calculations) and a control unit to organize and coordinate this data processing.

In almost all modern computers, data is handled and manipulated in digital form—that is, as numbers. In fact, most computers process data in the form of binary numbers, as sequences of the binary digits 0 or 1.

4.1.2. Analog versus Digital Computers

Analog quantities are signals or measurements that can assume a continuum of possible values. For example, physical quantities such as temperature, velocity, or position are normally assumed to change smoothly and continuously. *Digital* quantities, on the other hand, are inherently discrete with a magnitude represented by an integer number. Digital signals can therefore change only from one incremental value to another; there is no longer a smooth variation. Examples of digital variables include the number of units produced by a manufacturing plant on a given day and the time of day as measured by a digital clock.

It is possible to build a calculational device or computer using either analog or digital principles. For example, an electronic circuit can be designed that will produce an output voltage that is the product of voltages supplied at two separate inputs. This circuit will then function as a rudimentary analog computer. Its function will be to take two analog input quantities (voltages) and produce an analog output that is their product (the output voltage). Given this basic multiplier circuit, many other arithmetic operations can be carried out by properly combining such circuits. Modern analog computers often combine hundreds of these elements to produce an analog voltage that may vary as a complex function of several input variable voltages.

Example It is desired to compute the kinetic energy of an automobile as its speed changes. The input variable in this case can be the voltage developed by the speedometer of the car, which is an analog quantity. Using a proper combination of electrical circuits, this voltage can be multiplied by itself (or squared), producing a result proportional to kinetic energy ($\frac{1}{2}mv^2$). This result is an analog voltage output that varies continuously as the automobile speed changes. It could be supplied to a calibrated voltmeter mounted on the automobile instrument panel to give the driver an indication of the car's instantaneous kinetic energy.

FIGURE 4.3. A laboratory of analog computers. (Courtesy University of Michigan College of Engineering)

Although analog computers are still useful in some applications, there has been an overwhelming trend toward the use of digital computers in engineering. Here the input and output quantities are digital numbers that can be used to represent physical quantities. Therefore, most of this chapter will deal with the concepts of digital information together with the basic components and functions of digital computers.

SUMMARY

Analog quantities vary continuously, whereas digital quantities are restricted to a set of discrete values. Analog computers consist of electronic circuits or mechanical devices that stimulate the behavior of a system. Digital computers process data in the form of numbers. Modern trends in engineering have been toward the introduction of digital computation, sometimes as a substitute for older analog methods.

4.1.3. Evolution of Digital Computers

The first electronic digital computer, ENIAC, was developed at the University of Pennsylvania in 1945 and consisted of some 18,000 vacuum tubes. Subsequent machines of the 1940s and 1950s were also fabricated from discrete circuit components such as vacuum tubes, resistors, and capacitors. These computers were of necessity quite large in physical size and required careful environmental control (air conditioning). In order to provide a useful amount of computation capacity, such units would occupy a large room or perhaps an entire building. Because of their large size and expense, they were treated as central facilities. A user found it necessary to bring data encoded on punched cards ("do not fold, spindle, or multilate. . .") or magnetic tape for processing.

The introduction of transistor-based circuits in the late 1950s considerably increased capacity and speed, but the computer continued to be a large and expensive device servicing users from a central location. Such large, central computing facilities are still common today, but the substitution of modern miniaturized circuitry has allowed vastly greater capacity and computation speeds. These large devices are commonly referred to as *mainframe* computers (Figure 4.4), and each unit may cost as much as several million dollars.

The trend toward miniaturization in circuit design led to the introduction of the *minicomputer* in the early 1960s. Here the basic elements of a digital computer could be provided in a single electronic rack of convenient size. The entire computer (complete with input/output devices) would be about the size of a desk. Furthermore, costs were now sufficiently low ($10,000 to $100,000) to allow sepa-

FIGURE 4.4. A large mainframe computer. (Courtesy Control Data Corporation)

FIGURE 4.5. A tiny microprocessor, a computer on a single chip of silicon (in this case placed on the head of a paper clip for size reference). (Courtesy Bell Laboratories)

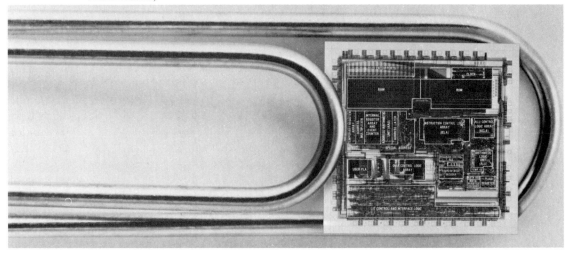

rate minicomputers to be dedicated to tasks in individual laboratories or at the site of their application. In this way data sources such as measuring instruments or sensors could be coupled directly to the minicomputer. This trend not only avoided the indirect transfer of data (carrying punched cards or magnetic tape over to the computer center), but it also permitted the direct involvement of the computer in the supervision and control of laboratory experiments and industrial processes. In this kind of *real-time* computation the signals from various sensors or instruments are used to derive a calculated result almost instantaneously. This result may then be used to control the future behavior of the system.

The next step in this evolution was the introduction of the *microprocessor* in the 1970s. Technical developments in large scale integrated circuits (IC) allowed the incorporation of the basic elements of a digital computer on a single chip of silicon (Figure 4.5). Tens of thousands of individual circuit elements could be etched onto the surface of a small wafer of silicon about the size of a fingernail. Such an integrated circuit could be designed to store, interpret, and execute instructions. When equipped with suitable input and output devices and further IC-based memory, the microprocessor became the central processing unit of a tiny computer. Such computers based on microprocessors and costing no more than a few thousand dollars are referred to as *microcomputers*.

While still considerably less versatile than large mainframe computers, microprocessors are continuing to undergo rapid evolution and improvement. Their

FIGURE 4.6.   Microprocessors have been built into "smart" instruments. (Courtesy Hewlett-Packard Company)

FIGURE 4.7. The personal computer has become a powerful tool of the scientist and
engineer. (Courtesy Hewlett Packard Company)

small size often allows their incorporation directly into the device or instrument
that actually generates the data. Thus we have seen the rapid evolution of
"smart" instruments in which some computational ability is provided within the
device itself (Figure 4.6). The microprocessor has already had a profound impact
on virtually all areas of instrumentation and engineering measurements. It has
penetrated into the consumer market with the appearance of multifunction
watches, telephone receivers, automobile instruments and controls, hand-held
electronic games, and so on, all based on microprocessors.

The impact of most immediate concern to the engineering student is the use of microprocessors in small pocket calculators or personal computers (Figure 4.7). Today we have small, inexpensive computers with the computational power of the large machines that were in use a decade ago. Already these devices have revolutionized engineering education and practice (Figure 4.8).

The use of computers has also changed over the years. Early mainframe computers were usually operated in a *batch* mode that required each user to bring input data to the facility for processing in the form of punched cards or magnetic tape. The entire computer was then dedicated to each job in turn as they were carried out in sequence. During the 1960s many computer facilities shifted to *time-sharing* systems (Figure 4.9). In this mode of operation many users are accommodated simultaneously from remote terminals connected to the central computer by phone lines. In a time-sharing system a small segment of the computer's time, say 0.01 second, is allocated to each user in sequence. If there are 100 users, then every second a typical user is given only one percent of the computer's time. Because the computer is so fast, a large number of calculations can be carried out in that short time segment. Very often the results needed by the user can be provided within a few such cycles, so that a response can be printed out at the terminal within a few seconds. In many cases the user is not even aware of the

FIGURE 4.8.   Personal computers are revolutionizing the activities of engineers and scientists. (Courtesy Apple Computer, Inc.)

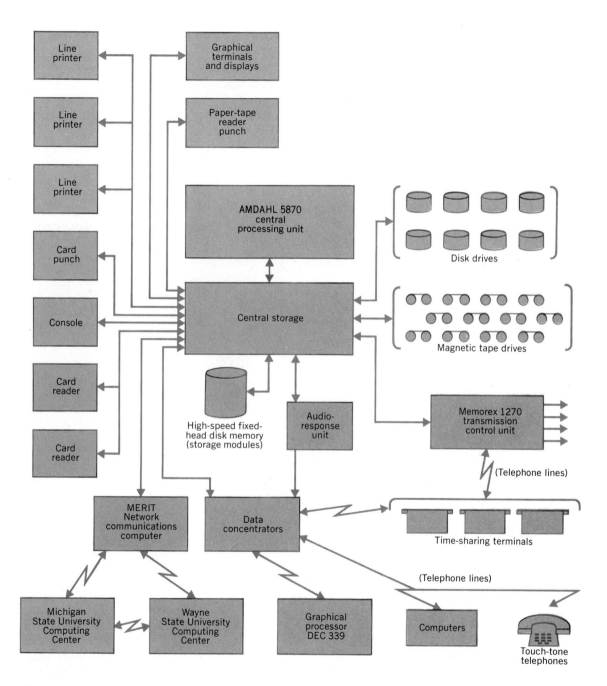

FIGURE 4.9. A schematic of a large time-sharing computer system (in this case the Michigan Terminal System). (Reprinted with permission from Brice Carnahan and James O. Wilkes, *Digital Computing, FORTRAN-IV, WATFIV, and MTS*, © 1979 by Brice Carnahan and James O. Wilkes)

FIGURE 4.10. Several of the more popular personal computers including the Apple-II, Radio Shack TRS-80 Model III, Zenith (Heath) Z-89, and the Hewlett-Packard HP-85. (Courtesy Apple Computer, Inc., Tandy Corporation, Zenith Data Systems and Hewlett-Packard)

delay. When the results of a calculation are available almost immediately, the input data or processing instructions can be changed to reflect the experience gained from earlier results. This mode of operation is known as *interactive* computing. In essence it allows the user to converse directly with the computer and allows a very rapid and efficient development of computing methods.

Another important aspect of interactive computing is the development of the personal or home computer (Figure 4.10). These microcomputers typically use a keyboard for input and a TV screen for output. Their large memory size coupled with inexpensive perhiperal devices such as floppy disk units and printers, provide the engineer with a powerful computer system that can be dedicated to individual needs, whether at office, home, or school. Home computers can also be converted rather easily into "smart" terminals to allow the user access to a large, central time-sharing system.

SUMMARY

The power of digital computers is based on their capacity to store large amounts of information and their ability to carry out the processing of this data at high speed. In addition to large scale mainframe computer facilities, the minicomputer and microprocessor have evolved to allow location of the computer at the site of its application. The increased flexibility of central facilities has allowed their use in a time-sharing mode of operation with rapid feedback of information. This increased use of interactive computing has also been stimulated by the development of the personal microcomputer.

Example The variety of modern computers is extraordinary. In Figure 4.10 we show several popular personal computers, including the Apple-II, the TRS-80, the TI-91, and the Heath 95. All of these units are based on 8-bit microprocessors with memory sizes of 8000 to 64,000 bytes and costing from $500 to $2500.

In contrast, Figure 4.11 shows the mammoth timesharing computer system at the Lawrence Livermore Laboratory. This system is based upon four large CDC-7600 mainframe computers and two CDC-Star-100 supercomputers. These computers are coordinated by several PDP-10 computers that act as dispatchers, routing data back and forth among the large computers, data banks, and an array of PDP-8 minicomputers that handle the some 2200 user terminals tied into the system. The Livermore system has been appropriately named "Octopus."

Exercises

What Is a Computer

1. Choose one of your own computers and identify and describe each of its major components.

2. Prepare a list of as many input/output (I/O) devices for computers as you can think of.

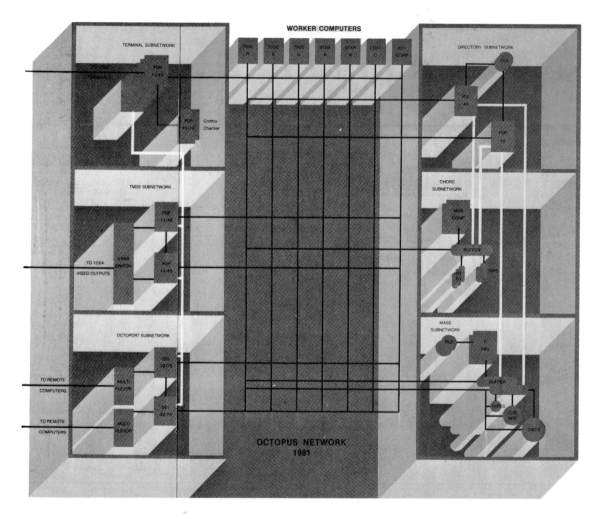

FIGURE 4.11. The Lawrence Livermore Laboratory Octopus computer system and data network. (Courtesy Lawrence Livermore National Laboratory)

3. How would you contrast the definition of a computer with that of a hand-held calculator?

4. Identify the major components of the "computer" of a human being.

Analog vs. Digital

5. In some circumstances both analog and digital devices are available to carry out somewhat the same function. One example is the watch, where the digital variety now coexists with the more traditional "big hand–little hand" analog design. List three additional pairs of analog

and digital devices that are also used for the same purpose in other applications.

6. The engineer's best friend of a decade ago, the slide rule, has been essentially replaced by digital pocket calculators. Using older texts as references, explain the principle of operation of the slide rule, and show how it functions as one type of analog computer.

7. Determine several applications of analog computers within your own college of engineering.

8. Identify and give examples of the replacement of analog by digital devices in automobile design.

Computer Evolution

9. Identify the role played by each of the following individuals in the development of computers:

William Babbage	Seymor Cray
Herman Hollerith	Frederick Fortran
John Von Neumann	The Countess of Lovelace

10. Visit your university computer center and obtain the following information:
   1. Type of mainframe computer(s)
   2. Computer size (main memory, peripheral storage, terminals)
   3. Computer speed (operations per second)
   4. Available language (FORTRAN, ALGOL, PL/1, etc.)

11. Identify how each of the following types of computers might be used in your particular field of engineering:
   1. Mainframe
   2. Microcomputer
   3. Microprocessor
   4. Analog computer

12. Trace the development of the personal computer (e.g., size, speed, support devices, I/O) by referring to back issues of microcomputer magazines in your technical library such as *BYTE* or *Personal Computing*.

13. Prepare a short shopper's guide comparison list of presently available personal computers (including primary characteristics and costs).

4.2 BASIC PRINCIPLES OF DIGITAL METHODS

To discuss the basic operation of a digital computer, we must first look at the way in which the digital information is handled within such devices. As we shall see, virtually any piece of information can be reduced to a sequence of numbers. A key element of our discussion is the way in which numbers can be represented by electronic signals within the circuitry of the computer. We shall therefore begin our discussion with a review of the different types of number systems, with particular emphasis on the binary system used by all basic logic circuits. We then show how these numbers can be combined or manipulated in arithmetic operations that are so important in the operation of any digital computer.

4.2.1. Number Systems

All of us have been trained to use the decimal number system. A specific number such as 359.8 is coded in terms of the base of the number system, in this case 10. The value of each digit depends on its position relative to the decimal point according to the following prescription:

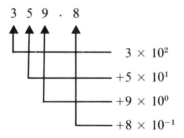

$$3 \times 10^2$$
$$+5 \times 10^1$$
$$+9 \times 10^0$$
$$+8 \times 10^{-1}$$

Of course there is nothing unique about the choice of 10 as the base. In most digital applications it makes more sense to deal in number systems with a base of 2, 8, or 16, since these are more compatible with the on/off behavior of most computer circuit components. These are called *binary*, *octal*, and *hexadecimal* systems, respectively.

In the *octal* number system, only the digits 0 through 7 are used. The digit represented as 8 in the common decimal system now is represented as 10 in the octal system. It is to be interpreted as

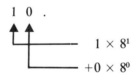

$$1 \times 8^1$$
$$+0 \times 8^0$$

Table 4.1 shows several of the first successive octal digits together with their decimal equivalents.

The *hexadecimal* representation requires six additional digits beyond those of the decimal system. These are given the designation A through F as illustrated in Table 4.1. In decimal notation these digits correspond to the numbers 10 through 15. Large numbers written in hexadecimal form will tend to have somewhat fewer digits than the decimal equivalent because each number position carries a greater range of values. The opposite is true for the octal system.

TABLE 4.1 Equivalent Number Representations in Different Base Systems

DECIMAL	BINARY	OCTAL	HEXADECIMAL
0	0	0	0
1	1	1	1
2	10	2	2
3	11	3	3
4	100	4	4
5	101	5	5
6	110	6	6
7	111	7	7
8	1000	10	8
9	1001	11	9
10	1010	12	A
11	1011	13	B
12	1100	14	C
13	1101	15	D
14	1110	16	E
15	1111	17	F
16	10000	20	10

The *binary* system is the simplest of all number systems and uses only the digits 0 and 1. The value of each digit position is therefore given by the following representation:

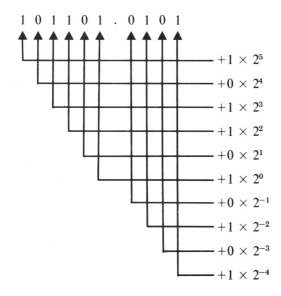

$$+1 \times 2^5$$
$$+0 \times 2^4$$
$$+1 \times 2^3$$
$$+1 \times 2^2$$
$$+0 \times 2^1$$
$$+1 \times 2^0$$
$$+0 \times 2^{-1}$$
$$+1 \times 2^{-2}$$
$$+0 \times 2^{-3}$$
$$+1 \times 2^{-4}$$

Each of these positions in the binary number is called a *binary digit* or *bit*. There can be only two possible values for a binary digit, 0 or 1. The counting sequence for binary numbers is also shown in Table 4.1. A given number requires many more digits in binary form than does its decimal equivalent. For example, the decimal number 35, which requires only two decimal digits has as its binary equivalent 10011, requiring five bits.

Any octal or hexadecimal number can easily be converted to its binary equivalent. Any of the octal digits 0 through 7 corresponds to a unique combination of three bits (see Table 4.1). Figure 4.12 illustrates the simple steps required to convert an octal representation to binary and vice versa. Each hexadecimal digit corresponds to a unique pattern of four binary digits and interconversion between these two systems is also direct and simple as shown in Figure 4.12.

Conversion between any of the above systems and the decimal is less direct. To convert a decimal integer into binary form, one must first find the largest power of 2 that is smaller than (or equal to) the decimal number. This power of 2 corresponds to the leading position (the leftmost digit) in the corresponding binary number. A "1" is entered in this position, and the corresponding power of 2 is subtracted from the original decimal number. If the remainder is larger than or equal to the next lower power of 2, a "1" is entered in the next bit position and the process repeated. If the remainder is less than the next lowest power, a "0" is entered into the next bit position. By repeating this process until the remainder is 0, all the rest of the bit positions are filled in. This scheme is illustrated in Figure 4.13. The octal or hexadecimal equivalent to the binary number can easily be obtained by grouping the binary digit into clusters of 3 and 4, respectively.

One binary digit can represent two numbers (0 or 1). Two bits can represent 4 numbers, while 3 bits can represent up to 8 numbers. In general, n bits will be able to accommodate 2^n decimal integers. To represent large decimal numbers, many bits are required. As shown in Table 4.2, for example, 12 bits are capable only of representing the decimal integers between 0 and 4095.

The familiar operations of addition, subtraction, multiplication, and division can be carried out directly using binary numbers. For example, addition requires the use of only the following simple rules:

$$
\begin{array}{ccc}
0 & 0 & 1 \\
+0 & +1 & +1 \\
\hline
0 & 1 & 10
\end{array}
$$

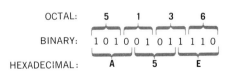

FIGURE 4.12.   The conversion between a given binary number and either the octal or hexadecimal system is quite simple. Starting at the right, each group of three binary digits corresponds to one octal digit. Each group of four binary digits converts to one hexadecimal digit. The digit conversions are shown in Table 4.1.

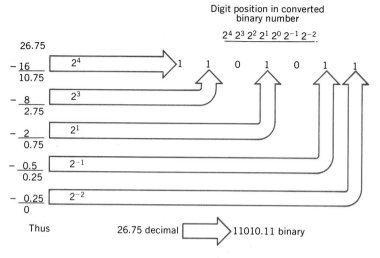

FIGURE 4.13. The conversion between decimal and binary systems is a bit more involved, as shown in this example.

TABLE 4.2 Number of Decimal Integers Corresponding to n Binary Bits

NUMBER n OF BITS	NUMBER OF CORRESPONDING DECIMAL INTEGERS (2^n)
1	2
2	4
3	8
4	16
5	32
6	64
7	128
8	256
9	512
10	1024
11	2048
12	4096
13	8192
14	16384
15	32768
16	65536

In the third example we see the case of "carrying" a digit to the next position to the left of the column in which the addition occurs. An example of the addition of two 5-bit binary numbers is shown in Figure 4.14. As in decimal addition, the process proceeds column by column, starting with the right-hand digit. Each step must correspond to one of the three cases shown above. Using equally simple rules, longhand subtraction and multiplication can also be carried out directly in binary representation.

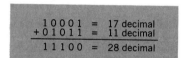

FIGURE 4.14. An example of the addition of binary numbers.

SUMMARY

Various number systems are defined by choosing different base values for the system. The binary, octal, and hexadecimal systems are defined with bases of 2, 8, and 16. All are used to some extent in the representation of numbers in digital computers. The binary system is of most general importance because it can be applied directly to describe the basic operation of logic circuits. Arithmetic operations such as addition can be carried out directly in any number system, including binary.

4.2.2. Representation of Information in Digital Form

Virtually any information can be reduced to digital form. That includes not only tables of numbers or other numerical data, but also books, pictures, or even a performance of Mahler's *Fifth Symphony* by the Chicago Symphony Orchestra. (Witness the recent development of "digital" recordings.) As a result, methods to store digital information are of paramount importance in many fields of endeavor, and the archives of the future may well be made up of collections of binary numbers.

Means for the storage (and retrieval) of binary information is consequently one of the most important topics of modern engineering. Since a binary digit can have only one or two possible values, it can be represented using pencil and paper simply by writing either a 0 or 1. Such written or printed numbers are easily recognized by humans, but not by machines. Ways of storing binary information have evolved over the years that are more compatible with the requirements of electrical or mechanical devices. For example, the bar codes on items at the supermarket can be sensed by an optical scanner and translated as a corresponding set of digital information. Because computers are electronic devices, common methods of representing digital information are the electrical signals produced by striking a key on a terminal keyboard or the signals from a pickup that senses the patterns on magnetic tape or disk. These media have become the means through which we can convey a wide variety of information to computers.

All digital computers are based on the storage and manipulation of information that is stored in the form of binary digits or *bits*. Most often this information is organized in the form of basic units called *words*. Each word consists of a fixed number of bits, commonly anywhere between 8 and 64, depending on the size and design of the particular computer. Most numbers that are stored or manipulated within the computer are coded in terms of the content of a single word. There are two distinct modes or conventions that can be used to code a particular number: *integer* mode and *floating point* mode.

Integer Mode

As its name implies, only integer numbers can be represented in integer mode. If we assume a 16-bit word for purposes of illustration (which is the common integer word length on most microcomputers), then all decimal integers between 0 and 65,535 could be coded by using each of the 16 bits to represent the equivalent binary integer. However it is likely that we shall also be interested in using or storing negative integers. In that case, the first bit (or that in the most left-hand position) can be given the special purpose of indicating the sign (0 will mean + and 1 will mean −) of the integer represented by the remaining 15 bits. In this way all decimal integers between −32767 and +32767 can be accommodated.

Example   In most digital computers, a system for representing integers known as *two's complement* is used in place of either of the above schemes. The two's complement of a binary number is defined as the binary number that, added to the original number, gives a sum of unity. For example, the two's complement of 101111000 is 010001000, as shown by the following addition:

$$101111001$$
$$\underline{010001000}$$
$$(1)000000001$$

The two's complement of a binary number is obtained simply by inverting the value of each bit (thereby obtaining what is known as the "one's complement") and adding 1 to the result. The advantage of the two's complement representation is that it simplifies the subtraction operation. Whereas the two's complement of the sum of two numbers is obtained by summing the two's complement of each, the standard binary representation of the difference between integers *A* and *B* is obtained by adding together the value of *A* (in conventional binary form) with the two's complement of *B*.

Example   Carry out the binary equivalent of the decimal subtraction: 51 − 29 = 22 in binary representation, using the two's complement of the number that is subtracted.

Solution: First we write

> 51 decimal ⟶ 110011 binary
>
> 29 decimal ▶ 011101 binary

Now, to obtain the two's complement of the second number, we first note that its one's complement is obtained by changing the value of each bit:

> 29 decimal ⟶ 100010 one's complement

Adding 1 to this result gives the two's complement:

> 29 decimal ⟶ 100011 two's complement

Therefore the original subtraction is equivalent to the following sum:

> 51 decimal ⟶ 110011 binary
>
> 29 decimal ⟶ <u>100011</u> two's complement
>
> (1)010110   22 decimal

Note that in forming the sum, we have assumed that the number length is limited to 6 bits. The 7th bit is therefore an "overflow" and is ignored.

The two's complement allows both addition and subtraction to be carried out on the same type of computer circuit. In fact, all arithmetic operations on binary numbers can be reduced to addition in a similar manner. A computer can get along without special circuits for subtraction, multiplication, and division provided it has the ability to add.

Floating Point Representation

In many situations it is necessary to represent numbers that do not fit within the integer format discussed above. Sixteen bits can represent only a limited range of values, and much larger numbers often arise in common calculations. Also, it is not possible to represent fractional values in integer format. For these reasons, the floating point representation has come into widespread use.

The floating point number representation is similar to the scientific notation discussed in Section 3.3. Here any number is represented by two parts: its mantissa and its exponent (normally assumed to be base 10). In most conventions, the mantissa is a number in the range 0.1000. . . to 0.9999. . . . The exponent then indicates the power of 10 by which the mantissa should be multiplied in order to give the represented number. The binary equivalent of the decimal point is allowed to "float" in relation to the bits stored in memory to achieve the most efficient representation of information. Some examples are shown below:

$$3647.1 \longrightarrow 3.6471 \times 10^3 \longrightarrow 0.36471E + 04$$

$$0.000\ 000\ 000\ 611\ 54 \longrightarrow 6.1154 \times 10^{-10} \longrightarrow 0.61154E - 9$$

Floating point numbers are generally represented by a 32-bit word length (on both micro and mainframe computers). One possible scheme is shown below:

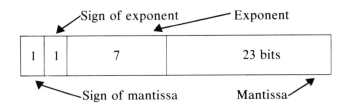

Sign bits must be reserved for both the mantissa and the exponent. The sign of the mantissa indicates the sign of the represented number, and the exponent can be either positive or negative to represent either large or small multiplying factors. In the 32-bit format shown, floating point numbers over the range of -2^{128} to $+2^{128}$ can be accommodated.

Example   There is one slight complication in this scheme for representing numbers in computers. Because each four bit binary sequence within a word can represent numerical values 0 to 15, it proves most convenient to store numbers (whether integer or floating point) in base 16 (or hexadecimal) form. For example, the representation of the integer 1 212 501 072 using a 32-bit word length would be

0	100	1000	0100	0101	0100	1100	0101	0000	Bit pattern
+	4	8	4	5	4	C	5	0	Hexadecimal

$$(48454C50)_{16} \longrightarrow (1\ 212\ 501\ 072)_{10}$$

while that of a floating point number $0.116\ 262\ 912 \times 10^{10}$ would be

0	100	1000	0100	0101	0100	1100	0101	0000	Bit pattern
+	4	8	4	5	4	C	5	0	Hexadecimal

$$(.454C5 \times 16^8)_{16} \longrightarrow (1\ 162\ 626\ 120.000)_{10}$$

In both instances the first bit is used to indicate sign. In floating point representation, the next 7 bits represent the exponent, while the last 24 bits represent the mantissa. An "express" notation has been used to eliminate the need for an exponent sign bit by assuming the exponent (in hexadecimal form) is 40 less than the 7-bit number in the binary string.

The difference between integer and floating point numbers is quite important for reasons other than convenience. Integer numbers are exactly represented, stored, and manipulated in a computer. There is never any loss of digital information due to rounding off. By way of contrast, the computer will automatically truncate or round off floating point numbers to maintain the required binary word length. This "roundoff error" can occasionally lead to problems in a long sequence of floating point calculations. The subject of inaccuracy and roundoff error is dealt with extensively in a field of mathematics known as numerical analysis.

Logical Information

Other types of information can be represented in digital form. Logical or Boolean variables (named after the British logician George Boole) can assume only the values "true" or "false." These can easily be represented in binary form ("0" or "1") and manipulated using logical expression such as "AND", "OR", and "=".

Character or String Data

In addition to representing numbers, there are many occasions in which we would like to store words and other text material in digital form. All that is needed is a prearranged code by which letters and other symbols are assigned a specific binary number. There are several different conventions in use, and Table 4.3 shows one of the most common, known as the 7-bit ASCII code (American Standard Code for Information Interchange). Seven bits will accommodate 2^7 or 128 different binary numbers. Therefore, 52 upper and lower case letters, 10 numerals, and a number of special symbols can be coded. A sequence of such characters is referred to as a *string*. String storage and manipulation is a key aspect in many computer applications such as word processing.

Example   Computer terminals with keyboards are a common means of communicating with digital computers. Entering the word "cat" into the computer memory requires a sequence of events. When the letter "c" is pressed on the keyboard, electrical circuits within the terminal must generate the 7-bit values (1s and 0s) of the corresponding ASCII code for the letter "c." Most terminals are connected to the computer by a phone line that can only carry audio signals. Therefore this bit pattern is translated by the terminal into a sequence of tones that is actually sent over the phone line. At the receiving end, these tones are reinterpreted as the corresponding bit pattern. The computer has been preprogrammed to recognize this pattern as the corresponding ASCII code for the letter "c." All

TABLE 4.3 The American Standard Code for Information Interchange (ASCII) Character Set

CODE		CHAR	CODE		CHAR	CODE		CHAR	CODE		CHAR
Dec	Hex		Dec	Hex		Dec	Hex		Dec	Hex	
Ø	ØØ	NUL	32	2Ø	SP	64	4Ø	@	96	6Ø	
1	Ø1	SOH	33	21	!	65	41	A	97	61	a
2	Ø2	STX	34	22	``	66	42	B	98	62	b
3	Ø3	ETX	35	23	#	67	43	C	99	63	c
4	Ø4	EOT	36	24	$	68	44	D	1ØØ	64	d
5	Ø5	ENQ	37	25	%	69	45	E	1Ø1	65	e
6	Ø6	ACK	38	26	&	7Ø	46	F	1Ø2	66	f
7	Ø7	BEL	39	27	'	71	47	G	1Ø3	67	g
8	Ø8	BS	4Ø	28	(	72	48	H	1Ø4	68	h
9	Ø9	HT	41	29	)	73	49	I	1Ø5	69	i
1Ø	ØA	LF	42	2A	*	74	4A	J	1Ø6	6A	j
11	ØB	VT	43	2B	+	75	4B	K	1Ø7	6B	k
12	ØC	FF	44	2C	,	76	4C	L	1Ø8	6C	l
13	ØD	CR	45	2D	−	77	4D	M	1Ø9	6D	m
14	ØE	SO	46	2E	.	78	4E	N	11Ø	6E	n
15	ØF	SI	47	2F	/	89	4F	O	111	6F	o
16	1Ø	DLE	48	3Ø	Ø	8Ø	5Ø	P	112	7Ø	p
17	11	DC1	49	31	1	81	51	Q	113	71	q
18	12	DC2	5Ø	32	2	82	52	R	114	72	r
19	13	DC3	51	33	3	83	53	S	115	73	s
2Ø	14	DC4	52	34	4	84	54	T	116	74	t
21	15	NAK	53	35	5	85	55	U	117	75	u
22	16	SYN	54	36	6	86	56	V	118	76	v
23	17	ETB	55	37	7	87	57	W	119	77	w
24	18	CAN	56	38	8	88	58	X	12Ø	78	x
25	19	EM	57	39	9	89	59	Y	121	79	y
26	1A	SUB	58	3A	:	9Ø	5A	Z	122	7A	z
27	1B	ESC	59	3B	;	91	5B	[	123	7B	{
28	1C	FS	6Ø	3C	<	92	5C	\	124	7C	|
29	ID	GS	61	3D	=	93	5D	]	125	7D	}
3Ø	1E	RS	62	3E	>	94	5E	∧	126	7E	~
31	1F	US	63	3F	?	95	5F	_	127	7F	DEL

this takes place in a fraction of a second. Therefore pressing the letters "a" and "t" at normal typing speed allows separate tone patterns to be generated and transmitted for these letters as well.

FIGURE 4.15.   The keyboard of the Apple-III computer provides an example of the ASCII format. (Courtesy Apple Computer, Inc.)

Example   In addition to the usual letter and numeral characters, the ASCII code assigns binary integer representations to 32 nonprinting *control* characters used to control input or output on terminals and printers. Examples of control characters include "RETURN" for carriage return (or linefeed), "RUB" for rubout, "EOT" for end of text, "ESC" for escape, and so on. Although there may be several special keys dedicated to such control characters on a computer terminal keyboard, such as the "ESCAPE" or "RETURN" key. Some control characters are transmitted using normal letter keys by simultaneously depressing a special "CONTROL" key analogous to the "SHIFT" key on a typewriter to enter such characters. For example, the "EOT" or end-of-text is entered by depressing the CONTROL key and the "C" key simultaneously—or in the jargon of computerese, typing a "CONTROL-C."

Machine Instructions

Instructions to the computer are also represented in binary form. A typical instruction based on a 32-bit word length is shown below:

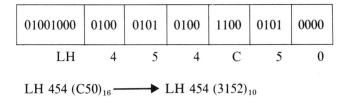

| 01001000 | 0100 | 0101 | 0100 | 1100 | 0101 | 0000 |

LH   4   5   4   C   5   0

LH 454 (C50)$_{16}$ ⟶ LH 454 (3152)$_{10}$

The first eight bits hold the command itself, while the remaining bits hold the memory location address and identify other computer components in the central processing unit to which the command applies.

SUMMARY

Virtually any type of information can be reduced to digital form. Devices such as computer terminals and tape readers are designed to convey this information to a computer in binary form. Numbers can be represented in computers using either integer or floating point mode. Logical data (true or false) can be represented in binary form. Through the use of the ASCII code, strings of characters and written text can also be represented in this form. Instructions for the computer are also coded in binary form and stored in computer memory.

4.2.3. Digital Signals

The concept of information is one about which we all have some intuitive feel. We all agree that a spoken sentence has the potential of transferring a great deal of information from the speaker to the listener. On the other hand, it is difficult to imagine that very much information would be carried by the static noise found between stations on the radio (or even *by* some stations on the radio). The science of *information theory* has been developed to provide a framework for quantifying some of these concepts and to provide analytical methods for analyzing the information content of signals of different types.

From a fundamental point of view, the most basic unit of information is a binary digit or bit. A single unit of information can be transferred from sender to receiver by simply indicating whether the value of this binary digit is either 0 or 1. Since no other values are possible, a further breakdown of the information content into smaller units is impossible, and we could consider that one unit or bit of information has been transferred. In principle, virtually any complex communication, be it words, numbers, musical tones, or pictures, can be represented to any desired accuracy by a collection of binary digits. The more complex the signal, the more digits are required. In principle, however, it is always possible to represent any piece of information to arbitrary precision if a sufficient number of binary digits can be used in its representation.

For example, suppose we are interested in the voltage produced by a pressure transducer that converts a measurement of pressure into an equivalent electrical voltage. This voltage is an analog quantity, which means that it may have a continuum of possible values as the pressure varies. At one specific time, we sample this voltage and call the result our signal. Depending on the degree of precision that is justified by the instrument or of interest in the measurement, we might represent the voltage as either 8.4 or 8.41362 volts. From our earlier discussion of Section 3.4, the number of significant digits we use reflects the estimated accuracy of the measurement. Both of these numbers have binary equivalents. The higher precision number, having a larger number of decimal digits, will also require a greater number of binary digits in its representation. Given enough binary digits, the voltage could be represented to any arbitrary degree of accuracy. Six bits will allow changes on the scale of one part in 2^6 or 64, whereas providing 12 bits will increase the precision to one part in 2^{12} or 4096.

The electronic devices that carry out the conversion of an analog signal (such as the voltage) to an equivalent binary number are generally called *analog-to-digital converters* or *ADCs*. Because most instruments and control devices in engineering provide an analog output signal, the ADC is an important first element when digital methods such as computer control are used with these instruments. Then all the powerful computational methods based on digital logic can be brought to bear on the problem of processing and interpreting the signal.

Another decided advantage in the digital representation of signals is the potential freedom from the effects of extraneous noise. Figure 4.16 is a sketch of a supposedly constant analog signal that shows the effects of random fluctuations called *noise*. Many sources of noise are possible, including random processes within the sensor itself, fluctuations added by electronic components, or interference along the line used to transmit the signal. An extreme example in which interference dominates would be the signal representing the temperature of a

FIGURE 4.16. The contrast between analog and digital signal transmission and reception.

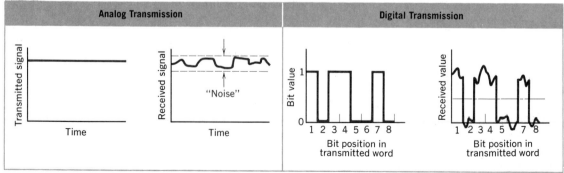

spacecraft after landing on the surface of one of the planets. Even a constant temperature signal would have the noisy appearance shown in Figure 4.16 after transmission over the vast distances back to the earth. If we were interested in the temperature at a specific time, our measurement would be uncertain because of the unknown component of the noise present at that instant.

A great improvement can result if the signal is digitized before transmission, as shown in Figure 4.16. Now the addition of a noise component to the binary signal does not detract from its information content, provided it is not so severe as to interfere with the distinction between the basic "0" or "1" states. Up to that point, the same digital number will be received, independent of the noise level. Therefore the noise addition does not interfere with perfect transmission of the information contained in the original signal.

SUMMARY

Virtually any type of information can be reduced to a series of binary numbers or sequence of bits. The larger the number of bits used in the representation of a quantity, the greater will be the precision possible in its representation. Analog to digital converters (ADCs) are common devices used to convert the output of many conventional instruments to digital form for subsequent processing by digital computers. One decided advantage of the digital representation of signals is the relative immunity from the effects of extraneous noise.

4.2.4. Binary Logic

All digital computers, whether they are at the level of microprocessors or large scale central computers, ultimately perform calculations in binary form. As we have seen, numbers are represented in such computers in the form of words made up of binary digits or bits. The awesome power of computers is really based on their ability to carry out relatively simple manipulations of these binary numbers. The most complex calculation ultimately is broken down into individual steps of addition, multiplication, incrementing, and so on, all of which are carried out at the binary level. As we have noted, a computer essentially consists of thousands of tiny electric circuits. These circuits perform the functions of storing binary digits and providing logical combinations of signals—the basic steps in any computation.

Only a few basic building blocks or *logic units* are necessary to construct the circuits that carry out these steps. We shall limit our discussion to three such logic functions: the NOT, AND, and OR functions. As we shall see later, these simple functions can be combined to carry out the basic process of binary addition. Since multiplication can be synthesized from a series of addition steps, we are already well down the path toward providing the necessary components for a digital computer.

Each of the basic logic functions can be described in terms of their effect on binary digits. Any such digit can have only two possible values, either 0 or 1. When represented within electronic circuits, these values correspond to two distinct voltage levels, usually called "high" or "low." No intermediate values are recognized, and ideally the voltage at any point in the circuit must assume either one of these two possible values.

The three basic logic functions are illustrated in Figure 4.17. We start our discussion with the first of these, called the NOT function. Its operation is to "invert" a binary digit, producing an output logical level that is the opposite of that applied to its input. Symbolically, the input level is represented as A and its inverse as $\bar{A}$. An input level corresponding to "0" is transformed by the NOT function into a "1", and vice versa. This operation is represented by the "truth table" also shown in Figure 4.17. Here a list is made of all possible input values (in this case, only 0 or 1), together with the corresponding output values. The truth table thus completely defines the operation of the logic function. It is particularly simple in this elementary case.

The second logic function is the AND function. It requires two separate input values and produces one output value depending on the state of the inputs. Its operation is also represented by the truth table shown in Figure 4.17. If we call the two inputs A and B, the output is called A AND B, and it is written symbolically as $A \cdot B$. The output is defined to be a 0 unless *both* A and B are 1, in which case the output then becomes a 1.

The OR function also operates on two inputs and generates a 1 at the output if either one or both of the inputs are at level 1. Otherwise the output is 0. If the inputs are again called A and B, A OR B is written $A + B$ and its value is shown in the corresponding truth table.

In digital electronic circuits these logic functions are carried out by units called *gates*. These are circuits composed of transistors, resistors, and capacitors

Basic logic functions:

NOT   AND   OR

Corresponding truth tables:

A	$\bar{A}$
0	1
1	0

A	B	$A \cdot B$
0	0	0
0	1	0
1	0	0
1	1	1

A	B	$A + B$
0	0	0
0	1	1
1	0	1
1	1	1

FIGURE 4.17. The basic logic functions and their truth tables.

that can be mass-produced in miniature on the surface of a silicon chip. Many complex digital operations can be built up on a single chip by properly combining such NOT, AND, and OR gates.

SUMMARY

Digital computation is based on a small number of basic logic functions. Most fundamental of these are the NOT, AND, and OR functions. The effect that each of these operations has on logic levels supplied to their inputs is indicated by the corresponding truth tables. Each of these functions has its electronic equivalent called a gate.

Example We shall now show how the most basic of all arithmetic operations, binary addition, can be carried out using a small number of these basic logic gates. The fundamental rules of binary addition were outlined in the previous section. In order to add two binary numbers, each with n bits, we must carry out n steps corresponding to the n separate columns in the addition process. We shall concentrate on just one such column, and realize that full addition of two n-bit numbers will require n of these units placed in series, one for each column (Figure 4.18).

Therefore, if we look at a typical column, our desired logic unit (which we will call a *binary adder*) must have three logic inputs: one each for the two binary digits that appear in the addition column, plus a "carry" input that may contain a bit

FIGURE 4.18. The summation of binary bits.

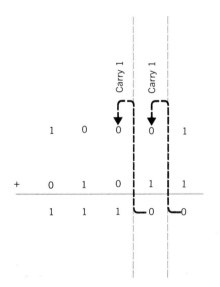

A	B	C	Sum bit	Carry bit
0	0	0	0	0
0	0	1	1	0
0	1	0	1	0
0	1	1	0	1
1	0	0	1	0
1	0	1	0	1
1	1	0	0	1
1	1	1	1	1

FIGURE 4.19. The truth table for a single bit binary adder.

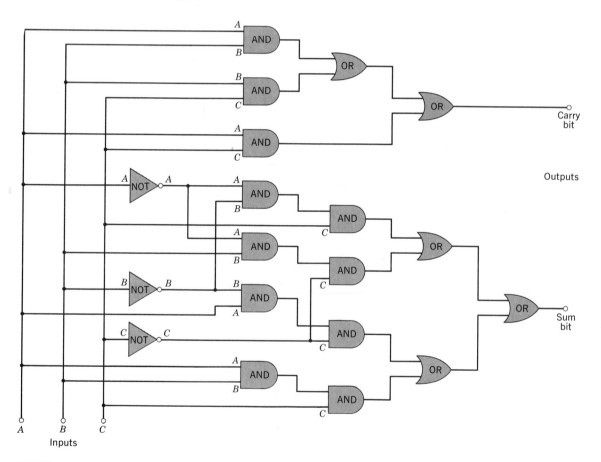

FIGURE 4.20. The logic circuit diagram for a binary adder.

carried over from the addition from the previous column. The binary adder will have two outputs: one corresponding to the sum bit or the value of the binary digit of the answer in the same column, plus a possible second bit that may be carried over to the next column to the left. All possible combinations of the input values are shown in the truth table of Figure 4.19, together with the corresponding values for the output sum and carry bits.

A logic diagram of a binary adder made up of 19 separate logic gates is shown in Figure 4.20. By following through each of the indicated logic functions, the reader can verify that this combination of elements exactly fulfills the conditions required in the adder truth table. For example, if all three inputs are held at 0, then 0s are also produced for both the sum and carry bit. The design shown is not reduced to the minimum possible components, but it is close to the minimum. Since the design deals with only one column in the addition shown on Figure 4.18, it is said to be a "one bit wide" adder. In order to build an equivalent adder that could accommodate words that are 16 bits wide, 16 such basic adders would have to be combined in cascade.

Exercises

Number Systems

1. Convert the binary number 101101001010 to octal and hexadecimal form.

2. Convert the decimal number 1015.5 to binary representation.

3. Convert the following numbers into their decimal equivalents:
   (a) 110.01_2 (d) 0.0011_2
   (b) 42.36_8 (e) $3D32.E_{16}$
   (c) $8B4.C_{16}$ (f) $BFFF_{16}$

4. Convert the following binary numbers into hexadecimal:
   (a) 101100   (d) 1.01101
   (b) 10.1101   (e) 110111.0
   (c) 110.1110   (f) 1001101011

5. Write the binary and hexadecimal equivalents of the following decimal numbers:
   (a) 36   (d) 1/3
   (b) 50.3   (e) π
   (c) 0.250   (f) e (2.718 . . .)

Digital Representation

6. What is the largest positive integer that can be stored in a computer of 32-bit word length?

7. A microcomputer has a memory size of 48000 K. How many bits are required to address each memory location?

8. We wish to store and then add 1000 three-digit numbers (001 to 999). How much memory will this require, assuming a binary number representation in the computer?

9. Estimate the memory required to hold the contents of this textbook (assuming the ASCII 7-bit character code is used).

10. Explain the bar code used on items in the supermarket.

11. Determine the two's complement binary form of each of the following numbers: 10, 7, 31, 28.

12. Subtract the following pairs of numbers in binary arithmetic using two's complement:
    (a) 10 − 7   (c) 6 − 2
    (b) 31 − 28   (d) 42 − 16

13. Determine the format used to represent floating point numbers in 16-bit word microcomputers.

14. Identify and describe the functions of five different CONTROL characters on your microcomputer or computer terminal keyboard.

Digital Signals

15. Contrast the number of bits required to represent the decimal integer 28 with the number required to represent the decimal 28.25.

16. If a temperature sensor is accurate to within 1° over a range from 0 to 100°C, how many bits should be provided by the output of an ADC to digitize its measurement adequately.

17. Locate three different ADCs that are part of devices that you frequently use.

18. Identify three different types of noise that are present in your everyday attempts at communication (oral, written, or graphical).

Binary Logic

19. Some compound logic functions are used frequently and have therefore been given special names. Supply a truth table showing the output values for all possible input values for the following compound functions:
    (a) The NOR gate consisting of an OR function and a NOT function in the sequence as shown in Figure 4.21.
    (b) The NAND function consisting of the AND and NOT functions in series.

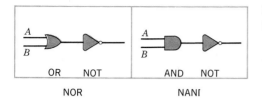

FIGURE 4.21. The logic diagrams for the NOR and NAND functions.

20. Using the three basic logic functions defined in the text, devise a logic circuit that meets the truth table in Figure 4.22

Inputs Output

A	B	
0	0	0
0	1	1
1	1	0
1	0	1

FIGURE 4.22. The truth table for Exercise 3.

4.3. DIGITAL COMPUTER HARDWARE

The basic organization of any digital computer is shown in Figure 4.23. This structure includes a central processing unit or CPU, input/output devices, a reasonable amount of high speed data storage or memory, cheaper low speed but high capacity storage, and a wide variety of peripheral devices. These elements are common to computers of all levels of sophistication, from the familiar pocket calculator to the most complex and powerful mainframe computer.

The *central processing unit* contains the control unit, the "brain" of the computer. This unit controls and coordinates all data processing activities. It retrieves instructions from memory, interprets, and executes these instructions in the proper order. The CPU also contains the arithmetic unit in which actual computations are performed. It must contain provisions to perform the basic arithmetic operations of addition, subtraction, multiplication, and division, as well

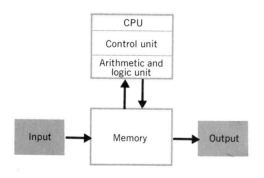

FIGURE 4.23. The basic components of a digital computer.

as logical operations such as comparisons and tests. The arithmetic unit is composed of very fast storage devices that temporarily hold the data or operands at the start of the calculation and the results at its completion. These storage units are called *registers* (or sometimes *accumulators*).

The main *memory* of the computer is composed of a large number of slower memory devices known as *memory locations* or *cells*. Each memory location is characterized by an address, typically 16 bits in length. The storage unit or memory of the computer provides a means of temporarily storing the input or output data as well as the necessary intermediate results. In most cases, however, the majority of the memory is commited to another task: to contain the set of instructions or program that determines the sequence of operations to be carried out by the computer. Computer systems also employ slower, high capacity memory devices such as magnetic disks and tapes.

The *input/output* (I/O) structure of the computer is the means by which data is communicated to and retrieved from the machine. It is the interface between the computer and the user. Common devices used to input data are a keyboard, punched cards, or magnetic tape or disk. Output devices are chosen to suit the application. They can provide printed records such as a typewriter or line printer, visual display as on a TV screen, or more permanent records such as on magnetic tape or disk.

SUMMARY

The organization of any computer, whether it is large or small, has basically the same elements: central processing unit (including control and arithmetic units), memory, and input/output devices.

Example   The CPU (in this case a tiny microprocessor) and the main memory of the Apple-II microcomputer are shown in Figure 4.24.

4.3.1.   The Central Processing Unit

Modern computers range in complexity from simple microcomputer devices to large scale complex central computing facilities. Where a particular system falls on this scale depends largely on two factors: the amount of fast memory that is provided and the complexity and variety of operations built into the central processing unit.

The CPU consists of two parts: A *control unit* to interpret, sequence, and process instructions, and an *arithmetic and logic unit* (ALU) to carry out particular operations on data (e.g., addition, multiplication, or logical comparison).

The ALU contains several very fast binary storage devices known as registers that can be loaded with the contents found in any specified address in the

FIGURE 4.24. The main circuit board of the Apple-II computer showing the principal components of a digital computer. (Courtesy Apple Computer, Inc.)

FIGURE 4.25. The basic components of a tiny microprocessor are identified. (The microprocessor is also shown in Figure 4.5.) (Courtesy Bell Laboratories)

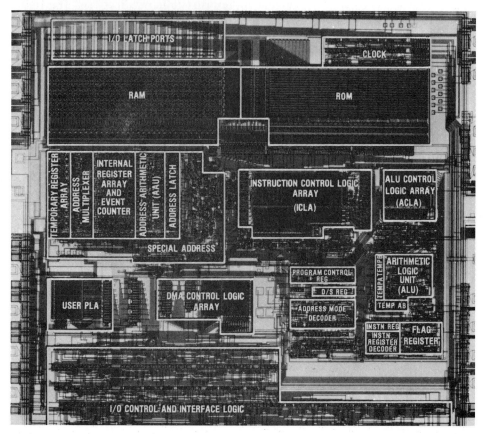

main memory. These registers can store data roughly 10 times faster than storage devices in the main memory (which are characterized by data storage and retrieval times ranging from 100 ns to 10 μs). The switching circuits in the CPU are even faster. For example, on most machines signals can pass through about 10 switching or logical devices in the time taken to store or retrieve information from the fast CPU registers.

A digital computer is designed with a set of built-in instructions, each of which is given a binary code. For example, all arithmetic units contain several registers that can be loaded with the contents found in any specified address in the memory. Thus, the code 0010 might represent the instruction "load register." If 4 bits are allocated to code the instruction as in this example, 2^4 or 16 different instructions could be coded. These might also include processes of addition, incrementing, transferring from one register to memory, and so on. Small machines can have an instruction set as small as 16 while larger machines have sets extending to 256 different instructions.

In addition to the coded instructions, some other information must also be provided to the ALU. In the case of our example, we must specify the address in memory from which the information will be obtained to load the register. If we wish to have access to a memory of 4096 locations (2^{12}), it will take 12 bits to specify all of these locations. Then the entire instruction would consist of a 16-bit combination: a 4-bit code specifying the operation (load register), followed by a 12-bit address of the location in memory whose contents will be transferred to the register. This 16-bit word then represents one step in a program that could be executed by the computer. One key to the success of modern computers is the fact that all programs are reduced to a sequence of such binary words that can be stored internally in the fast memory of the computer. When executing the program, the computer simply processes these instructions one at a time in the order in which they are stored within the memory. The entire program is really just a set of binary numbers, and only a prearranged code built into the computer allows it to be interpreted as a particular series of instructions. It is the function of the control unit to supervise the flow of data required by this set of instructions between the arithmetic unit, memory, and I/O devices of the computer.

SUMMARY

The arithmetic unit of a computer is provided with a basic set of instructions that allow it to perform elementary operations on data supplied to it. The instruction set is very limited for small devices such as microprocessors but may extend to hundreds of instructions for large scale computers. During its operation, the control unit supervises the flow of data to and from the arithmetic unit as required by each step of the computer program.

Example   The instruction to load a register does not directly lead to a useful result. However, it can be an important first step in carrying out obviously important processes such as addition. One of the prearranged instructions for the machine could be to add together the contents of the register with that of another location within the memory, and to replace the contents of the register with the sum. This operation could be carried out within the arithmetic unit using the binary adder discussed in an earlier section. Hence, if we wish to add together two numbers initially found in the memory, we could do so using a sequence of fundamental operations. The first of these will be to transfer the contents of the first of the numbers to the register using the command illustrated earlier. We then use the addition instruction to add the contents of an arbitrary location in memory containing the second number. A third instruction then transfers the new contents of the register, which now is the sum, to another location in memory for temporary storage. This value is then made known to the user by having the I/O portion of the computer seek out this new location in memory and supply its contents to an output device such as a printer or display screen.

4.3.2   Memory

Various devices are used to store binary data in a computer. We noted that the CPU contains very fast storage registers used in the arithmetic and logic unit. The main memory of the computer contains a large number of storage devices that can contain binary data. The memory is organized into individual units or cells that can be given a unique address. That is, the address of a memory cell is simply a number identifying the location of that particular cell within the memory as a whole.

Each memory cell can contain an amount of binary data known as a *word*. On large computers the word size can range as high as 64 bits. On small microcomputers the word size is more typically 8 bits. Several special terms have been introduced by computer engineers to characterize the size of a computer memory:

bit = one binary digit (0 or 1)

byte = 8 bits

nybble = half a byte or 4 bits

K or kilobyte = 2^{10} = 1024 bytes

Example   The word length for several of the more popular computers is given below:

microcomputers (Apple-II, TRS-80, PET): 8 bits (1 byte)

IBM 3081, Amdahl: 32 bits (4 bytes)

DEC VAX 11/780: 32 bits (4 bytes)

Control Data Cyber series: 64 bits (8 bytes)

Example   The main memory size of several popular computers is listed below:

Apple-II: 16 to 64 K

TRS-80: 4 to 48 K

Mainframe: 2 to 16 million bytes (2 to 16 megabytes)

The most general type of memory is called *random access memory* or RAM. It consists of integrated circuits into which data can be entered or retrieved very quickly. In a RAM cell with word length 2 bytes, 16 bits of data can be stored, recalled, or replaced with new data by specifying the appropriate address of the word in memory. Such memories typically can be read nondestructively—that is, the information at a particular memory address can be read many times without disturbing the information that is stored there. Writing into memory, however, normally destroys the information that was there previously.

At times there may be reasons to want to keep a portion of the memory without changing it. For example, it would be important to keep a specific set of instructions needed every time the machine is turned on in permanent form so that it is never inadvertently destroyed. Such units are called *read-only memories* (ROM) and cannot be changed once the original content is entered. (*Write-only memories* would be units into which information could only be deposited but not retrieved. For obvious reasons, this computer equivalent of a wastebasket serves no useful purpose.)

A related type of memory is *programmable read only memory* or PROM. A PROM is a ROM whose contents can be altered by electrical means. Both ROMs and PROMs are frequently programmed with special instructions and packaged in the form of a set of integrated circuit chips on a printed circuit board or card with contacts for electrical connections. These *firmware* cards can then be inserted into the computer when certain capability is required. For example, higher level programming languages such as BASIC are frequently supplied on ROM firmware cards for insertion into microcomputers.

A memory unit provides means for storing binary data. The smallest unit of such information is a binary digit or bit that can assume values of only 0 to 1. The physical system used to store information can be any element existing in either one of two states. For example, a common electrical switch could be used to store one bit of information. If the switch is open, it could represent the digit 0, while if the switch is closed, it could represent the digit 1. Early memories were made up of *magnetic cores* in which the direction of magnetization could assume either of two possible directions. The prevailing direction of magnetization could be sensed in order to nondestructively read out the existing data. The direction could be changed by applying an appropriate current pulse to write new information into the memory.

These magnetic core memories have all but disappeared. They have been replaced by all-electronic memories that can be packed to a much higher density on semiconductor-based integrated circuit chips (Figure 4.26). In early *semiconductor memories* the basic storage element consisted of what was known as a "flip-flop" circuit. Consisting of two transistors and a number of resistors, this circuit can exist in only one of two possible states: one or the other of the transistors conducting current while the other does not. These two states can also be remotely sensed or changed so one such circuit has the capacity to store one bit of information. Combinations of these elements can then be grouped to form words of 4, 8, or 16 bits within the memory units. Single semiconductor memory chips with as many as 64 000 bits of capacity were available.

More recent computer memory microcircuits store information in the form of charge on a capacitor. A metal-oxide silicon (MOS) transistor simultaneously serves both as a switch to allow the capacitor to charge or discharge and as the

FIGURE 4.26.   The rapid miniaturization of semiconductor computer circuits is evident in this comparison of the circuits used in the CYBER 205 supercomputer and the larger circuits used in the CDC STAR-100 computer (also considered to be in the supercomputer class). (Courtesy Control Data Corporation)

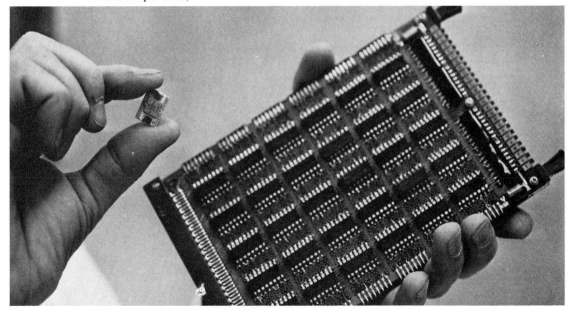

capacitor. Such one-device memory cells permit higher memory density. (Indeed, the charge stored in the capacitor memory cell of some present designs amounts to only about one million electrons.)

Even greater density of storage can be obtained through the use of *magnetic bubble memories* (Figure 4.27). These newly developed circuit chips are based on localized bubbles that can be created in a thin magnetic film. Each bubble is a tiny region of magnetization that can be readily created and detected in the film and that can be made to last indefinitely. The presence or absence of a bubble at a given point can indicate either a zero or a one value for the corresponding bit of information. The bubbles can be moved about on the surface of the film in a fixed pattern. In this way, a very large number of bubbles can be stored and moved into position when it is desired to read out their information. Because the bubbles can be made as small as one micrometer in diameter, as many as a million bits can be stored on a single chip.

Semiconductor memories allow quick access to stored information. They are used for the main memory of a computer. This memory might typically consist of anywhere between 8000 and several million bytes. It is used to supply the fast storage and recall needed in many computer operations. Since it consists of random access memory, it is possible to retrieve a byte of information in the time it takes the computer to recognize the address and sample that location. In many devices this time is well under a microsecond. Because of the fast access time, the

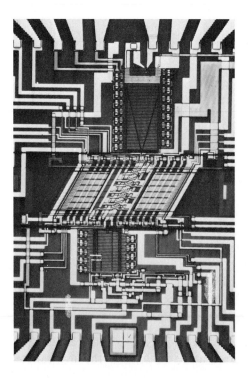

FIGURE 4.27. A microcircuit chip based on a magnetic bubble lattice that uses a hexagonal array of magnetic bubbles to read, write, and store digital information. (Courtesy IBM Corporation)

FIGURE 4.28. Magnetic disk memory units. (Courtesy IBM Corporation)

main memory is used to store the program being executed by the computer as well as data needed frequently in its execution. The speed is essential when carrying out arithmetic operations on a large scale. Other types of memory can be used that sacrifice some speed of access in return for a much larger capacity or lower cost than semiconductor memory.

One example of this auxiliary memory is the *magnetic disk* (Figure 4.28). Here the sequence of ones and zeros that make up each byte is represented by the magnetic pattern in the metallic oxide layer placed on the surface of a round disk. When in use the disk is rotated at high speed. A pickup head can move radially across the disk so that data written in the form of a circular track on the surface of the disk can be read back much in the manner of a conventional phonograph record, but at much higher speed. The pickup head can either read the existing pattern on the disk or can write new information in its place. The time needed to find any given location on the disk may be as long as ten to one hundred milliseconds. However, once the information is located, a large number of bytes can be transferred to main memory quickly. A standard disk unit can accommodate several million bytes of information on one interchangeable disk. Smaller units using a flexible base for the magnetic surface are known as "floppy disks" and have become quite popular in connection with small microcomputer systems.

Magnetic tape is another form of auxiliary memory (Figure 4.29). Here even larger capacities reaching tens or hundreds of millions of bytes can be achieved using tapes of convenient length. The access time is limited by the time required to spool the tape from one reel to another until the appropriate point along the tape is reached. This time can be many seconds or even minutes. As in the case of the disk, many bytes or words can be transferred at high rate once the appropriate

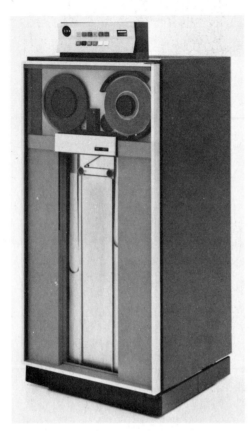

FIGURE 4.29. A magnetic tape drive unit. (Courtesy IBM Corporation)

position has been reached. Magnetic tape is most useful when large amounts of information must be stored and only infrequently accessed.

SUMMARY

The storage or memory of a computer is one of its most important elements. Random access memories allow for the quick storage or retrieval of information. Read-only memories have a content that cannot be changed, but their use eliminates the possibility of the inadvertent loss of stored information. Memories are usually organized into subunits of words, each of which is identified by an address. Semiconductor memories are the most common types used for the fast main memory of computers. Other media such as magnetic tape or disk have greater storage capacity but slower access time to the stored information.

4.3.3. Input/Output Devices

The user communicates with the computer through a variety of input/output (I/O) devices. In early batch mode facilities, input would typically be in the form of punched cards, prepared on a keypunch machine, and fed into the computer through a card reader. The output would be provided by a high speed printer.

With the introduction of time-sharing systems during the 1960s, the more common I/O device became a computer terminal. Early terminals looked quite similar to typewriters, with a keyboard for entering data and a printing device to provide the computer's response. Such terminals were either connected directly to the computer or equipped with cradlelike devices known as modems (for *mo*dulation-*dem*odulation) that would hold a telephone receiver and communicate with the computer over phone lines. Gradually terminals with television screens (or, as they referred to in computerese, CRTs or cathode-ray-tubes) for computer output became popular because they could be produced more inexpensively than mechanical printers. Keyboards and TV monitors have also become the most popular means for I/O with modern microcomputers.

Today's computers can utilize a wide variety of I/O devices. Input can be

FIGURE 4.30. A modern computer graphics system based on a small microcomputer. (Courtesy Hewlett-Packard Company)

provided not only by card readers or keyboard, but also from magnetic disk and tape, speech and visual pattern recognition devices, and analog-to-digital sensing devices. Output is provided on display screens (CRTs), teletype, high speed printer, magnetic disk or tape, speech synthesizer, and so on.

When one uses a computer (whether it be a large mainframe time-sharing system or a microcomputer) from a terminal, data or instructions entered at the keyboard will appear on the display screen. The user's location within this display is identified by a flashing marker known as a *cursor*. The user can specify a variety of different forms or *formats* in which to enter or receive data, including binary, integer, and floating point numbers, text or character strings, or graphical pictures. Data entered into or generated by the computer is temporarily stored in special memory locations known as I/O *buffers*. Data that appears on a display screen is sometimes known as ''soft'' copy since it cannot be carried away from the terminal (it disappears as soon as the screen is erased). If ''hard'' copy is desired, the computer output must be rerouted to a printing or graphics plotting device (Figure 4.31).

FIGURE 4.31. Many computer terminals or microcomputers are capable of providing hard copy. (Courtesy University of Michigan College of Engineering)

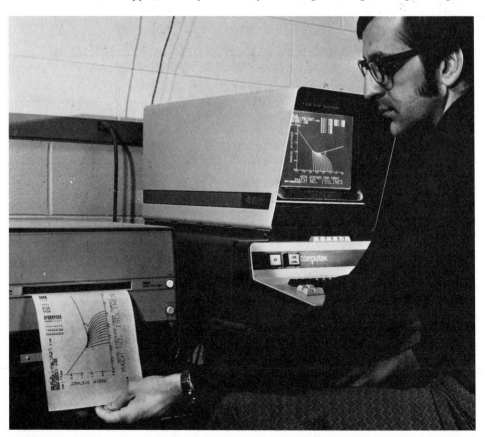

The development of inexpensive and reliable I/O devices has played a major role in the growing popularity of microcomputers. It has made possible the development of a truly portable and inexpensive computer system.

SUMMARY

Input devices include keyboards, card readers, magnetic disk and tape devices, and speech and visual pattern recognition systems. Output can be provided using teletypes and printers, CRT display screens, magnetic disk and tape, and voice synthesizers.

4.3.4. Pocket Calculators

The introduction of the pocket calculator made a major impact on engineering practice in the 1970s, just as it did on society at large. Almost instantly the slide rule became obsolete.

The early electronic calculators were essentially just very fast and convenient adding machines. They could perform only a limited number of operations or functions, albeit with a speed and accuracy far beyond that of mechanical devices.

With the introduction of the microprocessor and integrated circuit memories, the pocket calculator began to acquire more of the attributes of a computer. One could program in a sequence of instructions and store the results of intermediate calculations in memory. Yet the calculators of the 1970s were still limited to the display of numerical data and required that instructions be entered in a precise keystroke order.

One could distinguish among calculators by the type of keystroke logic that they employed. Most calculators employed an algebraic system, in which arguments and functions were entered in the same order as they would appear in an algebraic expression:

Example	Add 5 + 6	Key in first argument.	5
		Press the $\boxed{+}$ key.	$\boxed{+}$
		Key in second argument.	6
		Press the $\boxed{=}$ key	$\boxed{=}$
		Answer appears on display.	11

Such algebraic systems require some prescription to order operations in complex expressions with several operations. Usually a hierarchy was built into the logic (typically multiplication and division were executed prior to addition or subtraction). Parentheses were also used to assist in order operations.

An alternative system that avoided this difficulty was based on Reverse Polish Notation or RPN (developed by the Polish mathematician Jan Lukasiewicz). In this system the operator ($+$, $-$, $\times$, $\div$) was placed immediately after the operand(s), eliminating all ambiguity regarding the order of execution in compound expressions.

Example   Add 5 + 6   Key in first argument.   5

Press ENTER key to separate
first argument from second.   ENTER

Key in second argument.   6

Press the + key   +

Answer appears on display.   11

The more sophisticated pocket calculators based on microprocessors now allow the display of both numerical and symbolic data. In these devices one can enter a complex algebraic expression in its written form before execution, much as in the higher programming languages used on full-scale computers. In this sense,

FIGURE 4.32.   The evolution in hand-held calculators from simple four-function units ($+.-.\div.\times$) to microprocessor-based units approaching a personal computer in capability has been quite rapid. (Courtesy Hewlett-Packard Company)

the pocket "calculator" and the microcomputer are rapidly converging (Figure 4.32).

SUMMARY

The pocket electronic calculator rapidly replaced the slide rule during the 1970s. Since early calculators could accept only numerical input, keystroke logic systems such as algebraic or RPN logic were used to enter calculations. More advanced calculators are capable of receiving and displaying both numerical and symbolic information and are rapidly approaching the sophistication of microcomputer.

Exercises

Digital Computer Hardware

1. Compare the characteristics of the principal microprocessors used in personal computers.

2. Determine the number of basic machine instructions characterizing the CPU of your campus mainframe computer, a personal computer, a pocket programmable calculator.

3. Determine the types of memory devices available for your campus mainframe computer and a personal computer.

4. Using the ASCII code, how much memory would you estimate to be necessary for a 100 statement computer program (assuming at most 80 characters per statement)?

5. Define the following terms used with I/O devices:

Modem	Buffer	Handshaking
Baud	Cursor	Interrupt
File	Scroll	GIGO

Pocket Calculators

6. Describe the steps involved in performing the following calculation:

   $$55.78 \times 45.36$$

   using both RPN and algebraic logic.

7. Describe the steps involved in performing the following calculation:

   $$\frac{3.1416 \times (10.01)^2}{55.78}$$

   using both RPN and algebraic logic.

8. Use a pocket calculator to calculate a value for e using the series:

$$e = 1 + \frac{1}{1!} + \frac{1}{2!} + \frac{1}{3!} + \cdots + \frac{1}{n!} + \cdots$$

term by term up to 10 terms.

9. Use a pocket calculator to calculate the natural logarithm of 2 using the series expansion:

$$\ln x = 2\left[\left(\frac{x-1}{x+1}\right) + \frac{1}{3}\left(\frac{x-1}{x+1}\right)^3 + \frac{1}{5}\left(\frac{x-1}{x+1}\right)^5 + \cdots\right]$$

Perform the calculations to 4 terms.

10. Compare the features of the modern hand-held calculator with those of personal computers (e.g., cost, capability, programming).

4.4. COMPUTER SOFTWARE

Thus far our discussion has focused on the *hardware* of the computer systems: input devices for entering information into the computer, output devices for displaying results, memory units for storing information, and processing units for performing computations on the stored data. All data is stored and processed in the computer in binary form, as strings of binary digits or bits. The actual operations performed by the computer on this binary data are quite primitive and are performed using an array of simple electrical switches or gates.

But what makes the computer such a powerful device is its ability to store and interpret binary strings as *instructions* as well as data. The computer can control its own course of action if it is fed a suitable set of instructions, that is, a program. Hence the built-in hardware of the computer includes both circuits for performing arithmetical and logical operations on stored data, as well as circuits for interpreting and implementing stored instructions. These instructions provided as input to the computer are known as *software*. They serve to organize the various arithmetical, logical, and control functions built into the complex procedures needed to process information.

4.4.4. Machine Language Versus Assembly Language

In its most primitive form, a *program* is simply a sequence of instructions in binary form that can be loaded into the computer memory and then implemented by the central processing unit. Typically these instructions, which are referred to as *machine language*, consist of binary words of several bytes in length. These binary instructions control the processing of other binary data loaded into the memory of the computer.

Although machine language programs can be directly executed by the computer, the task of writing such programs is quite formidable. For example, in a typical computer each machine language instruction might consist of a 16-bit binary word. A typical program might consist of hundreds or thousands of such instructions. Each detailed operation within the computer would have to be specified by its binary code, together with proper addresses in memory and other details necessary for information transfer. "Hand coding" even a simple program in machine language is a herculean task, not to speak of the difficulties inherent in finding errors or making modifications to the program at a later time.

Hence it proves convenient to assign special symbols as a code to represent several machine language instructions. This development of a set of rules and special words and symbols to aid in the writing and implementation of computer programs is known as *assembly language*. It is the most primitive form of a programming language. A typical assembly language command to copy the contents of a memory location with address (in hexadecimal notation) F9C6 into a register might be "FETCH F9C6". Commands to multiply the contents of the register with the contents of location FE41 and then store this result in location C020 would be "MPY FE41" and "STORE C020", respectively.

After writing out a list of such assembly language commands, that is, an assembly language program, a special program known as an *assembler* can be loaded into the computer to translate these commands into machine language (binary form). The use of assembly language greatly simplifies computer programming over direct machine language coding. It is sometimes used today for situations in which very high execution speed is desired for a particular computation procedure.

However even assembly language programs are still very difficult to write. They require an intimate knowledge of the computer "architecture" (register characteristics, memory addresses). Furthermore assembly languages differ from computer to computer. An assembly language program developed for one type of computer will not run on another type.

When an engineer writes a program, the details of computer operation at either the machine language or assembly language level are usually of no interest. We might be interested in specifying that two numbers be added together, but we usually do not want to be concerned with the loading of registers or the transfer of data to and from memory that is required to carry out that addition. Consequently a number of higher level programming languages have been developed to ease the task of writing programs for computers. These languages use common English and mathematical expressions to represent machine language instructions. And in contrast to assembly language, these higher level programming languages are intended to be machine-independent. That is, a program written for one class of computers should be capable of straightforward adaptation to run on other machines.

SUMMARY

A program is a sequence of instructions that can be loaded into the computer and implemented by the central processing unit. At the most primitive level a program consists of binary words of several bytes in length known as machine language. To assist in programming, a set of rules and symbols known as assembly language is developed to represent the binary word instructions. A special program known as an assembler then converts these symbols into machine language. Assembly languages differ from computer to computer.

4.4.2.   Higher Level Programming Languages

A *higher level programming language* assigns common mathematical or English terms to represent a number of machine language instructions. Such languages also remove the dependence on machine architecture by referring to memory locations by names (called "variables") rather than specific addresses. At the present time there are over 150 programming languages in use throughout the United States for a variety of applications include scientific and engineering computation, simulation and modeling, process control, business and accounting, and so on (Table 4.4).

The most widely used programming languages are FORTRAN, Pascal, and BASIC. To illustrate how such languages are used, let us consider first the FORTRAN language. This is still the most commonly used language in science and engineering (at least in the United States). The original version of FORTRAN (a mneumonic for *For*mula *Tran*slator) was developed in 1954 for mathematical calculations. The language has evolved through several versions such as FORTRAN-IV, introduced in the 1960s and popular still today, and the more recent FORTRAN-77. FORTRAN closely resembles the algebraic manipulations the user would perform in attempting to solve a problem directly rather than program it for a computer. Although the language is easily understood by the programmer, it is totally incomprehensible to the computer until translated into machine language (binary) instructions.

This translation task is performed by a large and complex computer program known as a *compiler* or *translator*. The compiler does not simply transcribe a program into machine language (as an assembler transcribes assembly language into binary instructions). It also analyzes and reworks the program. It performs such complex tasks as reordering operations, choosing internal representations for data, eliminating redundant operations, and reserving memory locations. The compiler both checks the program to verify that all of the rules of the language

TABLE 4.4 Some of the More Popular Higher Level Programming Languages

LANGUAGE	YEAR INTRODUCED	ORIGIN OF NAME	PRIMARY USE
ALGOL	1960	*Algo*rithmic *l*anguage	Scientific
BASIC	1965	*B*eginner's *a*ll-purpose *s*ymbolic *i*nstruction *c*ode	Education, microcomputer
COBOL	1959	*Co*mmon *b*usiness-*o*riented *l*anguage	Business
FORTRAN	1954	*For*mula *tran*slator	Scientific
LISP	1956	*Lis*t *p*rocessor	Artificial intelligence
Pascal	1971	Blaise *Pascal* (famous French mathematician)	General, microcomputer
PL/1	1964	*P*rogramming *l*anguage 1	Scientific, business
Ada	1979	Augusta *Ada* Byron (Lord Byron's daughter and first computer programmer)	General

have been observed (the *syntax* of the language) and then optimizes the actual machine language translations of the program for efficient execution.

It is a difficult task to write a compiler program. One statement in the programming language may require many machine language instructions to control the transfers to and from memory, the arithmetic and logic operations, and the use of I/O devices. Fortunately, computer scientists have developed very sophisticated compilers over the past two decades for many higher level languages.

Example To illustrate the use of a compiler, let us consider how we might write and execute a FORTRAN program. We would first develop a set of instructions, a program, in the higher level language. (This is a major task in itself, and we shall return to consider it in greater detail in the next section.) This program is commonly called the *source* program, because it provides the source of instructions that will eventually be translated into machine language instructions by the compiler.

We would next give a command to load the FORTRAN compiler into the computer and then feed the source program in as data to be processed by the compiler program. The compiler will then process the source program, checking it for errors, translating it into machine language instructions, and optimizing this machine language program, which is referred to as the *object* program. The output from this phase known as *compilation* will consist both of the object program (or object "code") as well as diagnostic information provided by the compiler to identify any errors that might exist in the source program.

If the source program has been successfully compiled, that is, if the compiler detects no errors, then we would next load the object program (the machine language code) into the computer and execute it with suitable data provided as input.

While FORTRAN is probably the most common language used in scientific and engineering applications, in many ways it is a rather primitive language ill-suited to other applications. The listing of a FORTRAN program can prove quite incomprehensible to others (as well as to the original programmer after a short time away from it). It is sometimes difficult to write FORTRAN programs in a logical mathematical structure. In addition, FORTRAN is an awkward language to use for handling large amounts of non-numerical data (e.g., text material).

Perhaps the most popular language for nonscientific applications is BASIC (*Beginner's All-Purpose Symbolic Instruction Code*). This is an easy-to-use language especially suited to microcomputers. Even a novice can become reasonably adept at programming in BASIC with only a few hours of effort (conveniently provided by the numerous self-instruction modules written for this language). BASIC is designed as a conversational interactive language so that the programmer is in intimate and constant contact with the computer. A BASIC program is interpreted by a special program known as an *interpreter* each time it is executed, rather than translated or compiled once and stored in machine language (object) form such as FORTRAN. For this reason BASIC is an inherently expensive language to use in terms of CPU time required. Each execution of a BASIC program requires a new interpretation. Nevertheless, because of its simplicity, BASIC is frequently the language of choice for short programs (50 statements or less).

Other languages have been developed for special applications. For example, COBOL (*Common Business Oriented Language*) is a programming language suitable for commercial data processing work in business. SNOBOL is a symbol (string) manipulation language useful in text manipulation or matching patterns of characters. LISP is a symbolic language useful for artificial intelligence (robotics) applications. Recently there has been an attempt to introduce a standardized general purpose language known as Ada (after Lord Byron's daughter, Augusta Ada Byron, who served as the world's first programmer for Charles Babbage).

Probably the most powerful general purpose language available for microcomputers is Pascal (named after the famous French mathematician, Blaise

Pascal) developed in the early 1970s. Pascal is similar in many ways to an earlier scientific language ALGOL (for *Algo*rithmic *L*anguage). It is designed to take advantage of sophisticated programming methods. In contrast to FORTRAN or BASIC, a Pascal source program is compiled into a "p-code" language, somewhat akin to assembly language, which can then be interpreted by another program loaded into the computer. This feature makes Pascal probably the most machine-independent of all programming languages. Pascal is likely to replace BASIC and FORTRAN for many computer applications in the near future.

SUMMARY

Higher level programming languages assign common mathematical or English terms to represent several machine language instructions. The basic features of these languages are independent of computer type. The most popular such languages include BASIC (for small computers), FORTRAN (for scientific calculations), Pascal (for general use on both large and small machines), as well as special purpose languages such as COBOL (business applications) and LISP (artificial intelligence).

4.4.3. Operating Systems

Computer systems have become so complex and sophisticated that it would be impractical for one to depend on a human operator to run the systems. This has led to a class of computer programs known as operating or executive systems or simply as monitors. These programs sequence and monitor the progress of user computer programs in the system to take maximum advantage of the operating characteristics of both input and output devices and the central processing unit. In monitoring the progress of programs, the operating system protects other computer users (and the system itself) from user programming errors. It acts to limit total computing time and printing volume. Operating systems also perform the timing and bookkeeping operations necessary to assign appropriate charges for the use of the computer. On smaller microcomputers the operating system facilitates the use of peripheral devices such as disk drives. Among the more popular operating systems are the CP/M (Control Program for Microcomputers) system and the UNIX system.

In smaller machines the operating system or monitor may actually be hardwired into the system in ROM. However on large mainframe systems, the operating system must be loaded into the machine just as any other program. Individuals responsible for the care and feeding of the operating system are known as systems programmers. They play a vital role in modern computer applications.

Example Several years ago a major university computer center experienced a rather disturbing malfunction in its operating system shortly after one of the systems programmers had quit his job. One day during normal operation, all terminals tied into the system suddenly crashed and simultaneously printed out: "I WANT A COOKIE!" The system then resumed operation. This event would not have been so disturbing, had it not reoccurred once again several days later. At this point the systems programmers became concerned. But not matter how carefully they examined and tested the system, periodically the computer would stop everything and print out "I WANT A COOKIE!" To make matters worse, the frequency of its demands was gradually increasing. First every few days, then every day, and then every few hours, it would demand "I WANT A COOKIE!" Finally, in despair, they gave up and contacted the former systems programmer. Although he would not admit to tampering with the system, he did suggest that the next time the computer printed out "I WANT A COOKIE", they should type in the message "HERE IS AN OREO." And so they did. The computer then responded with "YUM, YUM!" and has behaved normally every since.

The operating system is responsible for running the computer, whether in batch or time-sharing mode. A set of special instructions is provided to allow the user to instruct the operating system to perform desired tasks, such as running computer programs or transferring data from one storage device to another or printing out results of a calculation. Operating system commands are frequently identified by a prefix character to distinguish them from other commands that might appear in a program.

Example Suppose that the prefix character for an operating system is "$". Then a typical job submitted by a user that would compile and run a FORTRAN program might look as follows:

```
$SIGNON ALONZO K. ZORCH ACT45 TIME=2S PAGES=20
$COMPILE FORTRAN

    {  FORTRAN program

$ENDFILE
$LOAD OBJECT
$EXECUTE

    {  Data cards for the FORTRAN Program

$ENDFILE
$SIGNOFF
```

In this case the user has first used the command "$SIGNON" to identify himself to the system, along with specifying time and printing limits. The "$COMPILE FORTRAN" command instructs the computer to compile or translate a set of programming statements, ending with the "$ENDFILE" command. The result of this compilation, the object program, is then loaded into the computer with the "$LOAD" command, and executed with the "$EXECUTE" command. When the job is finished, the user indicates to the system that there is no further work by the "$SIGNOFF" command.

SUMMARY

A set of special programs known as an operating sytem is used to control and monitor the use of computers. Most operating systems respond to a set of special instructions.

4.4.5. Programming

A computer program is a set of instructions that can be loaded into a computer to control its processing of information. In a more abstract sense, a program is a statement in some computer programming language of an *algorithm*: a step-by-step procedure for solving a problem. Computer programming is a term frequently used to describe the process of developing an algorithm to solve a problem of interest and then writing a computer program to implement this algorithm.

As we have noted, there are literally hundreds of programming languages available for different tasks and suitable for different types of computers. Such languages attempt to combine the precision of mathematical expressions with the flexibility of natural languages. Early programming languages such as FORTRAN were designed for specific tasks in mathematical analysis. They were not particularly well suited to the variety of computer applications ranging from data sorting to systems control to text processing that arise in practice.

There was yet another shortcoming with the early languages. It was hoped that these languages would be machine-independent. That is, no matter what machine a program had been written to run on, it would be possible to run it on any other machine with the appropriate compiler. Unfortunately this has not yet happened. Most programs cannot be moved easily from one machine to another without some changes in programming.

Yet another limitation of the early languages involved the difficulty they presented both in writing efficient programs and then tracking down errors that might be present in these programs. Computer programmers refer to an error in a program as a *bug*. Most programs contain many bugs when they are first written. (Indeed, many programs that have been used for years still contain bugs.) It can prove to be a tedious task to track down these errors, that is, to *debug* the program.

For these reasons, once a program is operating satisfactorily, there is a general tendency to continue its use, even though it may not present the best approach for a particular computer. Today many powerful computers are being used far short of their capacity because they are constrained by the programs or programming languages originally written for machines 10 to 20 years older.

There has been great interest in developing new methods of programming more suited to modern computers and the variety of tasks that they are employed to solve. Many of these methods involve entirely new approaches to problem solving as well as new computer languages.

Earlier approaches to computer programming were heavily influenced by the fact that computers can perform only one small instruction at a time. They tended to concentrate more on the order in which these steps were taken than on understanding how each step related to the solution of the problem as a whole. These early programming methods made extensive use of *flow charts* to diagram the logical sequence of processing that the program would follow. The sequence of operations in the flow chart could be followed as if it were a road map. Unfortunately, as the problem became more complex, so too did the flow chart become more complex and confusing.

Modern computer programming stresses a somewhat different approach known as *structured programming*. In this approach one uses an orderly approach that emphasizes subdividing large and complicated logical sequences into smaller and smaller modules, each of which performs a task that is conceptually separate and distinct from other parts of the program. Structured programming adopts essentially the reverse approach from the flow chart method. It leaves the order of processing to be determined until after the various major steps of the task have been defined. These steps are broken down into progressively finer and finer details. Each such component is then programmed as a separate part of the program.

Structured programming makes extensive use of subprograms or *procedures* (known as *subroutines* in earlier computer programming languages). Again the programmer makes extensive use of diagrams, but in this case the program is diagrammed as a hierarchic or *"tree"* structure. All logical connections branch away from the trunk of the tree, with no connections allowed to loop back on themselves. One starts at the trunk of the tree and follows the structure toward the branches, providing a progressive definition of more detailed levels. For this reason, structured programming is sometimes referred to as *top-down programming*.

Structured programming methods are actually quite similar to the approach we adopted in engineering problem solving in Chapter 2. Recall that we began there with a broad, detail-free definition of the problem. Then as we learned more about the nature of the problem, this definition became more and more precise, leading to more specific methods for its solution. The same approach is used in structured programming. One begins with a short list of only very rough and informal instructions. We then refine these instructions progressively until a satisfactory solution algorithm is achieved.

Many of the newer programming languages are now being designed to take advantage of structured programming methods. Notable among these are the Pascal, FORTRAN-77, and PL/1 languages.

SUMMARY

A computer program is a set of instructions that can be loaded into a computer to control its processing of information. Programs involve an algorithm, a step-by-step procedure for solving a problem. Early programming languages concentrated more on the order of instructions than on understanding how each step related to the solution of a problem as a whole. Newer languages employ structured programming in which one breaks up large and complicated logical sequences into smaller and smaller modules, each of which performs a task that is conceptually separate and distinct from other parts of the program. Structured programming is also known as top-down programming.

Exercises marked with an asterisk(*) require some degree of programming knowledge.

Exercises

Machine Language vs. Assembly Language

1. Find out, list, and define several assembly language commands for your university mainframe computer.

2. Define the following terms:
   Assembler   Opcode
   Disassembler   Operand
   Cross assembler   Editor

Programming Languages

3. Suggest appropriate variable names in FORTRAN, BASIC, or Pascal for each of the following:
   (a) Time
   (b) Velocity
   (c) Number of defective units

(d)   Result of a logical expression

(e)   Mass

(f)   User's response to the question: "What is your name?"

(g)   Number of terms to be retained in a series

(h)   Angle of inclination

*4.   Contrast the statements in FORTRAN, BASIC, and Pascal that accomplish each of the following tasks:

(a)   Sets the variable X equal to a value 1.57.

(b)   Adds together two variables, X and Y.

(c)   Calculates $e^{0.02}$.

(d)   Reads in data for the value of a variable X.

(e)   Writes out the value of a variable X.

(f)   Reads in a string variable such as a sentence.

(g)   Performs a given calculation 100 times.

*5.   Contrast the essential differences between FORTRAN IV (level G or H compiler) and FORTRAN-77.

*6.   Contrast the essential differences between BASIC and extended BASIC.

*7.   Roughly how much core memory is required for the BASIC interpreter program for a microcomputer? For the Pascal compiler and interpreter programs?

*8.   Write computer program statements (in any language such as BASIC, FORTRAN, or Pascal) for calculating the following algebraic expressions:

(a) $ax^2 + bx + c$

(b) $\dfrac{L + R/C}{L - R/C}$

(c) $\dfrac{e^{ax} - e^{-ax}}{e^{ax} + e^{-ax}}$

(d) $\dfrac{-b \pm (b^2 - 4ac)^{1/2}}{2a}$

*9.   Write a computer program for converting any decimal integer into binary form.

*10.   Write a computer program that will convert any binary number into a decimal integer.

Operating Systems

11.   Determine and list the commands on your campus computer operating system that correspond to each of the following:

(a) Signing on and off the system.

(b) Determining the status of your computer account.

(c) Creating a file in which data or program statements can be placed.

(d) Running a program.

(e) Compiling a source program in Fortran, PL/1, or ALGOL.

(f) Stopping program execution ("break").

(g) Listing the contents of a file on a printer.

12. Identify the operating system commands that will accomplish the tasks outlined in Problem 11 on a personal computer (e.g., the Disk Operating Sytem for the TRS-80 or Apple II computers).

*13. Write the sequence of commands analogous to those on page 284 that would compile and execute a FORTRAN source program on your campus computer.

Programming

Any suitable programming language can be used in the following exercises.

14. Develop an algorithm to calculate the number of elapsed days between any two dates.

*15. The Fibonacci sequence, 1, 1, 2, 3, 5, 8, 13, 21, . . . is characterized by the property that each number (beyond the first two) is the sum of the two previous numbers. Write and run a computer program to print out the first 30 terms in this series. At what point in the sequence will the size of the terms overflow the capacity of a 16-bit word?

*16. Write a computer program that will balance a checkbook. The program should first ask for the previous balance and then ask for each check written or deposit made.

*17. Write a program that will simulate dealing a 5-card poker hand from a 52-card deck by printing out the names of each card. This program can make use of the binomial distribution. Be careful to account for the change in the deck as each card is dealt.

*18. Write a program that will convert between English pounds and dollars using a specified exchange rate.

*19. Set up a string manipulation program that will write a technical paper. Begin by storing a variety of titles and phrases, and then use a random number generator to select out and combine various phrases to form the text.

*20. Write a program that will convert the measures used in a cooking recipe (e.g., cups, tablespoons, teaspoons, a pinch, . . .) into SI units (grams, cubic centimeters).

*21. Write a program that will take a cooking recipe and calculate the number of calories and nutrients per serving.

*22. Recalculate the Fibonacci sequence of Problem 15 to 100 terms using floating point arithmetic.

*23. The series

$$1 + \frac{1}{2} + \frac{1}{3} + \ldots + \frac{1}{K}$$

is known to diverge. That is, the sum increases without limit as K increases. However, if the sum is calculated for finite precision arithmetic, for example, 8-digit floating point, then for some value of K, K^{-1} will no longer contribute to the sum and the series will converge. For what value of K might you expect the sum to cease growing?

*24. Write a computer program to calculate the first N terms of the following series for any value of x:

(a) $\sin x = x + \frac{x^3}{3!} + \frac{x^5}{5!} + \cdots$

(b) $\cos x = 1 - \frac{x^2}{2!} + \frac{x^4}{4!} - \frac{x^6}{6!} + \cdots$

(c) $e^x = 1 + x + \frac{x^2}{2!} + \frac{x^3}{3!} + \cdots$

Test this program for various N by calculating the identity:

$$\sin^2 x + \cos^2 x = 1$$

*25. Write a program to play a game of TICKTACKTOE against the user. Use computer graphics if possible.

*26. Write a program to play a game of three-dimensional TICKTACK-TOE on a $4 \times 4 \times 4$ grid. Again use computer graphics if possible.

*27. Write a program that will interweave two text files line by line.

*28. Modify the file merging program of Problem 27 to weave files together on a sentence-by-sentence basis.

*29. Develop a course grading program. The program should handle the grades throughout the term of up to 30 students. It should combine these grades at the end of the term using some appropriate weighting and then perform a statistical analysis of the scores (computing average and standard deviation). The program should then assign letter grades (A, B, C, . . .) based on some appropriate "curve."

*30. Develop a computer program that will play Beethoven's *Für Elise* on a microcomputer.

*31. Develop a computer program that will "compose" melodies using a random number generator. Perhaps the simplest approach would be to write a Brownian Motion Boogie in which each note is selected at random to be equal to, one note above, or one note below the preceding note.

*32. Develop a program to generate "computer art" using a microcomputer with color computer graphics capability.

4.5. FUTURE DEVELOPMENTS

The technological trends in the rapid development of computers have stressed both their size and speed. Modern machines now vary in size from small microprocessors, with no more than a few hundred words of associated memory, to large mainframe computers with capacities that were unimaginable even a few years ago.

The speed of operation has also been an important factor in computer evolution. The number of operations performed per unit time for any given machine is inversely proportional to the time it takes to do one such operation. For many years progress was made by simply reducing the time required for each individual circuit to carry out its function. Ultimately, however, such improvements are limited by the finite time required to transmit signals from one point in circuitry to another. Even though these signals travel at near the speed of light, this transmission time can be a limiting factor in the very fastest machines. The Cray 1S computer shown in Figure 4.33 is built as a compact cylindrical arrangement of circuits just to minimize the length of interconnections required and therefore to reduce the necessary signal transmission times to a minimum.

High speed computation in digital computers is usually measured in terms of millions of floating point operations per second or MFLOPS. For example, large commercial machines such as the IBM 3033 or CDC Cyber 176 can achieve a speed of 1 to 10 MFLOPS. By way of contrast, supercomputers such as the Cray 1S or CDC Cyber 205 achieve speeds of 200 to 800 MFLOPs (Figure 4.34). Existing microcircuit designs are beginning to approach inherent speed limitations, both in the switching speed of their circuitry, and also because highly compact machines such as the Cray-1S dissipate so much heat that special precautions have to be taken to keep circuits within operating temperatures. (In the case of the Cray-1S, the computer has essentially been built inside an air conditioner.)

Perhaps the next phase of computer hardware development will be based not on semiconductors, but instead on superconducting devices known as Josephson junctions. These devices can act as switches, just as transistors. They are capable of miniaturization and incredible speeds. In fact, the Josephson junction is the

FIGURE 4.33. The Cray-1S supercomputer is a cylindrical circuit array to minimize the length of electrical connections. (Courtesy Los Alamos Scientific Laboratory)

fastest switch known, being capable of changing its state in as little as 6 pico-seconds (six-trillionths of a second). Moreover, since these circuits involve super-conductors with no electrical resistance, heat dissipation is no longer a serious limitation on computer size (although the circuitry must be kept immersed in a bath of liquid helium to achieve the superconducting operation). A high speed,

TABLE 4.5 Computer Speeds in
Millions of Floating Point
Operations per Second

COMPUTER	SPEED (MFLOPS)
IBM 3033	4.4
CDC Cyber 176	9.1
Cray 1S	240
CDC Cyber 205	800

high performance supercomputer based on Josephson junction devices would be capable of speeds up to thousands of MFLOPS, and yet would be packaged in a tiny cube several centimeters on a side (Figure 4.35).

The quest for faster operational speed has also led to the development of new types of computer architecture. Conventional computers carry out a program in sequence, moving from one instruction to the next as each is completed. This type of operation will always require a total time equal to the product of the average

FIGURE 4.34. The CDC CYBER 205 can perform up to 800 million operations per second and has four million words of central memory. (Courtesy Control Data Corporation)

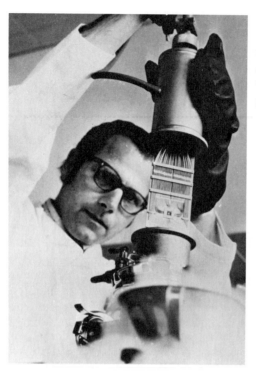

FIGURE 4.35. The experimental Josephson circuit computer will operate at supercooled temperatures at speeds orders of magnitude faster than semiconductor-based computers. (Courtesy IBM Corporation)

time required to carry out one operation multiplied by the total number of operations. Newer machines can reduce time using a technique known as *pipelining*. In this method, advantage is taken of the fact that certain circuit components may only be required for the early part of the computation and would normally be idle for the remainder of the operation. However, if results are required for a large number of different input parameters, these initial circuits may begin processing the next set of data before the first set has worked its way entirely through the computational process. If the sequence of operations is a long one, there could be, for example, as many as 10 independent operations going on at once, each operating on a different set of data. For a large collection of input data, the total time will therefore be only one-tenth that required if each set of data passed through the entire system before the next set were started. At any instant the pipeline will contain multiple data being processed according to the same set sequence of instructions. Pipeline architecture computers such as the Cray 1S or Cyber 205 are known as single instruction stream/multiple data stream (SIMD) or *vector* computers in contrast to conventional single instruction stream/single data stream (SISD) or *scalar* computers which act on only a single piece of data at a time. Even more advanced computer architectures are under development which utilize numbers of independent processors to achieve multiple instruction stream/multiple data stream (MIMD) operation.

Future developments in the area of microcomputers will also have a great impact on engineering practice. Today we already have small, inexpensive microcomputers that have the computing capability of relatively large machines of a decade ago. Just as the hand calculator quickly replaced the slide rule during the 1970s, the personal computer may well replace the hand-held calculator during the 1980s. Indeed, many "programmable" hand calculators are in fact microcomputers, although to date one must usually program them using a specially designed assembly language rather than a higher level programming language. Personal computers will almost certainly continue to address many of the needs that formerly required the use of large, mainframe time-sharing computer systems.

Perhaps the primary limitation on the role played by computers will occur not in computer hardware, but rather in the area of software development. To exploit fully the increasing power and sophistication of computers, we shall need to develop more flexible and powerful programming languages. It will be particularly essential in the area of personal computers, to develop languages (or I/O devices) that allow the layperson to "converse" naturally with the computer.

During the days of its early development, the computer was regarded primarily as a calculation tool—as simply a bigger and faster calculator. Only during the later stages of its development did it become apparent that the power of the computer could be directed at other applications such as the simulation and modeling of systems, data storage, organization, and processing, and systems control.

FIGURE 4.36. Computer programming can be an extremely frustrating experience. (Copyright, 1972, G.B. Trudeau. Reprinted with permission of Universal Press Syndicate. All rights reserved.)

It has been noted that the evolution of the computer has followed a path similar to that of the printed book. Early computers were "handmade," one of a kind items, expensive and complex to use, and available to only a few. Just as the introduction of the printing press led to the wide spread use of books, the implementation of time-sharing computer systems greatly expanded user access. And just as the Industrial Revolution made possible the personal book by providing inexpensive paper and mechanized printing, the microelectronics revolution of the 1970s has led to the development of the truly personal computer, inexpensive and portable, yet with the power of earlier machines.

However, there is one important difference between the development of the printed book and the computer. It took some 600 years for printed books to become available to the average member of society. It has taken just 40 years for the development of the personal computer. It is logical to expect that the pace of computer development and implementation will continue to be just as rapid in the future.

SUMMARY

Digital computers now cover a wide spectrum of complexity, ranging from microprocessors to large mainframe computers. Improvements in computational speed have been an important element in the evolution of all types of computers. As fundamental limits on speed are approached, attention has focused on new types of computer architecture. New concepts such as pipelining and vector processing have led to dramatic increases in computing power. At the other end of the spectrum, the development of the portable, inexpensive microcomputer has provided the engineer with a convenient and powerful tool.

Exercises

Future Developments

1. Determine the cost of your campus central computer, the number of hours it operates each week, and its speed of operation (number of operations per second). Then using this information, calculate the cost per million operations.

2. Repeat the calculation of computational cost outlined in Problem 1 for a typical personal computer (e.g., TRS-80 or Apple-II), making reasonable assumptions about usage.

3. An important question in engineering analysis involves the choice

between analytical (paper-pencil-calculator) and computer methods for problem solving. Which of the following problems would seem appropriate for "hand" calculation and which could make effective use of a digital computer:

    (a) Determine the geometrical shape that has the maximum surface-to-volume ratio.

    (b) Perform a statistical analysis (e.g., determine the average and standard deviation) for 5 temperature measurements.

    (c) Prepare an Internal Revenue Service tax return.

    (d) Prepare and place in alphabetical order words for the index of a book.

    (e) Analyze the air flow about a new hang glider wing design.

    (f) Prepare a technical report in which a number of revisions will be necessary at various stages.

4. Give examples of the following types of engineering problems:

    (a) A problem that can be solved by hand (without computer).

    (b) A problem that could be solved by hand, but might be more efficiently solved by using a computer.

    (c) A modest problem that requires the use of a computer.

    (d) A problem that would tax the speed and memory requirements of a personal computer.

    (e) A problem that would tax the speed and memory requirements of a large mainframe computer.

    (f) A problem that is clearly beyond the capabilities of present-day computers.

    (g) A problem that will *always* be beyond the capability of computers.

REFERENCES

General

1. Brice Carnahan and James O. Wilkes, *Digital Computing, FORTRAN IV, WATFIV, and MTS* (The University of Michigan, Ann Arbor, 1979).

2. William R. Bennett, Jr., *Scientific and Engineering Problem-Solving with the Computer* (Prentice-Hall, Englewood Cliffs, N.J., 1977).

3. Ivan E. Sutherland and Carver A. Mead, "Microelectronics and Computer Science," *Scientific American 237* (September, 1977), p. 210.

4. John G. Kemeny, *Man and the Computer* (Scribner's, New York, 1972).

5. Richard C. Dorf, *Computers and Man* (Boyd and Fraser, San Francisco, 1974).

6. R. J. Kochenburger and C. J. Turcio, *Computers in Modern Society* (Wiley, New York, 1974).

7. *Microelectronics*, a Scientific American Book (W. H. Freeman, San Francisco, 1977).

8. Joseph Weizenbaum, *Computer Power and Human Reason* (W. H. Freeman, San Francisco, 1976).

Programming

1. Jerrold L. Wagener, *FORTRAN 77 Programming* (Wiley, New York, 1979).

2. Daniel D. McCracken, *A Guide to FORTRAN IV Programming*, 2nd Edition (Wiley, New York, 1972).

3. Elliot I. Organick and Loren P. Meissner, *FORTRAN IV: Standard FORTRAN-WATFOR-WATFIV* (Addison Wesley, Reading, Mass., 1975).

4. J. G. Kemeny and T. E. Kurtz, *BASIC Programming*, 2nd Edition (Wiley, New York, 1971).

5. *The Applesoft Tutorial* (Apple Computer Company, Cupertino, Calif., 1979).

6. Kenneth L. Bowles, *Microcomputer Problem Solving Using Pascal* (Springer-Verlag, New York, 1977).

7. Jerome A. Feldman, "Programming Languages," *Scientific American 241*, No. 6 (December, 1979), p. 94.

CHAPTER 5

Experiments and Tests

As the needs of society have become more complex, the tools of the engineer have become more powerful and versatile. Modern science and mathematics have provided sophisticated methods of analysis and powerful aids such as the digital computer. Nevertheless, despite their power and versatility the tools of engineering analysis are incomplete until they are augmented by the experience gained by direct observation of engineering phenomena, that is, by experiment and testing. Experimental data and test results play an important role in all phases of engineering. These activities range from sophisticated experiments intended to investigate new scientific phenomena in the laboratory to product testing on the assembly line. Since all engineers will either perform experiments and tests or make use of the results obtained from such activities, it is important to consider experimental and testing methods in some detail.

In our discussion we shall distinguish between an experiment and a test. We define an *experiment* as an observation carried out under controlled conditions in order to discover an unknown effect or to test or establish a hypothesis. In this sense an experiment can be regarded as a planned observational process by which we increase our experience of the external world. The major objectives of an experiment are to discover and identify unknown effects and to prove or disprove theories. For example, experiments might be performed to determine the variation with temperature of the pressure of a gas in a container. Such an experiment would be designed to discover a new relationship, in this case between the temperature and pressure of the gas. One might attempt also to measure the deflection of starlight as it passes through the gravitational field of the sun in order to test Einstein's theory of relativity. These examples indicate that experiments can be performed on a small scale within the confines of the laboratory, or on the grander scale of nature itself, perhaps even on the scale of the universe. Scientific and engineering experiments represent an application of a procedure known as the scientific method, a procedure that includes the sequence of observations, analysis, and experiments that lead to new knowledge.

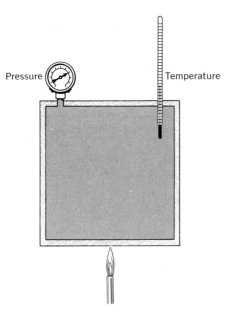

FIGURE 5.1. An experiment to measure the relationship between the temperature and pressure of a gas.

In contrast, an engineering *test* is a means of critical examination involving the controlled observation and evaluation of the results of engineering design, whether it be a product, a process or a component of a system. In this sense the purpose of a test is not to discover new phenomena, but rather to evaluate the characteristics of a design. Tests may be performed on a finished product, for instance, to determine whether an automobile meets federal standards for safety, exhaust emission, and fuel efficiency. Tests are also carried out before a product goes into production. For example, many different car radiator designs may be tested to determine the design that is most compatible with the tooling requirements of a new engine. Once the automobile goes into production, one radiator in a given batch (say, one in 100) might be tested to check the quality of the product.

Experiments and tests can provide valuable information only if they are properly designed and performed. A good experiment or test is one that yields the

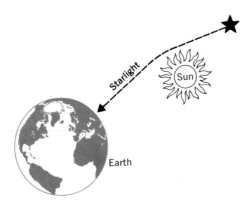

FIGURE 5.2. The deflection of starlight by a massive object such as the sun.

FIGURE 5.3.   Static testing of aircraft engines. (Courtesy United States Air Force)

required information with the minimum amount of time, effort, and cost. Unfortunately, experiments and tests often are run without proper care. The results of careless experiments and tests can give one a false sense of security without actually providing meaningful information.

The key to successful experimentation and testing is careful planning. Engineers should approach these activities with the same methods that they apply to any engineering problem. The application of the general procedure of engineering problem solving to the design, conduct, and analysis of engineering experiments and tests will be considered in this chapter.

SUMMARY

Engineers make extensive use of experimental data and test results. An experiment is an investigation carried out under controlled conditions in order to discover an unknown effect or to test or establish a hypothesis. A test is a means of critical examination aimed at evaluating the characteristics of engineering design. Both experiments and tests require careful planning prior to execution.

Exercises

Experiments and Tests

1. Give three examples each of experiments and tests that might be used in engineering practice in your field of interest.

2. Identify the unknown effect or hypothesis that was being investigated in each of the following famous experiments:
   (a) Galileo's measurement of the fall of stones from the leaning tower of Pisa
   (b) The Michelson-Morley measurement of the speed of light
   (c) The Stern-Gerlach measurement of the electron spin
   (d) Milliken's measurement of the ratio of charge to mass on an oil drop
   (e) Recent attempts to detect gravitational waves by sensing the vibration of a metal cylinder

3. What feature is being critically evaluated in each of the following tests:
   (a) Vibration tests of soil compaction at a construction site
   (b) Mechanically slamming a car door until hinge failure
   (c) EPA tests of automotive emissions
   (d) Measurement of radioactivity levels in nuclear power plant emissions

5.1. EXPERIMENTS

We have defined an experiment as a planned observational process by which we may increase our experience of the external world. Experimental investigations require careful thought, both in planning and in the analysis and interpretation of observed results or measured data. This combination of thinking and experimentation is commonly called the *scientific method*. Its elements consist of a sequence of steps that are followed, often unconsciously, in answering scientific questions. The steps involved in the scientific method can be identified as observation, hypothesis, and experiment.

5.1.1. The Scientific Method

Observation

The first step in the scientific method involves the *observation* of some new phenomenon. Frequently this observation is unplanned or by chance. For example, a mechanical engineer might notice that an automobile engine begins to vibrate noisily whenever its speed (rpm) exceeds a certain limit (Figure 5.4). Since

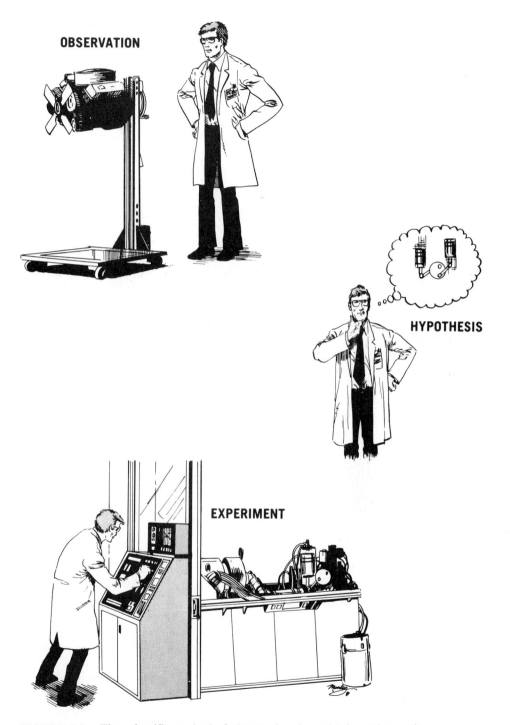

FIGURE 5.4. The scientific method of observation, hypothesis, and experiment.

the observed phenomenon is outside the range of previous experience, it provokes curiosity and stimulates speculation regarding the nature of the phenomenon. In our example, the engineer might suspect that the speed or frequency of rotation at which the vibration begins corresponds to the natural frequency of vibration of some component of the engine, such as the piston shafts. This speculation will generally stimulate further observations to confirm the presence of the phenomenon and determine its general features. However the observational phase of the scientific method remains quite qualitative and vaguely defined in most cases.

Hypothesis

After observing a new phenomenon, one attempts to explain or understand the observations by proposing tentative explanations that can be tested by further experimentation. These speculations or trial explanations are called *scientific hypotheses*. It is important to realize that such hypotheses or theories advanced to explain observations should be regarded as correct or truthful only to the extent that they are confirmed by further scientific observation or experiment. In a sense, scientists or engineers are just spectators, observing a complicated game played by nature without knowing the actual rules of the game. They hope that by watching long and carefully enough they can guess a few of the rules of the game: the laws of nature. These guesses then form the hypotheses advanced to explain a scientific observation.

To be of use, hypotheses or theories must be tested against the results of further observation. For example, our mechanical engineer might propose to test the hypothesis about piston shaft vibration by measuring the natural frequency of vibration of a single shaft and piston assembly. Only when confirmed by repeated observations do the hypotheses acquire the status of a theory or scientific law. The most useful approach is to formulate the hypothesis in such a way that it can be used to *predict* the results of further observations of a more careful and quantitative nature. It is this latter type of scientific observation, designed to test scientific hypotheses, that we call an experiment.

Experiment

The final stage of the scientific method is *experimentation* in which carefully planned and controlled observations are made to test a scientific hypothesis. The bits of information gathered in these observations are known as scientific or experimental *data*. This is the quantitative data that is used to confirm or disprove the predictions of a scientific hypothesis or theory. The experiment, then, is the acid test of a scientific investigation. Its goal is to establish new scientific knowledge. It is frequently difficult to distinguish between observation and experiment. As a general rule, however, we might characterize an experiment as an observation in which we interfere with nature to create conditions or events more favorable to our purpose.

It is important to realize that the scientific method of observation, hypothesis, and experimentation arrives at new scientific knowledge that may be only of a preliminary or temporary nature. That is, after the scientific hypothesis has been confirmed by careful experimentation, perhaps even acquiring the status of a scientific law, it may subsequently be contradicted or disproved by later observations or experiments. In this sense, scientific theories or laws should be regarded only as constructs of the moment, useful to the extent that they can help to explain the results of presently known experimental data, and doomed eventually to be discarded when contradicted by new observations.

SUMMARY

The scientific method involves the steps of observation, hypothesis, and experiment in order to determine new scientific knowledge. The first stage of observation of new phenomena is frequently unplanned. This stimulates the second stage of hypothesis in which a tentative explanation is given for the observed phenomenon. In the third stage of experimentation, carefully planned and controlled observations are made to test the scientific hypothesis. The information gathered in these observations is known as experimental data.

Example Galileo observed that there is a relationship between the time it takes a falling object to reach the ground and the height from which it was dropped. He hypothesized that this time was proportional to some power of the height

$$\text{time} \sim (\text{height})^n$$

According to legend, he then dropped balls from the leaning tower of Pisa and measured the time it took the balls to fall to the ground from different heights to deduce that $n = \frac{1}{2}$. (In fact, Galileo did not drop balls from the leaning tower of Pisa, but rather he rolled them down inclined planes.)

5.1.2. Design of Experiments

Experiments in science and engineering vary widely, both in scale and purpose. Nuclear engineers perform measurements on the microscopic scale of atoms and nuclei, while astronautical engineers measure properties of planets or stars or perhaps even the universe itself. Since practically every experiment is different, no one detailed experimental procedure will be applicable to all situations. Nevertheless we can identify some general features common to all experiments.

Some First Principles

Engineers must resist the temptation to rush into an experiment without adequate planning. The most critical aspects of experimental investigation are the preliminary analysis and planning of the experiment and the final analysis of the data. If the experiment is properly planned, its actual performance can frequently be reduced to a simple collection of data. Indeed, in many engineering experiments the data is actually taken by technicians rather than the engineer designing the experiment.

The nature of the experiment will be determined by several factors. Since the goal of any experiment is to test a conjecture or hypothesis about some new phenomenon, the engineer obviously should seek to acquire some understanding of the nature of the problem before beginning. This degree of familiarity with the subject of the experiment is occasionally called the *background* for the experiment. This preparation can vary widely, from the precise measurement of a quantity using a theory that is well-established (e.g., measuring the pressure, volume, and temperature dependence of a gas) to learning about a newly discovered phenomenon for which there is no directly relevant background information (e.g., measuring the surface composition of the Jovian moon Europa using a space probe). However, even though the familiarity with the subject of the experiment may vary widely, there is always some degree of reference to background ideas. That is, all measurements and experiments are conceived and interpreted in terms of current modes of thought.

A second factor is the degree of control that the experimenter has over the phenomenon. The physical properties of a new semiconductor material can be exhaustively studied using the sophisticated instruments in a solid state physics laboratory. In contrast, the production engineers investigating a large scale industrial process that is not functioning properly may be able to do little more than watch in the hope of making a guess at the fault. Such a limited degree of control is also characteristic of "experiments" in the social sciences in which observational surveys (e.g., product market surveys) play an important role. The planning phase is particularly important in such experimental investigations with limited control, for once the experiment is underway, the investigator may be able to do little more than passively watch its progress.

Another important factor is the influence of statistical fluctuations that limit the precision of any measurement. Since such statistical behavior is always present to some degree, the investigator should design the experiment and method of data analysis to account for these effects. The statistical concepts presented in Chapter 3 can be applied to the analysis of statistical variations in experimental measurements, as we shall demonstrate in the next section.

A final factor that will determine the conduct of the experiment is the required precision. For example, a very high degree of measurement precision is required when studying a phenomenon with a very small effect or when making a "benchmark" measurement intended to serve as a standard for further measurements. Of course, many experiments involving new or poorly understood phenomena assume essentially an observational nature in which precision is of no immediate concern. The skilled experimenter will design the procedure to yield the desired precision with the minimum investment of resources (equipment, time, and money).

SUMMARY

The design of an experiment involves many factors. The investigator should have acquired some degree of familiarity with the subject of the experiment before planning the observations. Important factors affecting the planning of experiments include the degree of control over the subject, the importance of statistical fluctuations in observed quantities, and the required precision.

Example An engineer is required to determine the power consumption of an internal combustion engine. The shaft of the engine is attached to an electric generator, and the power output of the generator is measured. The rpm of the engine is measured with an instrument called a tachometer. The rpm is varied during the tests by adjusting the throttle on the engine. These measurements establish a relationship between rpm and power output.

Variables

The basis of all scientific investigation is the assumption that similar events occur under similar circumstances or, put another way, that nature behaves according to some order. The similarities between events are usually restricted to a few features of particular interest. Then it is often found that similar circumstances can be identified by focusing attention on a rather small number of essential characteristics. These essential conditions that determine the occurrence of a given event are called *variables*.

FIGURE 5.5. An experiment to measure the dependence of engine power output on speed (RMP)

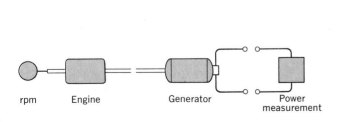
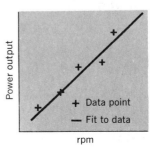

Therefore the first step in planning any experiment is to decide the kind of event to be studied and to identify the variables that prior experience suggests may be important. These variables may be conveniently classified into two groups. Variables under the direct control of the investigator are called *independent variables*. For example, in measuring the pressure–temperature relationship of a gas in a container, the experimenter might choose the gas temperature as the independent variable. This temperature could then be controlled by adding or removing heat from the container (e.g., using a heater coil or a cooling fluid).

The experiment will also involve variables that are not under the direct control of the investigator but are of direct interest in the outcome of the experiment. The most important of these uncontrolled variables are known as *dependent variables*. In our example above, the gas pressure would serve as the dependent variable. The experimenter will design the experimental procedure to measure values of the dependent variable for several values of the independent variables in order to learn as much as possible about the behavior of the system.

Of course many other variables are usually not under the control of the experimenter but nevertheless influence the outcome of the experiment. In fact many of these uncontrolled variables cannot even be identified or subjected to analysis. The degree to which mechanical friction or electrical resistance affect the transducer measuring the gas pressure and temperature are examples of uncontrolled and unknown variables. The experimenter should design an experiment to minimize the influence of these unknown quantities and to isolate the dependence of the variable of interest on only the few independent variables that can be controlled. Ideally the most desirable situation is to fix all but one of the indepen-

FIGURE 5.6.   Identification of dependent and independent variables in the measurement of the relationship between pressure and temperature of a gas.

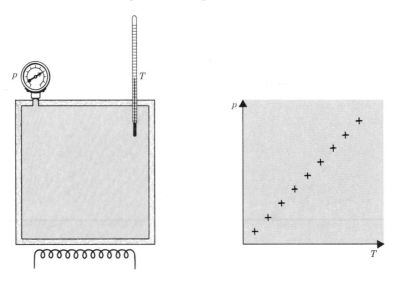

dent variables and then to study the variation of the dependent variable of interest with respect to only this single controlled variable.

A common axiom of good experimental procedure in the scientific laboratory is that we should hold all independent variables constant except that one under investigation. However this usually turns out to be a counsel of perfection rather than practice in more complex systems such as industrial processes. Here the variables may be so interdependent, so intertwined, that the effect of varying one may be to alter all the others that the experimenter had hoped to keep constant. That is, the supposedly independent variables are not really independent at all, but are rather strongly dependent on one another. The conduct of experiments for these situations becomes a matter of considerable skill and art (and frustration).

SUMMARY

The quantities characterizing an experiment are known as variables. They may be classified as independent variables (which determine the outcome of the experiment and are under the control of the investigator), dependent variables (which characterize the outcome of the experiment), and uncontrolled variables (which may influence this outcome, but are not under the control of the investigator). One should attempt to vary only one independent variable at a time when performing an experiment.

Example   An engineer sets out to measure the speed of sound through different gases. In this experiment the speed of sound is the dependent variable. It is hypothesized that the appropriate independent variables are the composition, temperature, and pressure of the gas. That is, the engineer designs the experiment to measure the dependence of the speed of sound c on these quantities:

$$c = f(\text{composition, temperature, pressure})$$

After the measurements are performed, it is found that near atmospheric pressure, the speed of sound does not change appreciably with pressure. Hence the original hypothesis is modified to indicate a dependence of sound speed on composition and temperature alone:

$$c = f(\text{composition, temperature})$$

Repetition and Reproducibility

Rarely does an experimental program consist of a single observation or measurement. Rather it is more common to repeat the measurement process several times to increase confidence in the results. For example, when electrical engineers measure the energy of the light beam produced by a pulsed laser, they fire the laser many times into a calorimeter and measure the energy deposited by each shot.

Of course even when an experiment is repeated under identical conditions, the measured results will usually be different. The experimenter is simply unable to control all of the variables affecting the outcome. The degree to which the uncontrolled variables influence the measurement will determine the *reproducibility* of the experiment. In most cases reproducibility is an important goal for experiment design.

There are situations, however, in which the results obtained in successive experiments performed even under identical conditions will continue to fluctuate. For example, an experiment in which a coin is tossed will yield heads or tails in almost a random fashion. This random behavior or uncertainty in the results of an experiment is present to some degree in nearly all measurements.

Methods have been developed to deal with the uncertainty or fluctuation in measurement. Later we shall apply the statistical analysis discussed in Chapter 3 to this problem. In fact, we shall find that the degree of fluctuation in the results of measurements can provide additional information about the phenomenon under investigation.

SUMMARY

Most experimental programs will consist of many observations or measurements performed under identical conditions. An important goal is the reproducibility of the results of experiments performed under similar conditions. There will always be some degree of variation in experimental results caused by uncontrolled variables or statistical effects that will appear as random fluctuations in measured quantities.

Example   On the microscopic level of atomic and nuclear physics, uncertainty in measurement occurs, which is explained by an important law of physics known as the Heisenberg uncertainty principle. This law states that the act of measurement itself will disturb the measured system in such a way as to introduce a random fluctuation in the measured quantity.

Relative versus Absolute Measurements

It is generally advisable to make comparative observations or measurements wherever possible instead of relying on absolute measurements. If only differ-

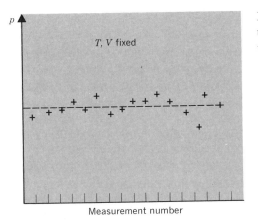

FIGURE 5.7. Measurements performed under apparently identical conditions will usually exhibit some fluctuation.

ences or ratios of a variable are used, then many measurement errors will be common to both cases and will cancel out in the result. Much of the skill in planning an experiment involves designing conditions under which such relative measurements are sufficient to test a hypothesis.

One way to proceed involves performing measurements on two different samples in which the only difference is caused by the quantity under investigation. In one sample the dependence between the dependent and independent variables would be measured as the latter is varied. These results would then be compared to measurements performed on a second reference or *control* sample in which variables are held constant. The use of a control experiment is particularly valuable when some perturbing influence is at work on both the sample and the reference.

An engineer might wish to measure the influence of static electricity on particulate emission from a power plant smoke stack. First a series of measurements is conducted to determine particulate flow and distribution under normal circumstances. This would be the control or background experiment. Then an electrical voltage would be applied across electrodes in the stack, and the measurements repeated under otherwise identical conditions. A comparison between the two classes of measurements should then yield the desired information.

Occasionally one can choose the controls as absolute standards whose properties have already been precisely measured. The experimenter then performs a measurement relative to the standard by comparing the data taken from the two samples. This is the general procedure used in all measurements, since the units in which one expresses the measurements of dimensions are defined relative to some standard (recall Section 3.2). A special case arises when the comparison value can be controlled so that the measuring instrument reads zero at the time the measurement is being made. In this type of "nul" measurement, the final reading is essentially independent of the characteristics of the detecting instrument and many other components of the system.

SUMMARY

It is generally advisable to make comparative observations or measurements between the experimental subject and a control or reference sample. It is particularly useful to choose the control sample as an absolute standard.

Example   An excellent example of a null measurement device is the Wheatstone bridge used to provide an accurate measurement of the resistance of a passive circuit element. The Wheatstone bridge circuit, shown in Figure 5.8, is arranged to balance reference resistances against the unknown resistance until the current in the central arm (between nodes I and II) is zero as sensed by a galvanometer placed in the arm.

The general approach is to adjust a variable but known resistance R until the galvanometer gives a null reading. Then, since nodes I and II must be at the same potential and

$$I_1 R_1 = I_2 R_2$$

$$I_1 R_x = I_2 R$$

we can divide the first equation by the second to find the unknown resistance in terms of the reference resistances as

$$R_x = R \frac{R_1}{R_2}$$

The Wheatstone bridge circuit allows very accurate measurements of resistances over a large range. It can also be used to measure small changes in resistance, as might arise, for example, in a strain gauge.

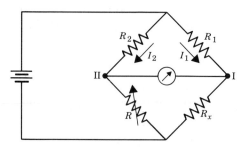

FIGURE 5.8. The Wheatstone bridge circuit.

Scientific Integrity

A fascinating phenomenon known as the "Cargo Cult" has been discovered on a number of islands in the South Pacific. These islands were inhabited by primitive, stone age cultures until their first contact with Western civilization through the voyages of Captain Cook. The natives were overwhelmed with Western material goods, which they referred to in their pidgin English as "cargo." During World War II their fondest dreams appeared to come true as the Allies developed several of the islands into military bases. Planes would land, and soldiers would come bringing with them enormous quantities of "cargo."

After the war, the soldiers left and the planes flew away, taking the cargo with them. Now the natives pray for the planes to return. They have developed a religion about the return of the cargo, a "cargo cult." They fashion bamboo shoots into primitive guns and march about like the soldiers did. On one island they have cleared off areas of land to look like a runway, and placed fires along side it to attract the planes. They even made a wooden hut out of old cargo crates for a native to sit in, with two wooden pieces on his head like headphones and bars of bamboo sticking out like antennas. They pray to this figure and wait for the planes to land.

The natives appear to be doing everything right. The form is perfect; it looks exactly the way it looked before. But something is wrong. The soldiers do not return. The planes do not land.

As members of a sophisticated culture, we find the cargo cult amusing. Yet, as the Nobel Laureate physicist Richard Feynman has noted, despite our sophistication, in many ways we tend to approach modern science and technology in ways remarkably similar to the cargo cult. Feynman has coined the term "cargo cult science" to refer to many approaches to science in our society today that appear to follow all the apparent precepts and forms of scientific investigation, but that are missing something very essential because they simply do not work—the planes do not land. We can all think of obvious examples such as astrology, the biorhythm fads, UFOs (close encounters of the third kind, the search for ancient astronauts), strange creatures (the Abominable Snowman, the Loch Ness monster, Bigfoot). Unfortunately, many aspects of engineering have also become infected with cargo cult science, particularly in those areas of technology that have become highly controversial and are debated vigorously in the public limelight.

But what is missing from these pseudosciences, these Cargo Cult Sciences? Feynman suggests that the missing ingredient might be identified as "scientific integrity," a principle of scientific thought that corresponds to an utter honesty, a leaning over backward to consider and present all aspects of a scientific investigation.

This principle of scientific integrity is particularly important in experimental investigation. If you are doing an experiment, you should report everything that you think might make your results invalid. You should present not only what you think is right about it, but also other causes that could possibly explain your

FIGURE 5.9. Some popular examples of cargo cult science. (Art and photography by Charles Mendez)

results; you should mention other ideas you considered but have rejected because of another experiment and why you have rejected them to make sure that others know it.

Details that could throw doubt on your interpretation must be given, if you know them. You must do the best you can—if you know anything at all wrong, or possibly wrong—to explain it. If you concoct a theory, for example, and advertise it or publish it, then you must also put down all of the facts that disagree with it, as well as those that agree with it.

In summary, the idea is to try to give *all* of the information to help others to judge the value of your contribution, not just the information that leads to judgement in one particular direction or another. We have learned from experience that the truth will eventually come out. Other experimenters will repeat your experiment and find out whether you were right or wrong. Nature's phenomena will agree or disagree with your theory.

You must be careful not to fool yourself, since you are frequently the easiest person to fool. After you have avoided fooling yourself, it is easy not to fool other scientists or engineers. You just have to be honest in a conventional way after that. But you also have an important obligation not to fool the public. It is this type of integrity, this kind of care not to fool yourself or others, that must be embraced to avoid excursions into cargo cult science.

SUMMARY

An essential ingredient of scientific inquiry is a kind of scientific integrity, an utter honesty in which the investigators present all available data and knowledge concerning a phenomenon, whether it supports their particular scientific hypothesis or not.

The Design of an Experimental Procedure

Experimenters should use their background knowledge of the phenomenon under investigation to design the experimental procedure. This knowledge may range from a mere suggestion about how the system might behave under various conditions to a well-established and highly developed theory of its behavior. Such considerations will assist in identifying appropriate dependent and independent variables, measurement procedures, and methods of data analysis and presentation.

The particular experimental design may vary considerably, depending on the various factors mentioned above. To illustrate the various considerations involved in designing an experiment, let us consider one example of an experimental procedure and then illustrate how this procedure might be applied to typical experiments:

Procedure

1. Sketch a diagram of the experimental problem, identifying all significant parameters.
2. List all significant variables (both dependent and independent) and their dimensions.
3. Determine which of these variables is treated as independent and under the control of the experimenter and which will be measured as dependent variables.
4. Determine the ranges of experimental variables (perhaps by making trial measurements).
5. Perform the appropriate measurements and record the data.
6. Evaluate the magnitude of errors or uncertainty in the results.
7. Analyze the results in terms of any available theory.

Example   Let us outline the design of an experiment to determine the relationship between the pressure p, temperature T, and volume V of steam.

Step 1

We begin by sketching a diagram of the problem, identifying all the significant parameters that might be relevant to the experiment (Figure 5.10).

Step 2

We then list the quantities of interest, along with other quantities that might influence our results. The dimensions of each such quantity are also noted.

pressure p (N/m³)
temperature T (K)
volume V (m³)
mass m (kg)

FIGURE 5.10.   Measurement of the pressure-volume relationship of steam.

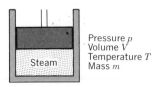

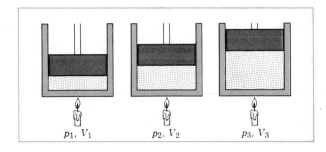

We include in this list all the quantities that we suspect will affect the measurement. It is also important, however, to exclude quantities that do not have significant effect.

Step 3

Notice that the above four quantities correspond to variables of the experiment. Let us choose volume as our dependent variable and identify the remaining three quantities as the independent variables under our control.

Step 4

We next estimate the ranges over which the measurements are to be made. In our example, the pressure, temperature, and volume may range from nearly zero to thousands of Pa, K, or m^3. It is impractical to perform the experiments for large numbers of values. The pressure, temperature, and volume ranges of interest must be decided upon before the measurements are performed. Suppose we use 1 kg of steam in our experiments and choose a temperature range of 200 to 1000°C and a pressure range of 0.02 to 0.60 MPa.

Step 5

We now perform the necessary measurements. During the experiment it is important to change only one variable at a time, while keeping all other variables constant. In our example, the temperature can be kept constant by using a heater, while the volume is increased step by step. For each volume the pressure is recorded as shown in Table 5.1. The data can then be plotted on a graph as shown in Figure 5.12. When the behavior of the data is completely unknown before the measurements, it is best to first take data at large intervals. This will establish the trend in the data. Once this trend is established, the gaps in the data can be filled in. It is also good practice to plot each data point as it is measured, since erroneous mea-

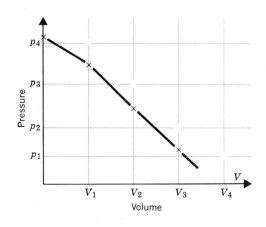

FIGURE 5.11. Plotting data taken in the pressure-volume experiment.

TABLE 5.1   Data Taken for the Pressure _p_ and Volume _V_ of 0.1 kg of Superheated Steam

TEMPERATURE: 200°C		600°C		1000°C	
_V_ (m³)	_p_ (MPa)	_V_ (m³)	_p_ (MPa)	_V_ (m³)	_p_ (MPa)
1.080	0.2	2.013	0.2	2.937	0.2
0.716	0.3	1.341	0.3	1.958	0.3
0.534	0.4	1.005	0.4	1.468	0.4
0.425	0.5	0.804	0.5	1.175	0.5
0.352	0.6	0.669	0.6	0.979	0.6

surements will often become apparent from the graph and can be checked or repeated immediately.

Step 6

We next estimate the error or uncertainty in the measurement.

Step 7

Finally, we establish, if possible, an analytical correlation for the data, which we interpret using our hypothesis or theory of the behavior of steam. We discuss error and data analysis in the next section.

FIGURE 5.12.   It is usually convenient to first take data at widely spaced intervals to establish a trend.

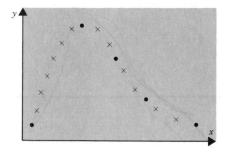

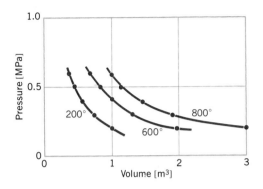

FIGURE 5.13. The pressure versus volume relationship for steam.

Example   Let us next design an experiment to determine the drag forces on a sphere moving with a constant velocity in a viscous fluid. (This is a classic problem in fluid mechanics.) Again we can follow our procedure step by step:

Step 1

First draw a diagram of the problem, labeling all of the forces acting on the sphere.

Step 2

Next, we identify the relevant variables of the problem:

Drag force F (N)
Velocity of the fluid v (m/s)
Speed of sound in fluid c (m/s)
Viscosity of fluid μ (N·s/m²)
Density of fluid p (kg/m³)
Diameter of sphere d (m)
Surface roughness ϵ (m)
Gravitational acceleration g (m/s²)

We have included in this list only those variables that may affect the drag on the sphere directly. For example, although the fluid temperature will have an effect on the viscosity, we shall not include temperature as a direct variable.

Step 3

We choose the drag force F on the sphere as our dependent variable, and represent its dependence on the remaining variables as a function:

$$F = F(v, c, \mu, \rho, d, \epsilon, g)$$

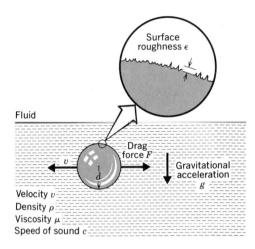

FIGURE 5.14. Measurement for the drag force on a moving sphere.

Surface roughness ϵ

Fluid

Drag force F

Gravitational acceleration g

v

Velocity v
Density ρ
Viscosity μ
Speed of sound c

But this suggests a bothersome complication. Our procedure in the previous experiment involved varying only one independent variable at a time. For example, we might vary only the velocity v while keeping all other variables unchanged. Then for each value of v we would measure the drag force F. Now suppose we were to repeat the measurements for four different velocities. We could then plot F as a function of v. We would repeat this procedure for the next variable, and so on, eventually measuring the drag force for all possible combinations of the independent variables. But, if we vary all seven variables independently for just 4 values, then we find we must perform $4^7 = 16\,384$ measurements! If we were to strive for more resolution and vary each variable 6 times, we would need $6^7 = 279\,936$ measurements. This is clearly an impossible task. We must find some means to alleviate the problem caused by a large number of independent variables.

Step 3′

We can reduce the number of measurements required by combining some of the independent variables. We have identified eight variables in the problem: F, v, c, μ, ρ, d, ϵ, and g. A result due to Buckingham (and known as the Buckingham Ⅱ-Theorem) lets us combine such variables into a number of dimensionless parameters or "groups" given by

$$i = n - k$$

where n is the number of variables in the problem ($n = 8$ in our example), and k is the number of basic dimensions that appear. In our example we have found three dimensions: s, N, and m, giving $k = 3$. (Notice here that kg is not a separate dimension since it is related to N by $N = \text{kg} \cdot \text{m/s}^2$.) That is, we may combine the variables into $i = 8 - 3 = 5$ dimensionless parameters—with the constraint that each original variable must appear at least once in the combinations.

The dimensionless parameters can be formed by various methods. The simplest way to proceed is just by inspection (fiddling with the variables). In our problem the dimensionless parameters can be identified as:

$$\frac{F}{\rho v^2 d^2} \qquad \left[\frac{N}{(kg/m^3)\ (m^2/s^2)\ m^2} = 1\right]$$

$$\frac{\rho v d}{\mu} \qquad \left[\frac{(kg/m^3)\ (m/s)\ (m)}{N \cdot s/m^2)} = 1\right]$$

$$\frac{v}{c} \qquad \left[\frac{(m/s)}{(m/s)} = 1\right]$$

$$\frac{v^2}{dg} \qquad \left[\frac{m^2/s^2}{(m)\ (m/s^2)} = 1\right]$$

$$\frac{\varepsilon}{d} \qquad \left[\frac{(m)}{(m)} = 1\right]$$

Of course these combinations are not unique. We could have formed other dimensionless parameters. Every combination is acceptable as long as it is dimensionless and each original variable appears at least once in one of the groups. Our particular choice was motivated by past experience with similar problems in fluid mechanics. In fact, four of the dimensionless parameters we have chosen arise so frequently that they have been given special names:

$$c_f = \frac{F}{\rho\,v^2\,d^2} \qquad \text{force (or drag) coefficient}$$

$$Re = \frac{\rho v d}{\mu} \qquad \text{Reynolds number}$$

$$M = \frac{v}{c} \qquad \text{Mach number}$$

$$Fr = \frac{v^2}{dg} \qquad \text{Froude number}$$

Note that by combining our variables into dimensionless parameters, we have reduced the number of independent variables from 7 to 4 and the number of measurements required from $4^7 = 16\ 384$ to $4^4 = 256$.

Step 4

We next determine the ranges of the independent variables over which the measurements are to be made. Notice that instead of varying each individual variable, we now have only to vary each group. For example, we need not vary both v and c, only their ratio v/c. Doubling v or halving c have the same effect, namely, the doubling of the ratio $M = v/c$.

We first inspect the dimensionless parameters to see if any of them are unimportant. If the sphere is smooth so that ϵ is much smaller than the diameter d, then the parameter ϵ/d plays no role in the problem. Similarly, if the flow velocity v is small compared to the speed of sound, then the Mach number $M = v/c$ does not influence the results. In this case we are left with a dependence of the drag or force coefficient on only two parameters, the Reynolds number and the Froude number:

$$c_f = c_f (\text{Re, Fr})$$

or

$$\frac{F}{\rho v^2 d^2} = f\left(\frac{\rho v d}{\mu}, \frac{v^2}{dg} \right)$$

If we were to vary each of the two variables, Re and Fr, 4 times, we would need to perform only $2^4 = 16$ measurements.

Step 5

The necessary measurements are now made. For example, we could vary the velocity v while keeping all other variables constant. Then for each value of v we would measure the drag force F. For each set of v and F we could calculate the ratios $F/v^2 d^2$, $\rho v d/\mu$, and v^2/dg and plot the results as shown in Figure 5.15.

Steps 6 and 7

If we had performed these measurements, we would have found that the ratio v^2/dg had no measurable effect on the results. In other words, for all values of v^2/dg the results would have been practically identical for the same Reynolds number. We thus conclude that for a smooth sphere (ϵ much less than d) in subsonic flow ($M = v/c \ll 1$), the drag force coefficient is a function only of the Reynolds number:

$$c_f = c_f (\text{Re})$$

The relationship between c_f and Re is provided from the measurements. Typical results are shown in Figure 5.16. The results in this illustration are not restricted to a specific sphere or a specific fluid. They apply to all

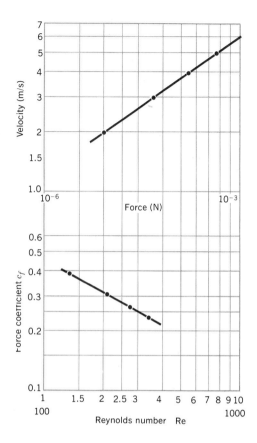

FIGURE 5.15. Velocity and force coefficient versus Reynolds number.

TABLE 5.2 Tabulation of the Data for an Experiment Measuring the Drag Force on a Smooth Sphere Moving at Subsonic Speeds in Air. Data Taken for a 1 mm Sphere moving in Air with Viscosity $\mu = 1.78 \times 10^{-5}$ kg/ms and Density $\rho = 1.23$ kg/m³

MEASURED		CALCULATED		
VELOCITY v (m/s)	DRAG F (N)	FORCE COEFFICIENT	REYNOLDS NUMBER	FROUDE NUMBER
2	1.92×10^{-3}	0.39	138	408
3	3.43×10^{-3}	0.31	207	917
4	5.31×10^{-3}	0.27	276	1631
5	7.38×10^{-3}	0.24	345	2548

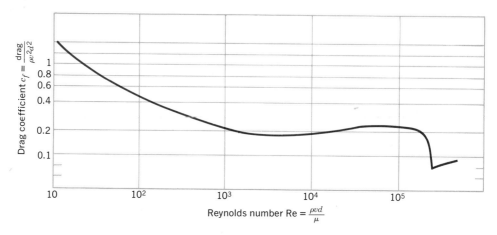

FIGURE 5.16.   The drag coefficient c_F versus Reynolds number.

spheres, large or small, and to all fluids, including both gases and liquids. The use of dimensionless parameters therefore enables us to compare the results obtained with different size objects. This feature is the reason why scale model testing (i.e., wind tunnels or ship towing tanks) are so useful in the study of fluid flow problems (airplane and ship hull design).

Example   A smooth ball of diameter 30 cm is placed in a river characterized by a flow velocity of 1 m/s. What force must be applied to the sphere to keep it from moving downstream?

Solution

We can apply our experimental results to answer this question if we recognize that the restraining force is equal to the drag force. The Reynolds number is given by

$$\text{Re} = \frac{\rho v d}{\mu} = \frac{(1000)\,(1)\,(0.3)}{(0.001)} = 3 \times 10^5$$

The corresponding force coefficient is found to be

$$c_f = 0.4$$

The drag force is therefore calculated as

$$F = c_f \rho v^2 d^2 = (0.4)\,(1000)\,(1)\,(013)^2 = 36 \text{ N}$$

SUMMARY

Although the detailed procedure for experiments will vary, a general approach might be:

1. *Diagram the problem.*
2. *Identify all significant variables.*
3. *Determine independent and dependent variables.*
4. *Determine ranges of experimental variables.*
5. *Perform appropriate measurements and record data.*
6. *Evaluate errors or uncertainty in reported results.*
7. *Analyze results.*

Exercises

Scientific Method

1. Identify the steps of observation, hypothesis, and experiment that would be appropriate in the investigation of the following phenomena:
   (a) Turbulent air flow about an aircraft wing
   (b) Energy content of a new blend of alcohol and gasoline (gasohol)
   (c) Change in tolerance of a part being machined as machine speed is increased
   (d) Strength of a prestressed concrete beam with various concrete compositions

2. Contrast the definitions of a hypothesis, a theory, and a law and give an example of each.

3. Give three examples of scientific laws, originally established by the scientific method, which have since been disproven by subsequent experiments.

4. Choose an observation in your everyday experiences (e.g., beer bubbles rising to the surface of a glass or the tendency of your instructor to lecture faster as students rustle about toward the end of class). Propose a hypothesis to explain this observation, and then perform an experiment to test your hypothesis. Write a brief summary of your results.

Design of Experiments

5. Rank the following experiments in order of decreasing background knowledge:
   (a) A new measurement of the speed of light
   (b) Measurements of X-ray sources in the universe
   (c) Measurement of switching speed of a new integrated circuit
   (d) Determination of atmospheric composition in the Los Angeles basin
   (e) Determination of voter preference in a political campaign
   (f) Measurement of failure strength of a new metal alloy
   (g) Search for lasing action at X-ray wavelengths
   (h) Soil composition measurements in construction site survey

6. Rank the experiments listed in Exercise 5 in order of decreasing degree of control.

7. Rank the experiments listed in Exercise 5 in order of decreasing presence of statistical fluctuations.

8. Rank the experiments listed in Exercise 5 in order of decreasing degree of required precision.

9. Identify the independent and dependent variables in each of the following experiments:
   (a) Measurement of heat resistance of new insulating material
   (b) Measurement of energy content of various blends of alcohol and gasoline (gasohol)
   (c) Loss of part tolerance in machine tooling with increases in machining speed
   (d) Strength of prestressed concrete beam with various concrete compositions

10. Identify possible uncontrolled variables in the following experiments:
   (a) Speed of crack propagation in material under tensile stress
   (b) Measurement of distance from earth to moon by bouncing a laser beam off a reflector left by the Apollo mission
   (c) Measurement of charge-to-mass ratio of an oil droplet (Milliken oil drop experiment)
   (d) Measurement of efficiency of conversion of light into electricity by a new photovoltaic cell

11. Suggest an experiment in which the independent variables are, in fact, highly interdependent in the sense that varying one will strongly influence another.

12. Rank the following measurements in order of decreasing degree of reproducibility:
   (a) Measurement of air temperature in a room

(b)   Measurement of the length of a meter stick using a standard

(c)   Measurement of noise level from airplanes passing overhead in a neighborhood near an airport

(d)   Measurement of vortex structure in water flow behind a ship propeller

13.   In measuring length, the unit of meter is used. Is this an absolute or a relative measure? Why?

14.   Is the measure of force (in newtons) an absolute or a relative measurement? Why?

15.   Suggest an experiment to perform a relative measurement of each of the following:

(a)   The amount of radioactivity in a sample

(b)   The degree to which level of education influences political attitudes

(c)   The degree to which the speed of a given chemical reaction depends on the concentration of a catalyst

(d)   The dependence of the rate of diffusion of a gas through a porous membrane on its molecular weight

16.   Give three examples of cargo cult science.

17.   Contrast the advocacy or adversary approach used in legal argument with the concept of scientific integrity used in scientific debate.

18.   Describe the experimental procedure you would use to establish the relationship between the resistance of a wire and its temperature.

19.   Describe the experimental procedure you would use to determine the relationship between the period of oscillation of a pendulum, the length of the pendulum, and the mass of the pendulum.

20.   Describe an experimental procedure to determine the relationship between the strength of a material and its temperature.

21.   Design an experimental procedure to measure the flow rate of water through a pipe for various pipe diameters and surface roughness.

22.   Design an experiment to measure the rate at which sand settles in a liquid.

23.   Design an experiment to measure each of the following quantities:

(a)   Effect of microwave radiation on biological organisms

(b)   Effect of radioactive krypton released from nuclear power plants on electrical conductivity of the atmosphere

(c)   Degree to which air bubble collapse can pit and damage ship propellers (erosion by cavitation)

(d)   Execution speed of a new computer algorithm

(e)   Effect of CO_2 concentration in the atmosphere on climate (or more narrowly, the degree to which CO_2 can act to trap heat radiation from the Earth's surface)

5.2. EXPERIMENTAL DATA

5.2.1. Data Recording and Documentation

Careful records should be kept of all experimental activities, whether they involve scientific research or engineering development. No matter how confident one feels at the time of an experiment, human memory is simply too fallible to be relied on. Every detail concerning the experiment should be put down in writing in some permanent form. This is true not only for lists of observations or data, but for all other aspects of description of the work as well. It is best to write down almost everything, since experimental work is essentially exploration, and one never knows what information may subsequently prove useful.

Laboratory notebooks should be permanently and strongly bound and of sufficient size, say roughly 20 by 25 cm, with numbered pages. Ruled pages are generally used, but this is a matter of personal taste, and some investigators may prefer unruled or cross-sectioned pages. One should begin by laying out the preliminary analysis of the experiment and the measurement and calculation program in the notebook. At this stage of preliminary work the notebook is particularly useful, since the average person can think more effectively on paper, writing down any given information, equations, circuits, etc., and working with them systematically, than by simply staring vacantly into space hoping for inspiration for the solution of the experimental problem.

Data should be entered directly into the notebook at the time of observation with a clear indication of the date. It should be recorded in ink (particularly if the notebook may be used as evidence to establish a patent). Sketches, drawings, and diagrams are essential, as is extensive documentation of all instruments involved in the measurements (e.g., serial numbers and circuit diagrams). Since so much observation is visual, it is important to record what is actually seen, including things not fully understood at the time. Bad or unpromising experiments, even those deemed failures, should be fully recorded. They represent an investment of effort that should not be thrown away, because often something can be salvaged, even if it is only a knowledge of what not to do in the future.

As a general principle, the experimenter should consider anything worth writing down as worth preserving. Completeness of description takes priority over conciseness of expression. In this sense, a laboratory notebook is quite different than a technical report, since it must serve to document the investigation rather than merely present the results.

Laboratory notebooks play a critical role in patent applications. For this reason, it may be desirable to witness and notarize notebook pages at various times. To maximize the legal worth of the record, each entry should be dated and no blank pages or spaces left. Many industrial and scientific laboratories have adopted precise standards governing the manner in which laboratory notebooks are kept.

FIGURE 5.17. An example of a laboratory notebook.

SUMMARY

Careful records should be kept of all experimental activities, including data, description of visual observations, experimental plan and data analysis, and any other information characterizing the experiment. Bound laboratory notebooks should be used; in certain applications, these notebooks should be periodically witnessed and notarized to serve as legal documentation of the experiment. All entries should be carefully dated.

5.2.2. Measurement and Accuracy

One important characteristic of scientific observation is its quantitative nature. This emphasis on numbers is an important aspect of all areas of technology. We shall define a *measurement* as a quantitative statement of the result of a human process of observation. It is important to recognize the human and fallible nature of the measurement process and its consequently limited validity. When one states that the speed of an automobile has been measured to be 100 km/h, this does not mean that the automobile is actually moving at this speed. Rather it means that the observer has used instruments (clocks and rulers) to measure the speed and concluded that it is 100 km/h. Of course the observer may have made a mistake in reading the instruments. Furthermore the experimental techniques are of limited accuracy. Hence a mere statement of the result of a measurement is usually insufficient, and some statement of the confidence one may associate with the measurement is very important. The definition of confidence in measurement is closely related to the concepts of significant figures, precision, and accuracy.

The number of digits that arise as the result of measurement are called *significant figures* (Section 3.3). When a number is written to represent the result of a measurement, it is always assumed that, unless otherwise stated, only the rightmost digit is uncertain. Significant figures are of considerable importance since they indicate to us the reliability of our measurements.

The term *precision* refers to how closely two or more measurements of the same quantity come to each other. In general, the more significant figures present in a given measured quantity, the greater the precision of the measurement.

The term *accuracy* refers to how close an experimental observation lies to the true value. Generally a more precise measurement will also be a more accurate measurement. The principal benefit of repeated measurements is not so much to improve the accuracy of the measurement as it is to provide an estimate of its precision.

Even with an ideally calibrated instrument under ideal conditions, there is a fundamental limit to the precision of the measurement by the instrument scale, since this is necessarily subdivided at finite intervals. In a sense, this is analogous to the situation in which a numerical value must be rounded off to some particular number of significant figures. Therefore great care should be taken to determine the precision or uncertainty in any measurement.

A common approach is to use the finest scale division on the instrument as a measure of the "maximum" range of uncertainty. But this definition is far too crude for most purposes. Rather the experimenter must actually *measure* the uncertainty in the results by repeating the measurement several times. Only in this way can the precision of the measurement be determined and a quantitative estimate of the possible error range provided.

SUMMARY

A measurement is a quantitative statement of the result of a human process of observation. The number of digits that are of importance in a given measurement are called significant figures. Precision refers to how closely two measurements of the same quantity come to each other. Accuracy refers to how closely an experimental measurement lies to the true value. It is important to determine the uncertainty in experimental measurements. Usually this must be done by repeating the measurement several times.

5.2.3. Errors

Data resulting from experiments always have inaccuracies or errors. To make the best use of such data, we need to know the magnitude of these errors. One can identify two general classes of errors.

Errors that influence all measurements of a particular quantity in a regular or equal fashion are classified as *systematic errors*. These are typically caused by inaccuracies in measurement instruments or experimental method. For example, an error in instrument calibration (e.g., a meter stick that is only 99 cm long) would give rise to systematic errors, since every measurement made by this instrument would be in error by the same amount (1%). Mechanical defects can influence instrument reproducibility in a systematic fashion. Other sources of systematic errors include limited investigator skills (such as in reading instruments), improperly designed apparatus, and so on. One must be aware of the presence of systematic errors and attempt to eliminate or minimize sources of such errors wherever possible, while quantifying and correcting for the errors that remain.

A second class of errors is characterized by statistical fluctuation or random behavior. Such random errors are due to the working of several uncontrolled variables, each of which has a small effect. As their name implies, successive observations of the same quantity will vary in a statistical or random fashion. Random errors can be caused by small changes in experimental conditions (temperature, pressure, illumination, vibration) or by the inability of the observer to read instrument scales precisely.

Our earlier definitions of accuracy and precision can be related to the type of error in the measurement. If the systematic error is small, the measurement is said to have a high accuracy. If the random error is small, the measurement is said to be of high precision.

In practice one should attribute random behavior to measurements only when no more plausible explanation of the observed pattern of results is available. Random errors can be analyzed using statistical methods.

SUMMARY

Errors that arise in a regular pattern to influence all measurements of a quantity in the same fashion are classified as systematic errors. Errors arising from uncontrolled variables that cause measurements to fluctuate in a statistical fashion are known as random errors.

Exercises

Data Recording and Documentation

1. Perform the following experiment: Sit down and rest for five minutes and then record your pulse rate. Then run up a flight of stairs twice. Record your pulse rate immediately after the run, then after a lapse of 1, 5, and 10 minutes. Record your observation in a proper laboratory notebook format.

2. Explain how you would present the results from any of the experiments proposed in Exercises 18 to 22 on p. 329.

3. Critically compare the method of presentation you are requested to use in your introductory chemistry or physics laboratories with that you might expect to be appropriate for an industrial laboratory.

Measurement and Accuracy

4. An engineer measures the wavelength of light emitted by a carbon dioxide laser and quotes this result as $\lambda = 10.613 \pm 0.002 \ \mu m$. How would this result be quoted with the appropriate number of significant figures?

5. Suppose the measurement in Exercise 4 had been written as $\lambda = 10.613 \pm 0.2 \ \mu m$. How would this result be quoted with the appropriate number of significant figures?

6. Suppose the measurement in Exercise 4 had been quoted as $\lambda = 10.613$ with a 0.5% relative uncertainty. How would this result be written in terms of absolute uncertainty and in terms of significant figures?

7. An engineer performs five different measurements of the temperature in a room and records this data as: 22.1, 23.0, 23.7, 24.1, and 24.5°C. It is known that the true temperature in the room is 23.9°C. Determine (a) the number of significant figures, (b) the precision, and (c) the accuracy of the measurements.

Errors

8. Calculate the percent error for the following measurements:
   (a) 10.613 ± 0.002 (c) 0.156 ± 0.002
   (b) 555.31 ± 0.4 (d) 100 ± 30

9. Calculate the absolute error for the following measurements:
   (a) $101 \pm 3\%$ (c) $5 \pm 50\%$
   (b) $15.672 \pm 0.02\%$ (d) $0.0065 \pm 10\%$

10. A meter stick, capable of being read to the nearest mm, is used to measure a length of 53 cm. What are the absolute and relative uncertainties in this measurement?

11. What is the smallest length that can be measured using a meter stick (read to 1 mm) if an uncertainty of 1% or less is to be achieved?

12. A stopwatch can be read to an accuracy of one-tenth of a second. What are the absolute and relative uncertainties that would be present in the timing of a 100 m dash using this watch?

13. A digital watch is guaranteed to be accurate to within 15 seconds per month. What is the uncertainty present in using this watch to time the results of a marathon run in which the winning time was roughly 2 hours and 10 minutes?

5.3   STATISTICAL ANALYSIS OF EXPERIMENTAL DATA

Suppose we follow the suggested experimental procedure and perform a number of measurements under identical circumstances. In most cases these measurements will not be in absolute agreement, but instead will exhibit some fluctuation, that is, random error. How do we analyze the results of these measurements to arrive at a quantitative statement of the result of the experiment?

The most commonly used procedure is to take the *average* or arithmetic *mean* of the measurements. If we make N observations and we obtain the data x_1, $x_2, \ldots, x_N$, then we could calculate the average of these observations as

$$\bar{x} = \frac{1}{N} (x_1 + x_2 + \ldots + x_N)$$

Clearly the average gives the number $\bar{x}$ about which the observations tend to cluster. The average is not necessarily the true value of the quantity being measured, but, in many cases, it is the best estimate for this true value.

It is also important to determine how the observations are spread about the average value. Do the observations all cluster closely about the average, or are they scattered widely about it? The most common measure of this spread in the data is provided by the *standard deviation*, defined by

$$\sigma = \left\{ \frac{1}{N} \left[(x_1 - \bar{x})^2 + (x_2 - \bar{x})^2 + \ldots + (x_n - \bar{x})^2 \right] \right\}^{1/2}$$

As the name implies, the standard deviation is a measure of the amount by which a typical data point differs from the mean.

Example   The diameter of a shaft is measured with a micrometer. Five measurements are taken, which yield the following data: 2.51, 2.49, 2.52, 2.50, and 2.48. What is the average of the measurements and their standard deviation?

Solution

Let us organize the data in tabular form:

x	$x_i - \bar{x}$	$(x_i - \bar{x})^2$
2.48	-0.02	0.0004
2.49	-0.01	0.0001
2.50	0.00	0.0
2.51	$+0.01$	0.0001
2.52	$+0.02$	0.0004
12.50		0.0010

The average is then

$$\bar{x} = \frac{12.50}{5} = 2.50$$

while the standard deviation is

$$\sigma = \left(\frac{0.0010}{5}\right)^{1/2} = 0.014$$

We could write the corresponding measurement as

$$x = 2.50 \pm 0.01$$

The use of concepts such as the average and standard deviation to characterize experimental data is an example of the application of methods of statistical analysis. By the term *statistics*, we generally mean those mathematical methods developed for describing and analyzing processes in which there is some degree of chance or random variability. Certainly the collection of experimental data is an appropriate example, since this data will usually contain random errors that cause a fluctuation in the results of measurements performed under identical circumstances. The engineer makes frequent use of statistical methods in the interpretation of experimental or testing data.

SUMMARY

Statistical methods are generally applied to analyze sets of repeated measurements that are characterized by the occurrence of random errors. In particular, the average or mean of the data is frequently taken as the best estimate, while the standard deviation is used to characterize the spread of the data.

5.3.1 Statistical Distribution Functions

We can apply statistical methods to analyze data in a more informative manner. Suppose our experimental "data" consist of a set of scores obtained on an examination given to a class of 100 students (Table 5.3). As we noted earlier, one way of characterizing these data is to calculate the average score by adding all of the scores and dividing by 100, the number of students taking the examination. For this example the average score is 77.4. We can also calculate the standard deviation of the scores to characterize the scatter about the average. Such a calculation yields a standard deviation of 10.2.

Statistics provides an alternative way to present and analyze the test data. As long as we are not interested in the exact sequence of scores, we can replace the data set by its corresponding *frequency* or *statistical distribution function*, which

TABLE 5.3 Scores for a Class of 100 Students

76	89	90	89	92
83	93	81	86	68
82	86	92	72	95
81	82	85	77	66
94	90	75	87	99
62	65	79	82	87
53	97	94	89	81
60	79	47	65	82
78	91	87	52	70
53	75	80	89	40
53	64	59	80	62
55	85	50	95	78
83	62	78	78	52
59	85	95	80	77
83	80	88	92	95
79	79	85	87	69
86	50	81	75	74
91	91	92	88	84
57	70	85	91	89
64	92	97	76	64

we shall denote by $F(x)$. The value of $F(x)$ is defined as the relative frequency with which the number x appears in the collection of data. In the example given by Table 5.3, the number 81 appears 4 times in the data set, so that the amplitude of the distribution function at this value, $F(81)$, is 4. Figure 5.18 shows the complete distribution function for the data set. Such a plot is known as a *histogram*. It is simply a graphical representation of the relative frequency with which a given quantity is observed versus its magnitude.

It is common to normalize the distribution function by dividing by the total number of scores. Hence, if there were 4 occurrences of the score 81, then $F(81)$ would become $4/100 = 0.04$. In this way, if we add up the values of $F(x)$ for all possible values of x, the total will come out to be unity.

Also shown in Figure 5.18 is a graphical representation of the 100 scores as a horizontal bar graph. By comparing the two plots, we can see that the distribution function tells us something about the way the data are distributed about the mean value. The scores range from a minimum of 40 to a maximum of 99. Therefore the distribution function has nonzero values only between these extremes. Its peak

FIGURE 5.18. A histogram for the class grade data.

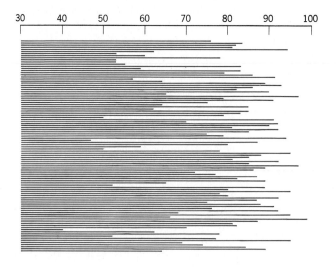

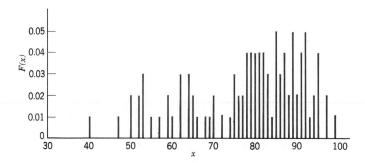

occurs at the most probable score (the score that appears most frequently), that is, at the *mode* of the distribution, which in this case is 78. We can also define the *median* of the distribution as that value for which there are an equal number of measurements whose values lie above and below the median value. The median for the data displayed in Figure 5.18 is 81. In general, the mean, mode, and median will not be the same.

We have defined the mean or average value of the data as the sum of all values of the measurements divided by the number of measurements. However we can also calculate this quantity directly from a knowledge of the distribution function $F(x)$. We need only multiply each value of the distribution function $F(x)$ by the value of the argument x and then sum over all values of x to find the average as

$$\bar{x} = x_1 F(x_1) + x_2 F(x_2) + \cdots + x_N F(x_N)$$

This is also known as the *first moment* of the distribution function.

The relative width of the distribution function is a measure of how much internal fluctuation exists in the data. If all the test scores in our example were very nearly the same, then the distribution would be narrow and sharply peaked. If, on the other hand, there were a great deal of variation of scatter among the individual scores, the distribution function would have a large width. This width can be characterized by a quantity known as the *variance*

$$\text{Variance} = (x_1 - \bar{x})^2 F(x_1) + (x_2 - \bar{x})^2 F(x_2) + \cdots + (x_N - \bar{x})^2 F(x_N)$$

A large variance implies a broad distribution, a small variance, a narrow distribution.

There is one technicality in this definition. It assumes that the mean value $\bar{x}$ is the true or exact mean characterizing all of the data that could be measured for the process of interest. However in practice, we can only take a limited sample of data and use this to estimate the mean. When this "experimental mean" is used to calculate the variance of the same limited set of data points, the derived value will tend to be slightly smaller than if the true mean from an unlimited data set were used. Therefore, the following formula should be used when calculating the variance of the same set of data used to derive the mean value:

$$\text{Variance} = \left(\frac{N}{N-1}\right)\left[(x_1 - \bar{x})^2 F(x_1) + (x_2 - \bar{x})^2 F(x_2) + \cdots + (x_N - \bar{x}) F(x_N)\right]$$

where N is the number of data points. The "minus one" in the denominator of the expression above increases the calculated value, compensating for the fact that we have used the experimental rather than the true mean.

From its definition we see that the variance is approximately the average obtained by squaring all the deviations of the data points from the mean value. By taking the square root of the variance, the standard deviation is obtained

$$\text{Standard deviation} = (\text{Variance})^{1/2}$$

The standard deviation is a measure of the amount by which a typical data point differs from the mean. Again, wide distributions will have large values and narrow distributions will have small values of the standard deviation (Figure 5.19).

Since the standard deviation is a convenient index of the amount of fluctuation that characterizes a given set of data, it has come to play an important role in statistics. In some cases it can be applied to represent the uncertainty that should be associated with a given measurement. The reasoning is as follows: Suppose we have a single measurement of a quantity that is subject to random fluctuations. It might be, for example, the number of people who are struck by lightning in the United States in a given year. However, data are available only for the latest year and show that 246 people fall in that catagory. We know that the number for the next year will undoubtedly be different. An insurance company would be interested in knowing the average number to be expected if data could be gathered over many years, so that they could properly set their premiums for lightning insurance. The company could suffer financial loss if the 246 was simply an abnormally low number, and the average were really 500 cases per year. If we had some idea of the standard deviation of the distribution from which the one sample

FIGURE 5.19. Distribution functions for two sets of data with differing degree of internal fluctuations.

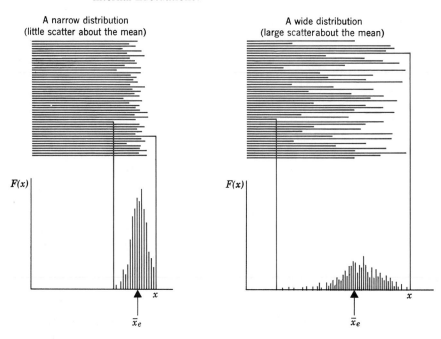

was drawn, we could then say that the difference between any typical sample and the true mean value should approximately be given by the value of the standard deviation. Therefore the standard deviation can be quoted as an uncertainty to be associated with that single sample as an estimate of the amount by which the true value might reasonably be expected to differ from the single value.

SUMMARY

A histogram is a graphical plot of the relative frequency with which a certain quantity is observed. When applied to a collection of data, such a plot is also called a frequency or statistical distribution function. The distribution function can be characterized by its mean, mode, and median values. The width of the distribution indicating the spread in the data is characterized by the mean-squared deviation of the data about the average or the variance. The square root of the variance is the standard deviation, a measure of the amount by which a typical data point differs from the mean.

5.3.2. Common Distribution Functions

Many physical phenomena can be described by well-known distribution functions. For example, sometimes we are interested in events that are known to occur with a given probability p. Then the distribution function characterizing the occurrence of x such events out of n repetitions of the selection process is given by the *binomial distribution*

$$F(x) = \frac{n!}{(n-x)!\, x!} p^x (1-p)^{n-x}$$

where $n!$ is the factorial function, $n! = n \cdot (n-1) \cdot (n-2) \ldots 3 \cdot 2 \cdot 1$.

Example The probability of rolling a given number on a Las Vegas style die is $p = \frac{1}{6}$. Suppose we use the binomial distribution to determine the probability of rolling exactly 5 sixes in 20 rolls of the die. Then we choose $x = 5$ and $n = 20$ find

$$F(5) = \frac{20!}{15!\, 5!} \left(\frac{1}{6}\right)^5 \left(\frac{5}{6}\right)^{15} = 0.1294$$

In Figure 5.20 we show the histogram of the binomial distribution characterizing

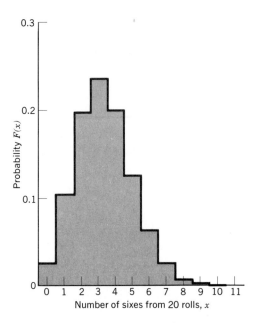

FIGURE 5.20. The probability distribution function for 20 rolls of a die as given by the binomial distribution function.

the probabilities of all possible results from 20 rolls of the die. For this example, the mean or average is 3.33, the median is 3, and the mode is 3.

One disadvantage of the binomial distribution in practical applications is that one needs to know both the number of trials and the individual probability of success in order to apply it. A useful simplification of this distribution can be used, however, when the individual probability of success is low, $p << 1$. The distribution function can then be rewritten as

$$F(x) = \frac{(p \cdot n)^x \, e^{-pn}}{x!}$$

However, since the product $p \cdot n$ is just the mean value $\bar{x}$, we can rewrite this as

$$F(x) = \frac{\bar{x}^x \, e^{-\bar{x}}}{x!}$$

This form is known as the *Poisson distribution*. Note that now only one parameter, the mean value $\bar{x}$, of the distribution needs to be known in advance in order to predict the probability that any result will be observed.

Example   As an example of the application of the Poisson distribution, suppose that a given engine factory produces 500 units per day. Experience has shown that, on a more or less random basis, one out of every 80 engines is defective and fails the quality control test. Suppose we wish to know the predicted probability distribution of the total number of failed engines over one day's productions.

Solution

In this case, $p = \frac{1}{80}$, and the necessary condition $p << 1$ for the application of the Poisson distribution is satisfied. We can therefore proceed immediately to calculate the expected mean number of failures as

$$\bar{x} = p \cdot n = \tfrac{1}{80} \, 500 = 6.25$$

This value can be used in the Poisson distribution to predict the probability of any given number of failures occurring. For example, the probability that exactly 8 failures occur on a given day is

$$F(8) = \frac{(6.25)^8 \, e^{-6.25}}{8!} = 0.111$$

Thus, about one day in every nine ($1/9 = 0.111$) we should expect eight failed engines.

SUMMARY

> *The binomial distribution can be used to predict the outcome of a given number of repetitions of a process for which the probability of success for any one trial is known. If the probability of success is small for any one trial, the binomial distribution may be replaced by the more convenient Poisson distribution. Here, the probability of observing any given number of successes can be predicted based only on a prior knowledge of the average number to be expected.*

Our previous example of a histogram characterizing the test scores of a class of students is useful for describing quantities that can take on only integer or discrete values (the number of successes or a definite count). Many other measurements involve continuous variables such as size, weight, or velocity whose value are not restricted to discrete possibilities. For example, the measurement might be that of the weight of each person in a large class. If we use a fine enough scale, no two weights will be exactly the same. In order to construct a histogram for this example, the weight scale must be arbitrarily divided into a number of intervals, and then the number of individuals whose weight lies within each of these intervals can be plotted as the histogram (see Figure 5.21).

The binomial and Poisson distributions introduced earlier apply only to discrete quantities or counts. In each case, the distribution gave the predicted probability of observing any discrete value of the variable. When dealing with continuous variables, distribution functions must be cast in a slightly different form. We can no longer talk about the probability of observing any specific value of a continuous variable, since that probability approaches zero as we examine a smaller and smaller interval about the value. Instead, we must now talk about the *probability density* characterizing the probability of observing a value between any two specified limits.

An example of a continuous distribution or probability density is the *Gaussian* distribution (also called the *normal* distribution) defined by

$$P(x) = \frac{1}{(2\pi\bar{x})^{1/2}} \exp\left[- (x - \bar{x})^2/2\sigma^2 \right]$$

In this expression, $\bar{x}$ is the mean value and σ is the standard deviation. A graph of this distribution for a mean value of 61.2 and a standard deviation of 8.6 is shown in Figure 5.22. We now interpret the area under the curve between two limits as the probability that a value of the variable x between these two limits will be observed. The example could represent the expected distribution of the size of particles formed in a given industrial process. The probability that we shall find a particle with a diameter between 40 and 55 μm is then given by the shaded area on the plot. The area under the entire curve must be equal to 1 since the probability is unity that the particle diameter must lie somewhere between the very small and very large limits.

One property of the Gaussian curve is that 68% of its area lies within limits that are one value of the standard deviation σ on either side of the mean. Therefore in any sample from a true Gaussian distribution, we would expect that the difference between a given sample and the true mean value will be less than one

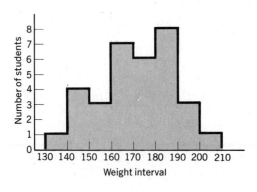

FIGURE 5.21. A histogram for the weight distribution of a class.

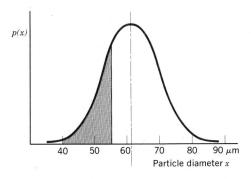

$p(x)$

40 50 60 70 80 90 μm

Particle diameter x

FIGURE 5.22. A Gaussian or normal distribution for particle sizes. The shaded area represents the probability that the particle size is between 40 and 55 microns.

standard deviation for 68% of all samples. A similar argument indicates that 97% of the samples will fall within two standard deviations, and 99% will fall within three standard deviations of the mean value.

Many experimentally observed distributions can be adequately represented by the Gaussian shape. Any population in which many factors combine to produce the resulting distribution is likely to appear Gaussian in shape because it can be shown mathematically that the combination of a large number of distribution functions, no matter how different they may be individually, will always tend to approach a Gaussian shape when combined (the so-called Central Limit Theorem).

The continuous Gaussian distribution is also often applied to situations in which the variable is really discrete, but the differences between adjacent values are small. For example, we might use the Gaussian distribution to describe the expected number of vehicles that cross a bridge each day, even though we know that the result must always be an integer value. As long as the mean value is large, however, differences between adjacent possible results are relatively small, and it is not a bad approximation to treat the variable as continuous.

For processes in which the individual success probability is small and the expected mean value is large, the binomial, Poisson, and Gaussian distributions are similar. One property of this common distribution is that the expected standard deviation is given by the square root of the mean value. Therefore, if the expected number of vehicles crossing a bridge in a day has an average value of 1000, we should expect that there will be an uncertainty associated with this statistic given by the square root of 1000, that is, an uncertainty of about 30. Thus on any given day, it should not be surprising if the traffic flow were to vary anywhere between 970 and 1030 just due to random fluctuations. In comparing the results of different samples, it is very important to bear in mind that the differences that are observed may be due only to these random variations and may not indicate any real differences or trends in the true quantity being sampled. Traffic planners would be wrong in interpreting a 3% gain in traffic flow across the bridge in two consecutive days corresponding to an increase from 970 to 1000 vehicles per day as significant, since this difference could easily be due to entirely to chance.

SUMMARY

Distribution functions can also be defined for continuous variables as well as for discrete variables. A common distribution function for continuous variables is the Gaussian or normal distribution. For such distribution, 68% of all measurements will lie within one standard deviation of the true mean value, 97% within two standard deviations, and 99% within three standard deviations. The standard deviation is therefore an indication of the uncertainty in any particular measurement.

5.3.3. Data Analysis and Correlation

Suppose we measure the electrical resistance R of a wire as a function of the wire temperature T and plot the data as shown in Figure 5.23. The question then arises as to what correlation may exist between the dependent and independent variables, R and T. The simplest such relationship between these variables would involve a *linear* dependence of the form

$$R = a_0 + a_1 T$$

where a_0 and a_1 are constants. Fortunately, linear relationships or correlations of this type are not only quite common in practice but can also be used to characterize more general correlations by using appropriate variable scales (e.g., log-log or semi-log).

Therefore it is good practice to first graph the data in several different forms to see if a linear correlation exists between variables x and y of the form

$$y = a_0 + a_1 x$$

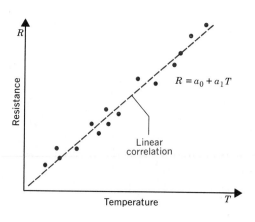

FIGURE 5.23. The correlation fitting resistance-temperature data for a wire.

To find the values of the coefficients a_0 and a_1 that yield the "best fit" to the data, one can use the *method of least squares*. If we represent each pair of observations using subscripts, x_i and y_i, then the method of least squares gives the coefficients as

$$a_0 = \frac{\sum_i y_i \sum_i x_i^2 - \sum_i x_i y_i \sum_i x_i}{n \sum_i x_i^2 - \left(\sum_i x_i\right)^2}$$

$$a_1 = \frac{n \sum_i (x_i y_i) - \sum_i x_i \sum_i y_i}{n \sum_i x_i^2 - \left(\sum_i x_i\right)^2}.$$

With this method we can fit a straight line to *any* data. Of course this does not imply that such a fit is appropriate. How well the data is represented by the straight line fit is expressed by the *correlation coefficient r* defined as

$$r = \frac{n \sum_i x_i y_i - \sum_i x_i \sum_i y_i}{\left[n \sum_i x_i^2 - \left(\sum_i x_i\right)^2\right]^{1/2} \left[n \sum_i y_i^2 - \left(\sum_i y_i\right)^2\right]^{1/2}}$$

The correlation coefficient r will range between -1 and $+1$, with a value $r = 1$ corresponding to perfect correlation. A value $r = 0$ implies no correlation. The r values obtained from the fitting procedure can be compared with those given by statistical tables to estimate the "goodness" of the correlation. The table gives the maximum values of r that can be expected to occur by chance alone. The calculated value of r should be larger than the value given by the table to ensure that it is indeed a correlation (Table 5.4).

FIGURE 5.24. The correlation coefficient indicates the degree of correlation.

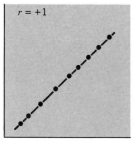 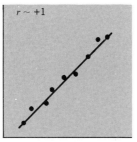 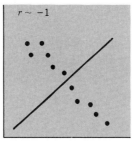 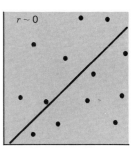

Perfect correlation   Good correlation (positive)   Good correlation (negative)   Poor correlation

TABLE 5.4 Value of the Correlation Coefficient r. (Here, n is equal to the number of data points minus 1.)

n	95% CONFIDENCE LEVEL				99% CONFIDENCE LEVEL			
	TOTAL NUMBER OF VARIABLES				TOTAL NUMBER OF VARIABLES			
	2	3	4	5	2	3	4	5
1	.997	.999	.999	.999	1.000	1.000	1.000	1.000
2	.950	.975	.983	.987	.990	.995	.997	.998
3	.878	.930	.950	.961	.959	.976	.983	.987
4	.811	.881	.912	.930	.917	.949	.962	.970
5	.754	.836	.874	.898	.874	.917	.937	.949
6	.707	.795	.839	.867	.834	.886	.911	.927
7	.666	.758	.807	.838	.798	.855	.885	.904
8	.632	.726	.777	.811	.765	.827	.860	.882
9	.602	.697	.750	.786	.735	.800	.836	.861
10	.576	.671	.726	.763	.708	.776	.814	.840
11	.553	.648	.703	.741	.684	.753	.793	.821
12	.532	.627	.683	.722	.661	.732	.773	.802
13	.514	.608	.664	.703	.641	.712	.755	.785
14	.497	.590	.646	.686	.623	.694	.737	.768
15	.482	.574	.630	.670	.606	.677	.721	.752
16	.468	.559	.615	.655	.590	.662	.706	.738
17	.456	.545	.601	.641	.575	.647	.691	.724
18	.444	.532	.587	.628	.561	.633	.678	.710
19	.433	.520	.575	.615	.549	.620	.665	.698
20	.423	.509	.563	.604	.537	.608	.652	.685
21	.413	.498	.552	.592	.526	.596	.641	.674
22	.404	.488	.542	.582	.515	.585	.630	.663
23	.396	.479	.532	.572	.505	.574	.619	.652
24	.388	.470	.523	.562	.496	.565	.609	.642
25	.381	.462	.514	.553	.487	.555	.600	.633
26	.374	.454	.506	.545	.478	.546	.590	.624
27	.367	.446	.498	.536	.470	.538	.582	.615
28	.361	.439	.490	.529	.463	.530	.573	.606
29	.355	.432	.482	.521	.456	.522	.565	.598
30	.349	.426	.476	.514	.449	.514	.558	.591
35	.325	.397	.445	.482	.418	.481	.523	.556
40	.304	.373	.419	.455	.393	.454	.494	.526

TABLE 5.4 (Continued)

45	.288	.353	.397	.432	.372	.430	.470	.501
50	.273	.336	.379	.412	.354	.410	.449	.479
60	.250	.308	.348	.380	.325	.377	.414	.442
70	.232	.286	.324	.354	.302	.351	.386	.413
80	.217	.269	.304	.332	.283	.330	.362	.389
90	.205	.254	.288	.315	.267	.312	.343	.368
100	.195	.241	.274	.300	.254	.297	.327	.351
125	.174	.216	.246	.269	.228	.266	.294	.316
150	.159	.198	.225	.247	.208	.244	.270	.290
200	.138	.172	.196	.215	.181	.212	.234	.253
300	.113	.141	.160	.176	.148	.174	.192	.208
400	.098	.122	.139	.153	.128	.151	.167	.180
500	.088	.109	.124	.137	.115	.135	.150	.162
1000	.062	.077	.088	.097	.081	.096	.106	.116

Example The resistance of a wire was measured as a function of temperature, and the following data were obtained:

Temperature x:	20	30	40	50	60
Resistance y:	1.0	1.13	1.25	1.38	1.51

Let us assume a linear correlation of the form

$$Y = a_0 + a_1 x$$

We can now apply the method of least squares to determine the fitting coefficients. It is convenient to tabulate the steps in our calculations as follows:

x	y	xy	x^2	y^2
20	1.0	20.0	400	1.00
30	1.13	33.9	900	1.28
40	1.25	50.0	1600	1.56
50	1.38	69.0	2500	1.90
60	1.51	90.6	3600	2.28
200	6.27	263.5	9000	8.02

Hence we calculate the fitting coefficients as

$$a_0 = \frac{(6.27)\ (9000) - (263.5)\ (200)}{(5)\ (9000) - (200)\ (200)} = 0.746$$

$$a_1 = \frac{(5)\ (263.5) - (200)\ (6.27)}{(5)\ (9000) - (200)\ (200)} = 0.0127$$

Thus the resistance-temperature relationship is

$$R = 0.746 + 0.0127\ T$$

The correlation coefficient can be calculated as

$$r = \frac{(5)\ (263.5) - (200)\ (6.27)}{[(5)\ (9000) - (200)^2]^{1/2}\ [(5)\ (8.02) - (6.27)^2]^{1/2}}$$

$$= 0.999$$

For 5 data points ($n = 5$), the Table 5.4 gives $r = 0.97$ at the 99% confidence level. Since 0.999 is greater than 0.97, it can be concluded with 99% confidence that the dependence between R and T is linear.

Many engineering problems are better described using relationships that are not linear:

$$y = 1 - e^{-a_0 x} \qquad \text{exponential}$$

$$y = a_0 + a_1 x^{1/2} \qquad \text{square root}$$

$$y = ax^b \qquad \text{power law}$$

$$y = a + b/x \qquad \text{inverse}$$

Each of these relationships can be transformed into a linear form. For example, consider the power law expression $y = ax^b$. If we take the natural logarithm of both sides of this expression, we find

$$\ln y = \ln a + b \ln x$$

or

$$Y = A + bX$$

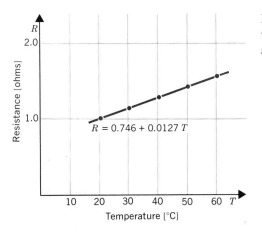

FIGURE 5.25. A fit of a linear correlation to resistance-temperature data characterizing a wire.

$R = 0.746 + 0.0127\,T$

where we have introduced the new variables

$$Y = \ln y$$

$$X = \ln x$$

$$A = \ln a$$

We can now apply the least squares method as before to determine the fitting coefficients A and b.

A similar procedure can be used to apply linear correlation methods to the other forms. The appropriate variable substitutions are given in Table 5.5

TABLE 5.5   Application of Linear Correlations to Nonlinear Relationships

NONLINEAR EQUATION	LINEAR EQUATION	Y	X
$y = 1 - e^{-ax}$	$Y = aX$	$\ln\left(\dfrac{1}{1-y}\right)$	x
$y = a_0 + a_1 x^{1/2}$	$Y = a_0 + a_1 X$	y	$x^{1/2}$
$y = ax^b$	$Y = \ln a + bX$	$\ln y$	$\ln x$
$y = a + \dfrac{b}{x}$	$Y = a + bx$	y	$\dfrac{1}{x}$

SUMMARY

Data analysis typically involves determining a functional form or correlation that characterizes the data taken in an experiment. The simplest such correlation assumes a linear relationship between dependent and independent variables. The least squares method can be used to determine the coefficients for such a linear fit. Linear fits can be applied to more general correlations by appropriate variable transformations.

Exercises

Statistical Analysis

1. The resistance of a resistor is measured with results: 5.12, 4.90, 4.95, 4.98, 4.10, and 5.05 ohms. What is the average of the measurements and their standard deviation?

2. The speed of cars passing through a certain point on a highway is measured with radar. The measurements give the following values: 85.5, 80, 82, 90, 95, 100, 102, 105, 89, 106, 103, and 104 km/h. What is the average speed of the cars and the standard deviation of the speeds?

3. The various ages of the players on a baseball team are 21, 20, 25, 35, 37, 28, 24, 26, 22, 24, and 35. Calculate the mean and standard deviation for these ages.

Distribution Functions

4. Find the probability that exactly 7 heads are observed after 10 tosses of a coin.

5. On the average, members of a certain poker group attend 3 out of every 4 sessions. What number of invitations maximizes the probability that exactly 7 players turn out on a given night?

6. Tires obtained from a given distributor have a 2.5% probability of being defective. What is the probability that all 5 tires issued with a new car are free of defects?

7. Records kept over a long period of time show that a municipal fire department makes an average of 4.5 runs per day. Assuming that alarms occur on a random basis, what is the probability that: (a) exactly three runs are made on a given day; (b) at least one run is made on a specific day?

8. A given industrial plant has compiled a record that shows that 2.5 reportable accidents happen in any year on the average. What is the probability that exactly 10 accidents occur over a four-year period?

9. An environmental engineer performs a series of measurements of the atmospheric concentration of carbon dioxide in a certain region and tabulates this data (in parts per million):

201	193	158	185
285	223	205	232
301	187	243	211
110	273	220	198
155	261	162	218

   (a) Draw a histogram of these observations, grouping them into intervals of 50 ppm.
   (b) Calculate the mean, mode, and median of the distribution of observations.
   (c) Calculate the standard deviation of the data.
   (d) If the distribution is assumed to be Gaussian in form, within what limit should 68% of the measurements fall? In actual fact, what percentage of measurements fall in this range?
   (e) Select two randomly chosen groups of five measurements from the data set and compute the mean and standard deviation for each of these subsets. How do they compare with each other and the larger sample?
   (f) If a single measurement of 392 ppm had been obtained, should this measurement be accepted or rejected?

10. Using the mean and standard deviations calculated from the data provided in Exercises 1, 2, and 3 on p. 352, plot the corresponding normal distribution for this data.

Data Analysis and Correlation

Suggest the variables or combinations of variables that should be plotted to determine the quantities in the following problems.

11. The position d of a body subject to uniform acceleration a is measured for various times. If it is known that

$$d = \frac{1}{2}at^2$$

how would one determine a?

12. The pressure p, specific volume v, and temperature T of an ideal gas are related by

$$pv = RT$$

If the pressure and temperature are measured for fixed v, how would one determine the gas constant R?

13. The electrostatic force beween two stationary charges q_1 and q_2 separated by a distance r is given by

$$F = \frac{q_1 q_2}{4\pi\epsilon_o r^2}$$

If the force between two charges of fixed magnitude is measured for various separations r, how would one plot this data to confirm the inverse square nature of the force law?

14. The mass m of an object moving at speeds comparable to the speed of light will vary according to the theory of relativity as

$$m = \frac{m_o}{(1 - v^2/c^2)^{1/2}}$$

If this mass is measured as a function of the speed v, how would one plot the data to determine the rest mass m_o and the speed of light c?

15. The number of disintegrations per second observed from a radioactive sample is known to decay exponentially as

$$R(t) = R_0 \exp(-0.693 \ t/T_{1/2})$$

How would you plot the data obtained from measuring the disintegration rate R versus time to determine the half-life $T_{1/2}$?

16. The resistance of a copper wire is measured as a function of temperature:

T(C)	10	20	30	40	50	60	70	80
R(ohms)	12.1	12.9	13.5	13.9	14.3	15.0	15.3	16.0

It is known that the resistance as a function of temperature is approximately linear

$$R = R_o (1 + \alpha T)$$

where R_o is the resistance at 0°C, and α is the temperature coefficient of resistance.

   (a) Using the least squares method, fit a straight line to the measured data.

(b) Using the slope and intercept of this linear fit, determine a value for R_0 and α.

17. A comparison of the number of students inquiring about admission to a given engineering school with actual school enrollment showed the following figures for a seven-year period:

YEAR	1	2	3	4	5	6	7
Enrolled	300	350	400	480	500	510	550
Inquired	1200	1450	1500	1670	1980	2010	2310

Using a least squares linear fit, estimate the number of students expected to enroll in a given year when the number of inquiries is 500.

18. In a chemical process the pH levels of a liquid were found to vary with the amount of liquid. Measurements gave the following results:

Liter	218	178	200	195	152	231	205
pH	5.5	2.9	4.6	4.4	3.8	6.0	4.6

Can a linear least squares fit adequately describe these results?

19. A vacuum pump is connected to a sealed chamber. The pressure is found to vary with time in the following manner:

Time (s)	1	2	5	10	60	120
Pressure (mm Hg)	752.44	744.95	722.93	687.68	417.10	228.81

Find a suitable expression correlating this data.

20. The fuel consumption of an automobile was measured while the car traveled at different speeds. These measurements gave the following values:

Fuel consumption (liters/100 km)	6.5	9.1	12.5	16.6
Speed (m/s)	40	45	50	55

Find an expression correlating this data.

21. The relationship between drag and velocity for an airplane wing was found to be:

Drag (N)	10	40	250	1000
Velocity (m/s)	10	20	50	100

Find an expression correlating this data.

5.4 TESTING

All engineering designs should be subjected to extensive testing. At each phase of the design process, tests are used to verify the design concept, control the quality of materials and production methods, and determine whether the final product or process meets design specifications. If the design involves a single item such as a building or a bridge, it is usually a straightforward task to design a suitable testing program. However, when many items are involved, such as integrated circuits produced in an electronics factory, the engineer is forced to draw conclusions about a large number of products based on tests performed using a limited number of samples. This would be a simple task if all the production items were completely identical. Unfortunately no two items are ever exactly alike. The test engineer must utilize a limited number of samples to infer information about a large and varied group of products. Since the results of these tests may be used to make important decisions, it is imperative that the tests be designed properly and the data analyzed correctly.

5.4.1 Testing Programs

Testing requires very careful planning. The engineer should specify the design characteristics that are most important and the methods to be used for measuring these characteristics. The conditions under which the tests are to be conducted must be defined. This might include start-up and shutdown conditions, operation under partial and full load, operation under the failure of auxiliary equipment, operator errors, material selection, and so on. Objectives of engineering tests include: (1) quality assurance of materials and components, (2) performance, (3) life, endurance, and safety, (4) human acceptance, and (5) environmental impact.

A good testing program might consist of the following sequence:

1. The engineer should determine whether a test is needed at all. Tests can be expensive, and unnecessary testing should be avoided.

2. The engineer should next decide what type of test is most suitable for the problem. For example, either destructive or nondestructive testing could be selected. The test could be terminated before the part ultimately fails, or the test can be run until failure. The test could be performed under the actual conditions or under conditions that only approximate the real life environment.

3. The engineer should design the tests so that the resulting data will be statistically significant. This involves ensuring that the samples tested are representative of the actual parts and that a sufficient number of samples are tested.

4. The test data should be properly analyzed and interpreted.

5.4.2   Analysis of Test Results

There are many statistical tools available to assist in the design of tests and in the analysis of test results. We shall demonstrate the use of statistical methods in test data analysis through the use of the normal distribution. Although such a distribution is not suitable for describing the results of every test, it is appropriate in many situations.

Let us recall the general form of the distribution of a random variable, as shown in Figure 5.26. This distribution has an average value $\bar{x}$ and a standard deviation σ. Suppose we are interested now in the probability that x is greater than some value of b. This probability can be denoted by the symbol $P(x > b)$ and is represented by the shaded area under the distribution curve. This area, and accordingly the probability $P(x > b)$, can be found by integrating the normal distribution from $x = b$ to $x = \infty$:

$$P(x > b) = \int_{b}^{\infty} \frac{1}{\sigma \,(2\pi)^{1/2}} \exp\left[-\frac{(x - \bar{x})^2}{2\sigma^2} \right] dx$$

Unfortunately this integral cannot be performed analytically. However it has been calculated numerically by mathematicians and tabulated for various cases. It is customary to introduce the variable

$$z = \frac{x - \bar{x}}{\sigma}$$

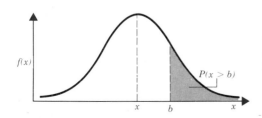

$f(x)$

$P(x > b)$

FIGURE 5.26. The probability that a given data point will be greater than b is given by the area under the distribution function curve.

so that the integral of the normal distribution can be presented as shown in Table 5.6. The use of this table is illustrated by the following example.

Example   Consider a manufacturing process in which plugs are produced for a bathtub. Plugs larger than 2.5 cm in diameter must be rejected. Measurements performed on 20 samples show that the plug diameters follow a normal distribution with a mean of 2.495 cm and a standard deviation of 0.01 cm. What percentage of the plugs can be used?

Solution

The problem is illustrated graphically in Figure 5.27. We can first calculate the reduced variable z:

$$z = \frac{b - \bar{x}}{\sigma} = \frac{2.51 - 2.495}{0.01} = 1.5$$

From the table, we can look up at the probability for $z = 1.5$ as

$$P(x > b) = 0.0668$$

Thus 6.68% of the plugs must be scrapped. The percentage of usable plugs is $100\% - 6.68\% = 93.32\%$.

5.4.3   Types of Testing

Performance Tests

One customarily thinks of engineering testing as a procedure directed toward determining whether products or processes meet design standards. Certainly such performance testing is an important feature of engineering. Whether one performs structural tests on a large building, engine tests on an automobile passing down an assembly line, or even tests of a new computer program to be used in inventory

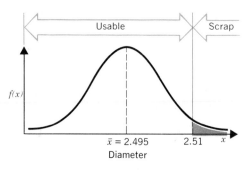

FIGURE 5.27.   The Gaussian distribution used to describe plug diameters.

TABLE 5.6 Tabulation of the Area under the Normal Distribution $P(x > b)$ from $x = b$ to $x = \infty$

z

b	.00	.01	.02	.03	.04	.05	.06	.07	.08	.09
0.0	.5000	.4960	.4920	.4880	.4840	.4801	.4761	.4721	.4681	.4641
0.1	.4602	.4562	.4522	.4483	.4443	.4404	.4364	.4325	.4286	.4247
0.2	.4207	.4168	.4129	.4090	.4052	.4013	.3974	.3936	.3897	.3859
0.3	.3821	.3783	.3745	.3707	.3669	.3632	.3594	.3557	.3520	.3483
0.4	.3446	.3409	.3372	.3336	.3300	.3264	.3228	.3192	.3156	.3121
0.5	.3085	.3050	.3015	.2981	.2946	.2912	.2877	.2843	.2810	.2776
0.6	.2743	.2709	.2676	.2643	.2611	.2578	.2546	.2514	.2483	.2451
0.7	.2420	.2389	.2358	.2327	.2296	.2266	.2236	.2206	.2177	.2148
0.8	.2119	.2090	.2061	.2033	.2005	.1977	.1949	.1922	.1894	.1867
0.9	.1841	.1814	.1788	.1762	.1736	.1711	.1685	.1660	.1635	.1611
1.0	.1587	.1562	.1539	.1515	.1492	.1469	.1446	.1243	.1401	.1379
1.1	.1357	.1335	.1314	.1292	.1271	.1251	.1230	.1210	.1190	.1170
1.2	.1151	.1131	.1112	.1093	.1075	.1056	.1038	.1020	.1003	.0985
1.3	.0968	.0951	.0934	.0918	.0901	.0885	.0869	.0853	.0838	.0823
1.4	.0808	.0793	.0778	.0764	.0749	.0735	.0721	.0708	.0694	.0681
1.5	.0668	.0655	.0643	.0630	.0618	.0606	.0594	.0582	.0571	.0559
1.6	.0548	.0537	.0526	.0516	.0505	.0495	.0485	.0475	.0465	.0455
1.7	.0446	.0436	.0427	.0418	.0409	.0401	.0392	.0384	.0375	.0367
1.8	.0359	.0351	.0344	.0336	.0329	.0322	.0314	.0307	.0301	.0294
1.9	.0287	.0281	.0274	.0268	.0262	.0256	.0250	.0244	.0239	.0233
2.0	.0228	.0222	.0217	.0212	.0207	.0202	.0197	.0192	.0188	.0183
2.1	.0179	.0174	.0170	.0166	.0162	.0158	.0154	.0150	.0146	.0143
2.2	.0139	.0136	.0132	.0129	.0125	.0122	.0119	.0116	.0113	.0110
2.3	.0107	.0104	.0102	.00990	.00964	.00939	.00914	.00889	.00866	.00842
2.4	.00820	.00798	.00776	.00755	.00734	.00714	.00695	.00676	.00657	.00639
2.5	.00621	.00604	.00587	.00570	.00554	.00539	.00523	.00508	.00494	.00480
2.6	.00466	.00453	.00440	.00427	.00415	.00402	.00391	.00379	.00368	.00357
2.7	.00347	.00336	.00326	.00317	.00307	.00298	.00289	.00280	.00272	.00264
2.8	.00256	.00248	.00240	.00233	.00226	.00219	.00212	.00205	.00199	.00193
2.9	.00187	.00181	.00175	.00169	.00164	.00159	.00154	.00149	.00144	.00139
3	.00135	$.0^3988$	$.0^3687$	$.0^3483$	$.0^3337$	$.0^3233$	$.0^3159$	$.0^3108$	$.0^4723$	$.0^4481$
4	$.0^4317$	$.0^4207$	$.0^4133$	$.0^5854$	$.0^5541^*$	$.0^5340$	$.0^5211$	$.0^5130$	$.0^6793$	$.0^6479$
5	$.0^6287$	$.0^6170$	$.0^7996$	$.0^7579$	$.0^7333$	$.0^7190$	$.0^7107$	$.0^8599$	$.0^8332$	$.0^8182$
6	$.0^9987$	$.0^9530$	$.0^9282$	$.0^9149$	$.0^{10}777$	$.0^{10}402$	$.0^{10}206$	$.0^{10}104$	$.0^{11}523$	$.0^{11}260$

[a] $.0^5541$ means $.00000541$.

control, performance testing is always necessary to provide assurance of the validity and achievement of design goals.

Quality Assurance Tests

Any engineering design is only as good as the components and workmanship used to realize it. Therefore a critical aspect in both construction and production engineering involves quality assurance, a continual testing to ensure that materials, parts, and labor used in engineering activities meet adequate standards. Quality assurance involves both testing of components and workmanship (e.g., welds), as well as complex management programs designed to monitor and ensure quality control. The latter activity becomes particularly critical in large construction projects and manufacturing.

Endurance Tests

It is important to determine how the performance of a product will change with time. It is not sufficient simply to verify that the performance of a new product meets design goals; the engineer must determine as well how it will hold up in use and whether it can stand up to overloading, misoperation, or other punishment without failure. Tests are performed to determine whether the product will perform adequately over its design lifetime. Moreover, product samples are usually also pushed beyond design loading or lifetime until failure occurs to determine their endurance (the degree of overdesign).

Safety Tests

Engineers should always be concerned about the safety of a design. A thorough and carefully planned testing program may be required not only during the construction (or manufacturing) stage, but also during operation. Safety testing is becoming increasingly regulated by various government agencies. In many cases, the ever-changing and occasionally confusing array of safety regulations generated by government bureaucracy represent a serious challenge to effective engineering design.

Environmental Impact

Another aspect of testing is to determine the impact of a design on its natural environment. This occurs in both product testing (e.g., measurement of exhaust emissions from automobiles) and construction (e.g., soil runoff from excavation). As with safety, environmental impact has come under the increasing control of government regulatory agencies (such as the Environmental Protection Agency). The rapid growth in regulations governing this aspect of engineering design has had a major impact in many areas.

Human Acceptance Tests

Engineers should be concerned with the interaction of their designs with the people who buy or use them. Tests should be performed to determine whether the product meets the physical, mental, and emotional requirements of the average person for whom the design is intended. This frequently requires extensive surveys conducted on large groups of users.

Exercises

Testing Programs

Suggest a testing program for each of the following situations:

1. Test the integrity of radioactive-waste shipping casks under accident conditions.
2. Test the failure strength of pressure vessels for boilers.
3. Test the lifetime performance of flashlight batteries.
4. Test the comfort of new commercial aircraft seat designs.
5. Monitor the quality of concrete being used in highway construction.
6. Ascertain whether EPA (Environmental Protection Agency) regulations governing effluents from waste treatment plants are being met.
7. Monitor quality of welds on piping in nuclear power plants.

Analysis of Test Results

8. The acceptable value of resistors produced in a factory is 0.900 ± 0.005 ohm. The standard deviation of the resistances measured in the plant is 0.003 ohm. What percent of the resistors must be discarded?
9. The concentration of a pollutant in the exhaust gas stack of a power plant was measured periodically. The measurements yielded the following values (in ppm): 14.1, 14.9, 14.8, 15.1, 15.6, 14.4, 14.9, 15.2, 15.3, 15.2, and 15.9. Estimate the number of measurements when the concentration level can be expected to be below 15.5 ppm.
10. The flow rate in a pipe is measured over a period of time: 30.1, 7.2, 10.1, 5.1, 1.3, 6.1, 16.2, 6.5, 24.1, 3.9, 17.2, 8.4, 13.2, 14.9, and 18.9 m^3 per minute. Determine the percentage of time the flow rate would be greater than m^3 per minute.

REFERENCES

The Scientific Method

1. E. B. Wilson, *An Introduction to Scientific Research* (McGraw-Hill, New York, 1952).
2. Russell L. Ackoff, *The Scientific Method* (Wiley, New York, 1962).
3. Marshall Walker, *The Nature of Scientific Thought* (Prentice-Hall, Englewood Cliffs, N.J., 1963).

Experiments

1. D. C. Baird, *Experimentation: An Introduction to Measurement Theory and Experimental Design* (Prentice-Hall, Englewood Cliffs, N.J., 1962).
2. E. B. Wilson, *An Introduction to Scientific Research* (McGraw-Hill, New York, 1952).

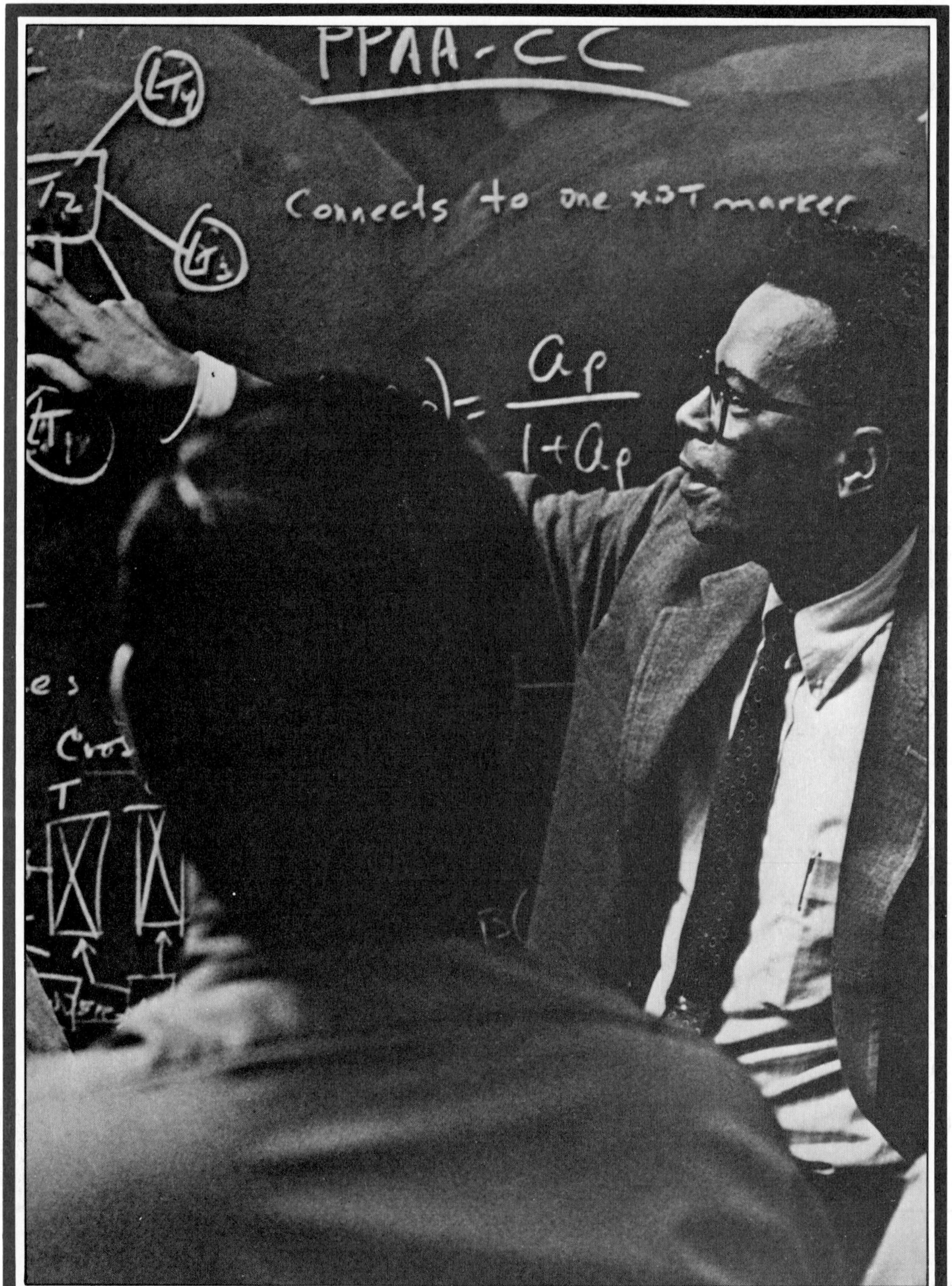
PPAA-CC
Connects to one

CHAPTER 6

Communication

There is a natural tendency to consider an engineering task completed once the technical activities of design and analysis have been finished. Sometimes the quality of technical achievement is emphasized to the exclusion of other engineering activities that may have a comparable impact on the success of the effort. In particular, both students and practicing engineers alike frequently tend to underestimate the importance of communication in engineering. An engineer's efforts are incomplete until they have been communicated in an appropriate form to those who will use them. In fact a significant portion of an engineer's activities are spent not in actual technical work but, rather, in communicating the results of that work to others.

The engineer's audience might consist of fellow engineers, supervisors, or skilled craftsmen who will implement the design. Engineers may even be called upon to describe their activities directly to members of the public at large. Sometimes the information engineers must convey is of a highly technical nature. They might wish to present the results obtained from a complex computer program or prepare an operating manual for a new machine. Sometimes the information is more qualitative and descriptive, such as would be the case in the sales description of a product. An engineer might even be faced with writing an article explaining a project in a manner suitable for a layperson.

Regardless of the intended audience and purpose of the communication, the presentation must be clear and precise so that the message comes across. Unless engineers can accomplish this, their technical achievements will be of little use. For even the most brilliant engineering design will lay dormant, gathering dust, if those who must implement it remain unaware (or unconvinced) of its merits.

6.1. GENERAL ASPECTS OF EFFECTIVE COMMUNICATION

The importance of effective communication in engineering cannot be overemphasized. In some situations the engineer's efforts may even be judged more on the quality of presentation than on the merit of the work itself. A final product may be able to stand on its own, of course. But before any idea or design can be translated into reality, it must first be sold successfully to others many times, as it passes from primitive design concept through development to production and marketing. No matter how good the idea is, it will not be accepted unless the engineer can describe it in a clear and convincing fashion. As disappointing as it may sound, the First Law of Madison Avenue sometimes applies to engineering just as it applies to advertising: A great presentation of lousy work has a better chance of acceptance than a lousy presentation of great work.

The engineer should master a wide range of communication skills. Foremost among these are fundamental verbal skills in writing and speaking. Written communications, for example, might consist of brief informal notes or memos (memoranda) to other engineers within a large organization (perhaps even to oneself!) Engineers will frequently need to write letters. To the chagrin of many engineers, longer technical reports or proposals will usually be required, both for internal distribution within an organization or for wider distribution outside. Sometimes the engineer may even wish to write a paper or a book suitable for publication.

In a similar manner, oral communication may occur on several different levels. The engineer will have daily, informal discussions with colleagues to exchange thoughts and explore new ideas. Instructions to subordinates or craftsmen may be required. Also, sometimes the engineer must give formal presentations or lectures describing technical activities.

There are other important modes of communication in addition to the verbal skills of writing and speaking. Engineers make frequent use of graphical materials such as drawings, charts, pictures, or models. In fact, engineers are frequently singled out for their ability in graphical communication. Most reports or lectures presented by engineers will make frequent use of visual materials. Engineers would do well to become familiar with computer graphics, which has revolutionized the design and preparation of graphical materials.

The computer has influenced communication in other ways. Certainly an important aspect of engineering communication is the presentation of computer materials, including both computer software (programs) and computer-generated output detailing the results of a particular computational task. The computer has also revolutionized the manner in which the engineer approaches the more mundane aspects of written communication by making available powerful wordprocessing (text editing) systems. Today the microcomputer is rapidly replacing the typewriter (and to a more limited extent, the secretary) as the medium through which the engineer translates ideas into verbal form.

Unfortunately, while communication activities occupy a large fraction of practicing engineers' time and play an important role in determining their effectiveness, engineers are often poorly equipped to handle communication tasks. Many engineers are poor writers and speakers. All too true is the well-worn joke about the senior engineering student who observed, "Four years ago I didn't even know how to spell 'engineer,' and now I *are* one!"

FIGURE 6.1. Communication plays a vital role in engineering. (Ken Karp)

Many engineering students are the products of a natural selection process operative in our educational system that has reinforced their analytical skills (mathematics and science) at the expense of their verbal development. Unfortunately, some engineering curricula compound this situation by placing insufficient emphasis on the development of verbal skills in composition and speaking. All too frequently students (and later, practicing engineers) must acquire verbal communication skills mostly on their own, often through hard-earned work experience.

Many engineering students have problems with the basic mechanical skills of communication: grammar and spelling. Grammatical and spelling mistakes can make a written report incoherent (and certainly embarrassing), and poor grammar detracts from an oral presentation.

But these are not the primary sources of difficulty in verbal communication. The grammar and spelling in a written report can always be corrected by an editor. Grammatical errors in an oral presentation may be overlooked if the talk is audible, the accent understandable, and the material well organized. Many foreign-born engineers and scientists are better speakers, even with their thick accents, than their American counterparts. We need only recall brilliant engineers such as Werner von Braun and Theodore von Karman who conveyed their ideas with absolute clarity despite their heavy accents. When someone once asked von Karman what language he used to communicate his ideas, he is said to have responded: "In the universal language of science and engineering, bad English." Foreign-born engineers and scientists have played a significant role in this country's technological development. They could express their ideas clearly, though admittedly in less than perfect English.

It is not our intent to downplay the importance of good grammar and correct spelling in engineering communication. However, these mechanical aspects of verbal skills are not the major reasons for poor communication. The most significant deficiencies are improperly organized material and a corresponding lack of clarity in expression of thoughts and ideas. Grammatical and spelling errors can be corrected in a written report or overlooked in an oral presentation. But no verbal communication will be successful without careful organization and clarity. Attention to these points is the engineer's responsibility.

In college students are required to write many reports and even give an occasional oral presentation. However in most cases they are concerned less with communication that with "performing" for their instructor. Their goal is to impress the professor with their knowledge. The situation in engineering practice will be quite different. Unlike the professor, the audience will usually be less informed on the subject than the engineer. Furthermore, it may initially be quite uninterested in the engineer's activities and require some stimulation. Hence the engineer's goals are to inform, instruct, and often to excite and persuade. To achieve these goals, a presentation must be clear, coherent, and concise.

For many people communication becomes a form of entertainment or a means of creative expression. For engineers, communication is far more important because it is the means by which they convey the results of their labors and persuade others of their value. Without a high degree of skill in this endeavor, engineers would become cloistered and ineffective. Their technical achievements would lay fallow, unrecognized and unappreciated by those for whom they are intended.

SUMMARY

Communication plays an essential role in engineering, for without adequate presentation, an engineer's work will remain unrecognized and unappreciated. The engineer should master a wide range of communication skills, including the verbal skills of writing and speaking, as well as other skills in mastering graphical and computer methods. In all forms of communication, stress should be on careful organization and clarity of presentation.

Exercises

Communication

1. Rank the various roles or functions of engineers (recall Section 1.3) to the degree to which they depend on effective communication.

2. Identify three examples of poor communication in engineering that have led to serious consequences.

3. Estimate the number of papers and talks you will be required to prepare during your college education.

4. Give three examples in which a very effective presentation has led to the acceptance of a poor product (or idea).

5. Give three examples in which ineffective communication has hindered the public acceptance of a new technology.

6.2 WRITTEN COMMUNICATION

Written communication plays an important role in engineering. One usually thinks first of technical reports detailing the engineer's activities. But written communication assumes other forms as well. Engineers frequently find it useful to keep a written record of their activities, sometimes using only informal notes, at other times using dated and witnessed entries in bound notebooks (such as the laboratory notebooks used in experimental work). Communication by brief memorandum or letter is a common task. Occasionally the engineer must prepare more elaborate documents such as technical reports describing engineering activities and results or proposals seeking to persuade others of the need for or usefulness of a particular project. An engineer may be required to provide the detailed description or technical specifications of a given design, or to write an operating manual for the benefit of those who will implement or use the design. Some engineers find themselves working with marketing people to write sales descriptions for a product. Most engineers will be called upon at some point to provide a written description of their activities suitable for the layperson. Obviously a wide range of writing skills are required in engineering.

6.2.1. Notes and Records

Since the memory capacity of the human brain is limited, engineers find it useful to record in written form various aspects of their day-to-day activities. They may write informal notes to remind themselves of a given calculation or conversation with a colleague. Their notes may also be in a more formal arrangement, such as in the bound laboratory notebook in which entries are made in permanent form and carefully dated (see Chapter 5). This notebook may serve later as a legal document in a patent application.

It is wise to approach the task of keeping written notes or records of one's activities with some care. Even informal notes used as reminders should not be treated like a grocery list, to be scribbled on scrap pieces of paper and taped at strategic locations about the office (or pinned to the engineer's shirt). Rather a carefully systemized approach should be used to record information and then file this in such a way that it can be easily retrieved at some later time. Bound notebooks are one excellent approach. We have already noted the importance of laboratory notebooks in experimental work and in recording the results of calculations. Such notebooks can also be used to enter notes of meetings or conversations.

Many companies and laboratories have established careful procedures for keeping records of the engineers' activities. Whether or not these formal procedures exist, the engineer would be well advised to develop the habit of keeping careful written records of professional activities.

SUMMARY

> *Engineers should develop a habit of keeping careful written records of their professional activities. They should use a systemized approach to record and then file information in an easily retrievable form. Bound notebooks are one excellent approach.*

MECHANICAL ENGINEERING LABORATORIES
THE UNIVERSITY OF MICHIGAN

TITLE OF TEST: CALIBRATION OF ORIFICE METER (WATER)

DATE: APRIL 10, 1982   10 AM TO 5 PM

TEST PERFORMED BY: G. S. SPRINGER

NUMBER OF TEST	PRESSURE DROP ΔP (mm)	PRESSURE P (mm)	TIME t (s)	REMARKS

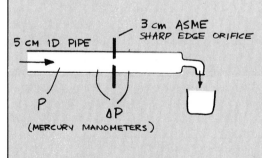

5 CM ID PIPE

3 CM ASME
SHARP EDGE ORIFICE

P

ΔP

(MERCURY MANOMETERS)

TIME t REQUIRED TO
COLLECT 5000 cm^3
OF WATER

WATER TEMP. = 23 °C

6.2.2. Memoranda and Letters

Engineers should become skillful at using short memoranda or letters to communicate with others both within and outside their organization. Memoranda such as those shown in Figure 6.3 provide a useful vehicle for communications within a company. The general format for such memoranda contains the date, name, title, and address of the receiver (TO) and the sender (FROM), a short statement of the topic (SUBJECT), and a brief text concerning this subject. The sender usually initials or signs the memorandum after his or her name and retains a copy in a correspondence file. If copies are provided for other individuals, these are usually indicated at the bottom of the memorandum with the notation "cc" (for carbon copy).

Communication by letter is also an important aspect of engineering activity. Most courses on English composition treat the business letter in some detail, and this format suffices for most engineering applications (Figure 6.4). As with all business letters, the text of the letter should be brief and to the point. Clarity of presentation is all-important, as it is in any form of engineering communication.

SUMMARY

Short memoranda and letters are a valuable communication tool of the engineer. The text of these documents should be brief and to the point.

6.2.3. Technical Reports

The difficulty of writing increases with the length of the written material. While the preparation of a one-page memorandum or letter may be quick and easy, the writing of a long report is painful for everyone, even for professional writers.

The first difficulty that must be faced and overcome is the dreaded "blank page terror"—sitting at your desk with newly sharpened pencils and a clean, white expanse of blank paper before you—and drawing a mental blank. How do you begin to write a technical report? How do you overcome the initial mental block? A novelist might fall back on the six-pack-of-beer (or Cognac-and-cigar) approach. A more productive approach for the engineer is to make a detailed *outline* of the report.

Most technical reports follow a fairly standard organization structure shown in Table 6.1. Although a technical report should present information in roughly this order, the actual preparation of the report follows almost the reverse order. For example, one frequently prepares the illustrations and tables first. The title page and abstract are usually the last parts to be prepared.

Department of Mechanical Engineering

The University of Michigan

Memorandum

December 10, 1980

TO: All technicians, Grades 8-12, in G. G. Brown Laboratory

FROM: Professor George S. Springer *GSS*

2282 G. G. Brown Laboratory

SUBJECT: Christmas Vacation

The machine shops in Rooms 1-121 and 1-122 in the

G. G. Brown Laboratory will be closed from 4 pm on

December 24, 1980 until 8 am on January 2, 1981.

cc: Mr. W. Marlatt

Administrative Officer

mhl/GSS

FIGURE 6.3.   Memoranda provide a brief, informal means of written communication.

January 23, 1981

Professor Alonzo K. Zorch
Department of Mechanical Engineering
The University of Michigan
Ann Arbor, Michigan 48109

Dear Professor Zorch:

   In response to your inquiry concerning the
availability of aircraft flight simulator programs
for microcomputers, we would note three of our software
packages of particular interest:

   SSS-12: Piper Super Cub 150

   SSS-14: Sopwith F.1 Camel

   SSS-15: Boeing 767

Each of these programs is designed to run (with full
three-dimensional, high resolution graphics display)
on either the TRS-80 or Apple-II microcomputers.

   More detailed program descriptions are enclosed.

                              Sincerely yours,

                              Susan K. Smith

                              Susan K. Smith
                              Sales Engineer
                              Software Simulations, Inc.

mh/SKS

Enc.

FIGURE 6.4. Technical letters written by engineers follow a form similar to those of business letters.

TABLE 6.1   Outline of a Technical Report

Title Page
Abstract
Table of contents
List of figures and tables
Nomenclature
Acknowledgments
Text:
  Introduction
  Experiments and tests
  Analysis
  Results and discussion
  Summary and conclusions
References
Appendices
Illustrations, graphs, and tables

Let us consider each of these components in the order of preparation (rather than presentation).

Graphical Material

We shall return to discuss the preparation of graphical materials such as technical illustrations, graphs, and tables in more detail in the next section. However we should note here that the quality of the graphical material used in a report is very important. Many readers will judge a report from the graphs and illustrations it contains and will only scan the text. These graphical materials are often copied and distributed without the main body of the report. Hence it is extremely important that all figures and illustrations be as self-explanatory as possible.

Text

Once the illustrations, graphs, and tables are completed, the task of writing the main body of the report begins. As mentioned, one first begins with an outline. Depending on the length of the text, several outlines may be required. The first outline should be general and cover only the major points. Successive outlines should include more and more detail.

As we noted earlier, organization is the key to a successful report. Thus each outline should be reviewed, analyzed, and rearranged until the sequence of presentation becomes logical. An important observation should be made here. Students and novice engineers frequently attempt to present material in the order in which the work was accomplished. However this "historical" approach is rarely suitable for a technical report since it drags the reader through all of the false starts, mistakes, and faulty logic of the original work. Rather one should emphasize logic and clarity of presentation in an outline. While the report must contain all the important facts (and, needless to say, only the facts), these need not be presented in the same chronological order in which they were discovered.

Having completed a detailed outline, one can proceed to write the text material. It is best to be simple. The writing should be clear enough to be understood by anyone, including people unfamiliar with the subject matter. It is therefore good practice to ask a friend or colleague to read the draft and comment on its clarity.

The *Introduction* section introduces the subject matter of the report: what was done, why it was done, and how it was done. This section provides a background for the work, relating it to similar studies performed in the past. Frequently an extensive review of similar work (sometimes known as a "literature survey") is included in the introduction to provide a context within which to judge the importance of the work.

Many readers will read only the introduction. Therefore special care must be taken to write this part lucidly. The purpose and scope of the work should be introduced and described in the first page or two of the introduction, relegating to later pages additional discussion and references to past work. In this way the reader can learn immediately whether the topic is of sufficient interest to warrant further reading.

The introduction is followed by a *Description of the Work*. Here the major decision facing the author is just how much detail to include. Most writers tend to include too much material. Generally it is sufficient to outline the method of approach and the solution technique in just enough detail for the reader to understand what was done. If necessary, details of algebra, apparatus, computer programs, and such can and should be relegated to appendices.

One can distinguish between the presentation of engineering analysis (mathematics) and experimental work. The detailed outline for the presentation of analytical work follows that given in Section 2.1. We have provided similar details for the description of experimental work in Section 5.2.

The results are given in *Results and Discussion*. This section should contain not merely the graphs with the data and verbal statements of the information given on the graphs, but it should also critically evaluate these results. In this section the author may provide an interpretation of the results and opinions regarding their usefulness and significance.

The final section is the *Summary and Conclusions*. Although many of these conclusions may have already been given in the results and discussion section, they are usually scattered throughout the text and intermingled with the author's opinions. In the summary and conclusions section only the facts (and not opinions) are given. These facts are best presented using short, concise statements. For emphasis, it is sometimes useful to number the conclusions consecutively. It is important to remember that many readers will only scan the abstract, introduction, and conclusions sections of the report.

Abstract

The abstract is a concise summary of the work that is usually written after the main text. Abstracts are usually less than a page long and in some cases may be only a paragraph in length. In the abstract the author states the problem that was attacked, the method of solution, and the major conclusions. In a sense, the abstract is a summary of the introduction and the conclusions sections. The purpose of the abstract is to provide enough information to enable the reader to assess whether or not the report is of interest. Thus the title provides the first clue to the contents of the report, while the abstract gives more detailed information on its nature.

Abstracts are of particular significance since they are frequently compiled as an independent summary of the report. There are journals and computer-based data banks that compile titles and abstracts of reports. The reader may selectively scan and obtain copies of any abstracts of interest from these sources. Therefore abstracts should provide sufficient information to allow the reader to decide whether or not to obtain and read the entire report.

Appendices

Details of the calculations, the experiments, and the data should be included in the appendices. As a rule of thumb, the appendices should contain information that is of interest only to a person who might want to repeat the work.

References

The references are grouped together following the main text. Each reference should provide sufficient information so that the reference could be easily located or obtained. A variety of standard forms are used for references, including books:

G. S. Springer, *Sex and the Singles Tennis Player* (Adult Book Press, Ann Arbor, 1982)

journal articles and symposium proceedings:

G. F. Knoll, "Knitpicking for Fun and Profit," *National Inquirer*, Vol. 23 (August 1982), pp. 45–49.

and reports:

G. S. Springer and G. F. Knoll, "Cardiovascular Development in Sedentary Engineering Faculty," Report No. 82-5, College of Engineering, The University of Michigan, Ann Arbor, Michigan 48109 (1982)

Title Page

The following information is included on the title page: the title of the report, the name and title of author(s), the name and address of the organization issuing the report, and the number and date of the report.

Table of Contents

The table of contents lists the titles of all sections and subsections and their page numbers.

List of Figures and Tables

This includes the titles of all figures and tables and their page numbers.

Nomenclature

All symbols used in the report are listed and defined in the nomenclature section, along with the page number where the symbol first appears in the text.

Once the text has been completed, the task of revision and editing begins. The report must be read and reread many times. The chapters and sections must be divided into subsections; appropriate titles and subtitles must be added. Superfluous parts must be deleted, and ambiguous parts must be clarified. The grammar and spelling must be corrected (making frequent use of style manuals). It is likely that the author will revise the entire manuscript several times.

An engineer who prepares a report usually "lives" with it for several days or weeks. It is common to become so familiar with (and perhaps so sick of) the report that one may develop blindspots to errors that remain in it. One useful technique (if time permits) is to let the report rest or "incubate"(cf. Sec. 2.2) for a few days while other unrelated activities are pursued. Coming back to the project will often bring a fresh insight and perspective. Even an overnight break will help.

It is also useful to have the final draft of the manuscript read by someone else. In many organizations professional editors trained in technical composition are used to edit the final draft. A good editor will not only correct grammatical mistakes, but will also point out where the text is unclear. After this editing, the report is finally ready for typing. The typed copy needs to be proofread again before it is ready for reproduction.

Writing is a long and arduous process, requiring continual revision and polishing to achieve a satisfactory result. All too often insufficient time is allowed for this critical engineering activity. Many a technical report has been hurriedly thrown together under the pressure of impending deadlines. The resulting document may become so flawed and confusing that it undermines the technical merit of the project being reported.

SUMMARY

The primary components of a technical report include title, abstract, front matter (tables of contents, list of figures, nomenclature), text, references, and appendices. One begins by developing a detailed outline of the body of the text, including the introduction, procedure, analysis, results and discussion, and summary and conclusions. Next the abstract is written and appendices and references are added. Finally the front matter (title page, etc.) is prepared. Successful report writing requires time for careful organization and revision.

6.2.4. Manuals and Technical Specifications

A different style of writing is required for instruction manuals or technical specifications and documentation that may accompany a given design. Operating manuals are frequently directed toward individuals without extensive technical background. The pitfalls of writing such instruction manuals are all too familiar. (This is particularly apparent to those of us who have been reduced to a state of utter confusion late on Christmas Eve after attempting to follow the assembly instructions that accompany most modern toys—"Plug tab A into slot B . . . light fuse and retire quickly . . . '") Writing manuals is somewhat akin to writing textbooks. Although engineers are a necessary part of this process, they should realize their own limitations and blindspots caused by their intimate familiarity with the device or process involved. Perhaps the most useful approach in writing operating manuals is extensive testing of the material on typical users.

A somewhat different tack is taken in the writing of the documentation or technical specifications for a design. Here one writes a reference intended for experienced users. Such technical specifications frequently assume a legal significance since they may be used to document the details of the design. Here accuracy and completeness should be emphasized above all else. Only the engineer who is intimately familiar with the design can prepare the technical specifications.

SUMMARY

Operating manuals are frequently directed toward individuals without extensive technical background and therefore should be tested on a trial basis before implementation. Technical specifications are intended as a reference for experienced users and therefore must be accurate and complete.

6.2.5. Proposals and Sales Literature

As we noted earlier, engineering communication is not only a means of conveying information but of persuasion as well. Nowhere is this more apparent than in the technical proposal. This is a document written by engineers to "sell" an idea, that is, to propose a certain project or activity. It may take the form of a brief letter or memorandum to the engineer's supervisor requesting permission for a certain activity. It may also assume the form of a lengthy technical document, prepared by hundreds of engineers in a company, that will be routed to an external agency (such as the federal government) for funding consideration.

Technical proposals are similar in many ways to technical reports. Their general outline is provided in Table 6.2. Just as with technical reports, clarity is paramount. But, in contrast to reports that are intended for a general audience, technical proposals are aimed at only one person or organization: the prospective client or sponsor. The concerns of this reader must be kept uppermost in mind during the preparation of the proposal.

A related type of technical writing involves sales literature. Although much of this material is prepared by professional marketing or advertising staff, the engineer may be called upon to provide technical information. The facts provided by the engineer must be sufficiently clear and detailed so that the sales information can be accurately prepared.

SUMMARY

Engineering communication is a means of persuasion as well as information exchange. Technical proposals are documents prepared by engineers to promote a given project or design. They are prepared in a manner similar to a technical report except that they are targeted specifically to the interests of the client. Engineers also become involved in preparing sales literature.

Exercises

Notes and Records

1. Set up a written record showing your daily activities.
2. Design a sheet for recording the electric power consumption of a typical home.
3. Set up a record showing your weekly expenditures.
4. Design a logbook for recording the gasoline consumption of an automobile.
5. Design a logbook for recording the weight of a person over a one-month period.

**TABLE 6.2 The Outline of a
Technical Proposal**

Cover material
Abstract
Table of contents
Text
  Introduction
  Objectives
  Background
  Method of approach
  Timetable
Documentation of qualifications
  Participating engineer biographical data
  Facilities of supporting institution
Other required documentation
  Cost estimates and budget
  Patent agreements
  Security provisions

---

6. Design a logbook for recording the data taken in an experiment to measure the relationship between the pressure, volume, and temperature of a gas.

Memoranda and Letters

7. Write a memorandum to announce the cleanup schedule of a laboratory.

8. Write a memorandum announcing the office hours during the Christmas holidays.

9. Write a memorandum designed to encourage energy conservation in your college.

10. Write a memorandum specifying who may use the student machine shop.

11. Write a memorandum introducing a new secretary to the staff.

12. Write a letter to the dean requesting a postponement of your final examination.

13. Write a letter providing a government agency with a statement of your project's monthly expenditures from a federal contract.

14. Write a letter to the Environmental Protection Agency explaining why your plant failed to comply with federal water pollution standards.

15. Write a letter to the local traffic violations court contesting a parking ticket.

16. Write a letter to a book publisher requesting permission to reprint material from a book.

17. Write a letter to a local manufacturer complaining about the quality of parts you have been receiving.

18. Write a letter summarizing briefly the technical accomplishments of an aircraft design project during the last month.

Technical Reports

19. Write a 500-word report describing how you would recommend that a student prepare for a final examination.

20. Visit a supermarket and then write a technical report on improvements you feel could be made in their service toward customers.

21. Prepare a report comparing the lifetime earnings of plumbers with those of engineers with B.S., M.S., and Ph.D. degrees.

22. Examine a house and then write a report with recommendations on how energy conservation measures could be implemented.

23. Write a technical paper discussing the impact of television on the election of the President of the United States.

Manuals and Technical Specifications

24. Write a user's manual for (a) an electric drill and (b) an electric toaster.

25. Write the technical specifications and user's manual for a corkscrew.

26. Write a user's manual for the novice on how to sign on to your university's computer system.

Proposals and Sales Literature

27. Write a sales brochure for
    (a) A bicycle
    (b) A digital watch
    (c) Plastic drawing aids (rulers)
    (d) A freshman engineering text

28. Prepare a proposal to your dean requesting funds for a set of tools for the laboratory.

29. Prepare a proposal to the Department of Energy to support the development of a new energy system that will make use of organic materials in chicken manure.

30. Prepare a proposal to the mayor of your town for permission to hold a rock concert on the steps of City Hall. Also propose that the city provide some form of financial assistance.

31. Prepare a personal resume that you could submit to prospective employers.

6.3   ORAL COMMUNICATION

Another important means of communication in engineering is speaking. Although the goals of oral and written communication are essentially the same, namely to convey information and to persuade, the actual techniques involved are somewhat different.

For many engineers, giving an oral presentation is even more difficult and painful than writing a report. Writing can be done in seclusion without anyone else being aware of the author's trials and tribulations. The writer can tear up pages, cross out paragraphs, and start over again. A manuscript can be revised again and again until it meets the author's satisfaction. A speaker has neither the luxury of privacy nor the opportunity to correct a speech, at least not when finally in front of an audience. Then, stagefright must be overcome, the sinking feeling in the pit of the stomach suppressed, and the talk must be given.

Even the most experienced speakers have butterflies before important speeches. It is sometimes said, in fact, that some degree of anxiety is helpful when giving a speech, just as it is in athletic competition. Furthermore, anxiety and nervousness can be overcome to a large extent by preparing the presentation very carefully.

Organization plays just as important a role in oral communication as it does in written communication. A technical talk should begin with an introduction describing the objectives of the work and the reasons for undertaking it. This should be followed by a description of the work and the results. Finally, the conclusions are given, and the principal results and conclusions summarized again. Hence we can identify three primary components of an oral presentation: (1) the introduction, (2) the description of work and results, and (3) the conclusions and summary. Within this general framework the speech should be subdivided into smaller segments, each of which is similar to the subsections of a report.

Although the organization of written and oral reports are similar, they do differ significantly in several details. The reader of a written report can control the pace of the communication. Difficult passages can be read more slowly; parts of the report can be reread if necessary. In contrast, the audience of an oral presentation must follow the pace set by the speaker. It is the speaker's responsibility to ensure that the audience remains awake, interested, and follows the subject matter of the presentation.

One should never read a speech from a prepared text. Except for highly skilled actors, few people can read a text without lulling their audience to sleep.

FIGURE 6.5. The engineer is frequently called on to make an oral presentation. (Courtesy Xerox Corporation)

Fortunately most speeches can be delivered without reading. If the speech is long, cue cards or visual aids can be used. The major points of the speech can be written on index cards, to which the speaker can refer during the lecture. Major statements, results, and conclusions should also be presented in visual form using slides or viewgraphs. By asking for the "next slide please" (or pressing the remote slide advance control), the speaker can trigger a reminder of the next topic in the speech. In this sense, visual aids not only serve to convey ideas to the audience, but they can also be used to prompt the speaker's memory.

Several recommendations concerning the use of visual aids are appropriate here. One should avoid putting too much information on any given slide or viewgraph. It is also important to avoid using too many slides. Usually four to five slides for every ten minutes of presentation will suffice. One should make certain that the lettering on the slides is sufficiently large. For example, typewriter lettering photographed with normal cameras will usually not project well onto a screen.

Visual material such as slides or viewgraphs may contain statements, mathematical symbols, or charts. The use of mathematical symbols and equations should be minimized. It takes time for any audience to absorb the meaning of mathematical symbols. It is best, wherever possible, to replace equations and formulas by verbal statements. After all, the main purpose of the presentation is to get the central ideas across to the audience, unencumbered with details. Any details can be provided in written reports that accompany the talk.

The presentation should be simple and to the point if the speaker is to keep the interest and comprehension of the audience high. A special effort should be made to hold the audience's attention during the talk, for once people lose the speaker's train of thought, they "tune out" and stop listening. To avoid this, the talk must be on a level where everyone in the audience can understand it. Research has shown that only three or four major ideas can be absorbed from a 30 to 50 minute talk. Therefore the speaker must restrict the subject matter accordingly.

Many novice speakers try to present too much information too rapidly. It is better to proceed slowly and to repeat major ideas several times throughout the presentation. Slides should not be flipped on and off but should be left on the screen for several minutes.

Before a speech is given, it is advisable to practice it alone or before a friendly

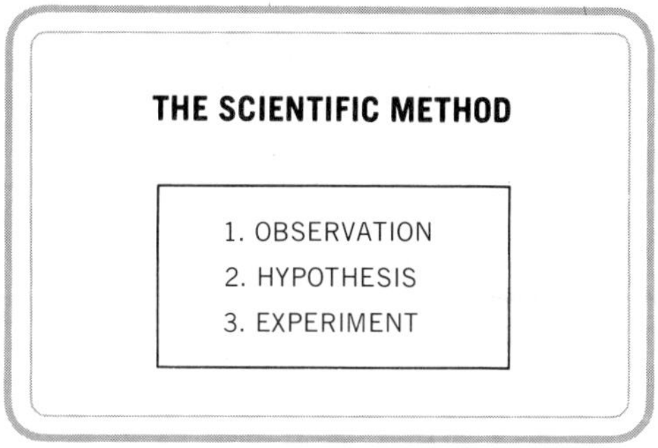

THE SCIENTIFIC METHOD

1. OBSERVATION
2. HYPOTHESIS
3. EXPERIMENT

FIGURE 6.6. This simple slide illustrates the limited amount of technical information that should be transmitted on one slide.

audience several times. During practice, the length of the speech can be timed, the effectiveness of visual aids tested, and the delivery refined. Remember the old saying in boxing, "Train hard, fight easy!"

SUMMARY

Although the basic goals of oral and written communication are similar, there are some differences in technique. Careful organization is the key to both. However a speaker has the additional concern of keeping the audience interested. Speeches should never be read from a prepared text. Extensive use should be made of visual aids such as slides or viewgraphs. Only a limited number of key points should be made in any talk.

Exercises

Oral Communication

1. Prepare a 10-minute talk on the effect of an oil embargo on the life of a United States citizen.
2. Prepare a 15-minute talk on the impact of recent integrated circuit technology on our everyday life.
3. Prepare a 5-minute talk on the importance of pocket calculators and computers to engineering students.
4. Prepare a 30-minute talk on the energy consumption in the United States and the various means by which energy can be conserved.
5. Prepare two 10-minute talks, one in support of and one in opposition to nuclear power.
6. Prepare a talk on the effects of government regulation on the automotive industry.
7. Prepare a 1-minute talk on the impact of word processors on office work.
8. Prepare a 5-minute talk explaining the importance of L'Hôpital's rule to your United States congressman.

6.4 GRAPHICAL COMMUNICATION

Graphical displays play an important role in conveying technical information. More than just the familiar saying, "A picture is worth a thousand words," is the

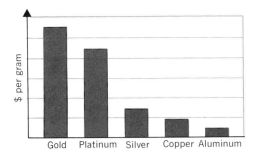

FIGURE 6.7. Bar charts are a convenient way to represent the general features of engineering data.

fact that many technical concepts simply cannot be conveyed with words alone. Illustrations, charts, graphs, figures, diagrams, and drawings are necessary to communicate technical information.

In general *graphics* refers to methods for transmitting technical information pictorally or symbolically by means of images or illustrations as opposed to text or mathematical symbols. Graphics is an essential tool for communicating engineering information. In a sense, graphics is the language of design, just as mathematics is the language of engineering analysis. The introduction of powerful new methods of computer graphics into design has expanded and enhanced the importance of graphical communication.

6.4.1. Graphs

Graphs are the mechanism most commonly used by engineers to present data. Engineers can choose among a wide variety of graphical representations. For example, pie charts illustrate the relative magnitude of various components of a system; bar charts or histograms illustrate data of a discontinuous nature; and line

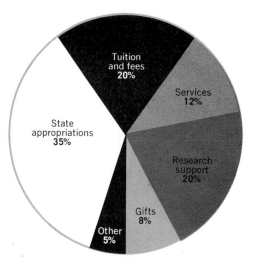

FIGURE 6.8. A pie or circle chart can be used to indicate the division of a particular quantity (in this case, college revenue).

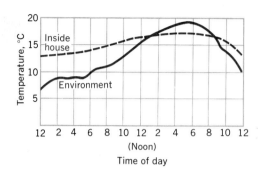

FIGURE 6.9. The line chart or graph provides a more detailed presentation of quantitative data.

graphs characterize more complex functional dependence. The engineer can also select a graphical scale most appropriate for the problem at hand. A linear or logarithmic scale might be chosen; the grid spacing could also be chosen to emphasize points of interest.

In general we might characterize the process of constructing graphs by the following steps:

1. Select the scale to be used.
2. Identify the scales and the parameters on each axis.
3. Plot the points on the graphs using proper symbols.
4. Connect the points (perhaps using a fitting or interpolation scheme).
5. Identify and label each curve and symbol.
6. Add all necessary legends.
7. Add captions, including figure number and title.

Graphs should be self-contained, including all information relevant to their understanding. This information may be provided on the graph itself or included in the caption. Since graphs taken from reports are frequently reproduced in journals after some reduction in size, one should take care that the lettering is sufficiently large to remain legible after photoreduction.

SUMMARY

Engineering data are usually presented in graphical form, such as pie charts, bar charts or histograms, and line graphs.

		Redwood			Birch		
		Thickness	Width	Length	Thickness	Width	Length
Type of Test	Mass loss	1.91	12.7	12.7	3.3	12.7	20.
	Tensile test	2.79	12.7	215.9	3.8	12.7	228.
	Shear test	3.81	12.7	22.9	3.5	12.7	20.

FIGURE 6.10. Tables can be used to present actual data for further reference.

6.4.2. Tables

The use of tables should be discouraged. Numbers in tables are difficult to comprehend. They fail to illustrate trends in the data. In fact, tables are sometimes used to disguise poor data. Results should be presented in graphical form wherever possible. Tables should be used only if it is necessary to present actual numbers for future reference. Even in this case, the data should also be presented in graphical form.

As with graphs, tables should be titled and contain sufficient information to make them self-explanatory.

SUMMARY

> *The use of tables should be avoided wherever possible. Data can be more easily understood when presented in graphical form. Tables should be used only when the actual numerical values of the data are necessary.*

6.4.3. Technical Illustrations

Illustrations are frequently used to depict a device or process. The illustration might be a precise mechanical drawing of the object of interest. A simple line drawing is frequently sufficient to illustrate the basic concept. The illustration might also take the form of an abstract diagram.

Technical illustrations play an extremely important role in engineering. Through graphical means the engineer communicates with colleagues and conveys the precise instructions needed for the production and assembly of engineering designs. In years past, engineering students spent one or more semesters learning the techniques required to prepare proper engineering drawings. Such courses, generally referred to as "engineering graphics," would teach the proper use of mechanical drawing instruments, lettering, engineering drawing standards, and how to prepare and understand technical drawings.

However for the past two decades, the tasks of preparing detailed design drawings have been assumed by draftsmen or engineering technicians. Engineers have become less concerned with the details of the preparation of engineering drawings, and courses in engineering graphics have gradually disappeared from many engineering programs.

Nevertheless there is still a need for graphics material in an engineering education. Engineers should be able to understand and communicate through the use of graphics. They should be able to set ideas down in graphical form and to understand and interpret the drawings prepared by others. They should also be thoroughly familiar with the various types of graphics used in engineering practice.

Interestingly enough, the introduction of computer graphics into engineering practice has intensified this need for a background in graphics (although graphics taught merely as "mechanical drawing" is of limited relevance to the modern engineer). As we shall discuss in greater detail in the next chapter, modern computer-aided design (CAD) systems have replaced the draftsman with computer graphics and, in so doing, have brought the engineer back into direct contact with preparation of engineering graphics. For example, the engineer might sketch a rough design directly on a computer terminal display screen with a light pen. The computer system could then be instructed to refine the sketch in an interactive mode, reforming it into a more precise shape, adding appropriate dimensions, performing geometrical analysis, and generating alternative graphical representations of the design. When the engineer is finally satisfied with the result of this engineer-computer interaction, instructions can be given to the computer to produce finished design drawings. In a very real sense computer graphics has reemphasized the importance of graphics education in the engineering curriculum.

A variety of types of graphics are used in engineering practice. For convenience we shall divide technical illustrations into several classes: block diagrams, schematic line diagrams, pictoral drawings, assembly drawings, and multiview drawings.

FIGURE 6.11. Block diagrams are commonly used to illustrate the major components of a system such as a hi-fi set.

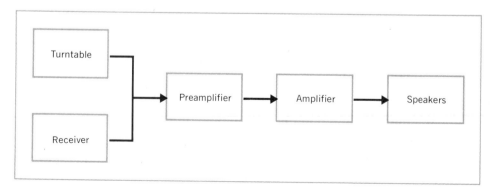

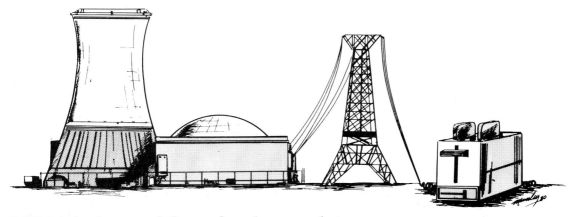

FIGURE 6.12. A conceptual diagram of a nuclear power plant.

Block Diagrams

Block diagrams are commonly used to illustrate the major components of a system and the relationship among these components. For example, the block diagram of a high fidelity system identifies components such as amplifiers, tuners, turntables, and speakers. It shows how these components are related. Note that block diagrams do not contain sufficient information on how to assemble (or even connect) such components, but they do describe their function.

Schematic Line Diagrams

Schematic line diagrams also show the components of a system, but in greater detail than block diagrams. Line diagrams are sufficiently illustrative that they can be used to assemble the system.

FIGURE 6.13. Line drawings provide a somewhat more detailed representation of system components.

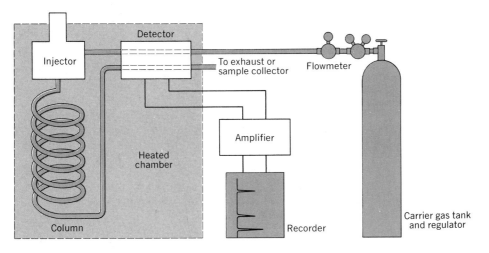

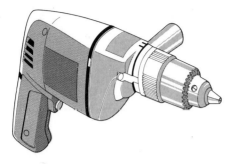

FIGURE 6.14. Pictorial drawings present a view of an object.

Pictorial Drawings

Pictorial drawings are similar to photographs in that they depict a view of the object. The disadvantage of this type of illustration is that all details cannot be shown. This is particularly true for complex objects. For example, while the picture of a typewriter might be helpful to a mechanic assembling typewriters, the assembly could not be accomplished on the basis of a pictorial drawing alone.

Pictorial drawings can be difficult to draw, particularly for those who are not artistically inclined. Nevertheless, even students less talented in art can draw freehand sketches by following some simple rules and by practicing sketches.

There are two major types of pictorial drawings: isometric and perspective drawings. In isometric drawings, lines parallel on the object remain parallel on the drawing. In perspective drawings parallel lines recede off to a point in the distance. Isometric drawings are easier to make but create the improper perspective of the object. In these drawings receding lines appear to be spreading apart. The top surface of objects appears to be shortened. Perspective drawings provide a more realistic picture of the true object, but are somewhat more difficult to construct.

FIGURE 6.15. Pictorial drawings can be either isometric (with parallel lines) or perspective (with lines receding to a point).

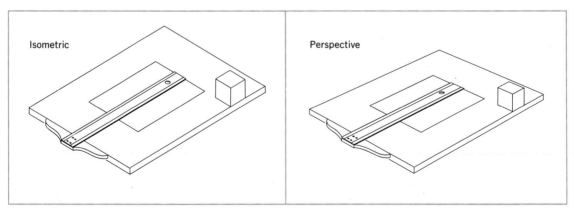

Isometric

Perspective

FIGURE 6.16. An example of an isometric sketch.

Each pictorial drawing (be it isometric or perspective) shows the object only as viewed from one direction. Thus the orientation of the object in the drawing must be selected carefully so that the resulting picture will illustrate the most significant features of the object. Sometimes it may be necessary to draw pictures from several different directions to bring out all the necessary details.

Assembly Drawings

Assembly drawings are somewhat similar to pictorial drawings. However, while the purpose of pictorial drawings is to illustrate the appearance of an object, assembly drawings show how individual parts and components are fitted together. On assembly drawings, the individual parts are identified and major dimensions are included.

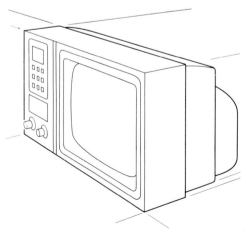

FIGURE 6.17. An example of a perspective sketch.

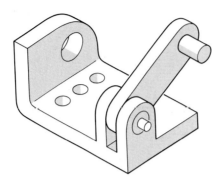

FIGURE 6.18. Assembly drawings are similar to pictorial drawings except that they illustrate how individual components fit together.

FIGURE 6.19. Multiview drawings show several views of an object.

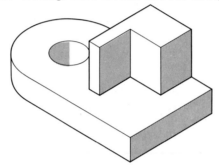

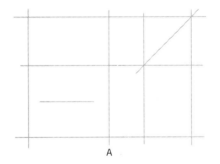

A

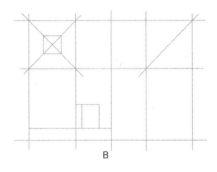

B

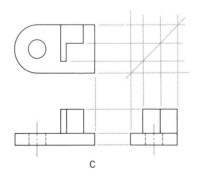

C

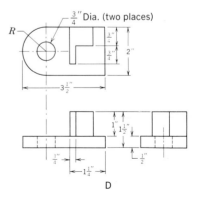

D

Assembly drawings are used to provide instructions on how to assemble a product from the various parts. Engineers make frequent use of assembly drawings to illustrate their ideas. The draftsman prepares the detailed multiview drawings of the object and its components on the basis of assembly drawings.

Multiview Drawings

Multiview drawings must be used when an accurate, detailed description of a part is needed. Multiview drawings are used when an object or product is to be fabricated from the drawing. These illustrations thus contain complete descriptions of the part, including all dimensions, tolerances, and surface finishes.

Multiview drawings show several simple orthographic projections of an object. Thus each view appears as a projection on a plane (say, a plate of glass) perpendicular to the line of sight of the viewer. These planes are then placed "flat" on the paper to provide the multiview presentation of the object.

The views are chosen so as to describe the object in sufficient detail. For single objects, two or three views are generally sufficient. For more complex objects, more views may be necessary.

Interior details do not appear on projections made of the outside contours only. Simple interior details such as a hole can be illustrated by dotted lines. To show more complicated details, sectional views must be used. To prepare sections, the object is sliced through with an imaginary plane, and the projection of the object on this plane is depicted on the drawing.

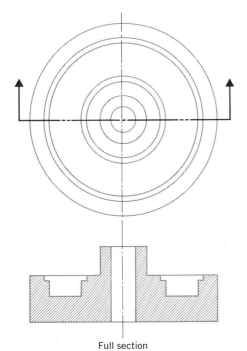

Full section

FIGURE 6.20. A full section drawing slices an object with an imaginary plane and then depicts the projection of the object on this plane.

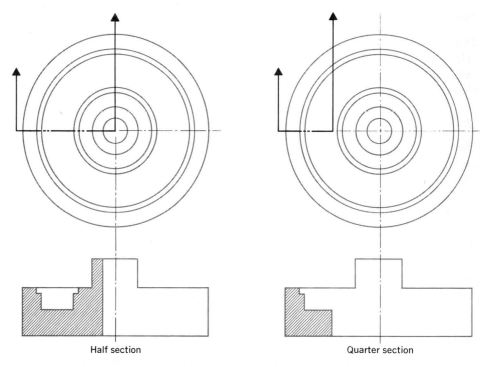

Half section   Quarter section

FIGURE 6.21. Half and quarter section drawings.

In the case of a full section the imaginary plane cuts through the entire section. Half sections show only the interior of half of an object. Half sections are useful for symmetric objects. Offset sections cut through the entire object along a staggered line. Partial sections are used to show internal details of pictorial drawings. In sectional views solid parts cut by imaginary planes are indicated by cross-hatched lines.

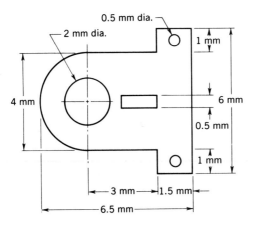

FIGURE 6.22. Dimensioning of an engineering drawing.

Dimensions

Dimensions must be added to all the different types of drawings to show the relevant sizes. In pictorial assembly drawings often a few major dimensions are sufficient. In multiview drawings the dimensions of all details must be given clearly.

Photographs

Photographs are difficult and expensive to reproduce. Unless absolutely necessary, photographs should not be included in a technical report. The information contained in a photograph can usually be conveyed in a line drawing.

We have classified engineering drawings by type, but it is also possible to classify drawings by function. For example, *construction* drawings are used for the construction of a particular project. Orthographic projections are used, and dimensions and tolerances are kept relatively simple. *Production* drawings include detail and assembly drawings for mass produced items. Again orthographic projections are used, but in this case detailed dimensions and tolerances are included. *Systems* drawings used in either production or construction use block and

FIGURE 6.23. A photograph can usually be replaced with a drawing. (Courtesy Chevrolet)

schematic line diagrams. *Concept* drawings are similar, but concerned mainly with the presentation of ideas.

Engineers should become adept at graphical communication. We shall return to consider this subject in further detail in the next chapter when we discuss computer-aided design.

SUMMARY

Technical drawings are extremely important in communicating engineering information. They include block diagrams, line diagrams, pictorial drawings, and multiview drawings. Although it is not essential that engineers be adept in rendering such drawings, they should be able to understand them and use them in communicating with others.

Exercises

Graphs

1. The annual expenditures for a company are given as

Salaries	$1,750,000
Rent	250,000
Supplies	50,000
Materials	600,000
Miscellaneous	9,000

   Present these expenditures in the form of (1) a bar graph and (2) a pie chart.

2. A company manufactures shirts in four size ranges. The number of shirts produced each month in each size range are:

Small	800
Medium	1600
Large	1300
Extra Large	800

   Present this data on bar graphs and pie charts.

3. The emission for an industrial plant are broken down into the following components:

Unburned hydrocarbons	4.0%
Oxides of nitrogen	0.5%
Carbon monoxide	0.3%
Carbon dioxide	0.6%
Other	95.6%

   Present this data in a pie chart.

4. The age distribution of workers in a plant is

	MALE	FEMALE
20–30 years	625	428
30–40	710	809
40–50	350	375
50–60	250	295
other	475	612

Present this age distribution data in pie and bar chart form. The two pie charts should be side by side. Combine the male and female distributions on a single bar chart.

5. Plot the weight of steel and wooden spheres as functions of their radius. The density of steel is 7703 kg/m^3 while that of wood is 705 kg/m^3.

6. The relationship between the Celsius and Fahrenheit temperature scales is given by

$$C = \frac{5}{9}(F - 32)$$

Plot this relationship over the temperature range 0°F to 250°F.

7. A certain machine part is removed from a temperature bath. The temperature of the part is found to vary with time t according to the formula:

$$T(t) = 300\,e^{-2.45t} + 20 \ °C$$

where t is the time in seconds. Plot the variation of the part temperature with time on both regular and semi-log paper.

8. Plot the pressure versus volume for saturated steam. Use the data given in the steam tables.

9. The drag on a sphere was measured and found to vary with velocity as follows:

Velocity (m/s)	1	2	4	8	10
Drag (N)	100	400	1600	6400	10 000

Plot the variation of drag with velocity using both regular and log-log scales.

Technical Illustrations

10. Draw a block diagram of the following systems:
    - (a) An automatic car wash
    - (b) The computing process in a computer center
    - (c) The power train of an automobile
    - (d) The process of television transmission and reception

11. Draw a line diagram of the following items:
    - (a) The heating system of a single family home
    - (b) An automatic coffee percolator
    - (c) A toaster
    - (d) A process that can be used to produce desalinated water

12. Draw both isometric and perspective drawings of the following objects:
    - (a) A pencil
    - (b) A chain
    - (c) A desk
    - (d) A hammer
    - (e) A paperclip
    - (f) A shoe
    - (g) A car
    - (h) A bicycle
    - (i) A crane
    - (j) A bridge

13. Make assembly drawings of the following items:
    - (a) A ballpoint pen
    - (b) A telephone receiver
    - (c) A filing cabinet drawer
    - (d) A bicycle pedal
    - (e) A faucet

14. Draw multiview diagrams of the objects listed in Exercise 13. Include sections wherever necessary.

6.5 COMPUTERS AND COMMUNICATION

6.5.1. Impact on Communication Activities

For many years the engineer has approached technical reports, articles, or letters by writing draft after draft. Typically notes written or dictated would comprise the first or rough draft. In some cases the engineer would edit (correct and modify) this written draft directly. In other instances a secretary would tediously translate the engineer's frequently illegible handwriting, correcting the more glaring spelling and grammatical errors, and type a clean copy of the material to serve as a draft for further editing. This process might then be repeated several times, revising and polishing the draft manuscript until an acceptable form was reached.

A similar procedure was followed in preparing graphical materials. The engineer would prepare a rough hand sketch that could be used by a draftsman to produce the desired illustrations or drawings. Once again several iterations might be required until an acceptable drawing was obtained.

Many of these activities have changed dramatically with the introduction of the computer into both office and home and the development of sophisticated wordprocessing systems. Today many engineers compose their first draft while seated at a computer terminal or microcomputer. As they enter text on a keyboard, the material appears before them on a monitor screen. The computer can then be used to edit this material, adding, deleting, or correcting text with the touch of a key. Entire sections (sentences or paragraphs) can be instantly shifted about. Most wordprocessing systems can aid in editing and composition in many other ways, such as by conducting word or phrase searches, adjusting and justifying margins, and even correcting spelling. They allow the engineer to compose and manipulate written material on a TV screen, almost as a sculptor would mold and form a lump of clay. Working drafts are stored in digital form on floppy disks and are available for instant retrieval. When the engineer is finally satisfied with the draft, simple commands instruct the computer to print out the text typewriter or high speed printer. Or, the engineer may simply send the draft, while it is still in digital form (on magnetic disk or tape), directly to the publisher for computer composition.

FIGURE 6.24. Wordprocessing systems have revolutionized the preparation of written reports. (Courtesy Texas Instruments Inc.)

The computer can be used in a very similar manner to prepare graphs, charts, and drawings. For example, the engineer can sit before a microcomputer or computer terminal and build up a graphical display on a screen using either the keyboard or a light pen (or electric tablet). The computer can assist in this task in a way that would have been beyond the imagination of draftsmen or artists several years ago. It can instantly modify color or shading, connect even the most complex diagram with properly placed lines, superimpose carefully constructed shapes, and so on. Once again the graphical information is stored in digital form on disk or tape until it is desired to produce "hard copy" using a printer or plotting device. In some cases the images produced on the display are reproduced directly using photographic methods.

In some sophisticated computer networks there is a capability to generate simultaneous displays on graphics terminals that are geographically remote from one another. This opens up an entirely new form of communication, since engineers can transmit complex graphics images of designs directly to one another.

Already the computer has had an enormous impact on the routine (and often tedious) tasks involved in written and graphical communication. This impact is certain to become even more pronounced with the introduction of voice interpreters to receive spoken commands and translate these into text or graphical materials.

SUMMARY

The computer has changed dramatically the manner in which both written and graphical materials are prepared. Modern wordprocessing systems have greatly improved the ease and efficiency of composition and editing. Likewise, computer graphics is rapidly replacing mechanical drawing (drafting) as the primary graphics tool of the engineer.

6.5.2. Computer Software

Just as the computer has revolutionized the conventional means of communication, it has also introduced a need for new kinds of communication in the area of computer "software." As we discussed, this term refers to the computer programs written to perform certain tasks. These programs represent the result of creative intellect and achievement, just as much as any device or process produced by the engineer. As computer software becomes more varied and complex, it becomes more important to provide these programs with suitable documentation and instructions for their use.

Standards are now being established for writing and documenting computer programs. For example, the top-down or structured programming methods de-

FIGURE 6.25. A modern computer graphics system based on a microcomputer. (Courtesy Hewlett-Packard Company)

scribed in Chapter 4 are now commonly recommended for all software applications. Sufficient comment (or REM) statements should be included to explain the program logic. This is particularly important in the older computer languages such as BASIC of FORTRAN as compared with the more literal languages such as Pascal. It is also important to recognize that computer software is now regarded as another form of publishing. Computer programs can be copywrited and protected from plagiarism. This point should be kept in mind when using or writing computer software or communicating it to others.

SUMMARY

A new area of communication has grown up involving computer programs or software. Standards are now being established concerning program structure and documentation.

6.5.3. Computer Results

One advantage of the computer is that it can process enormous quantities of information very rapidly. This also means that it can churn out huge stacks of output with ease, whether the output is relevant or meaningless rubbish. Sometimes computer users are tempted to overwhelm colleagues with the sheer magnitude of results, perhaps calculating a number to 16 decimal places when only two are sufficient, or including page after page of numerical results generated by the computer.

Therefore a premium must be placed on data reduction and interpretation when presenting the results of a computer calculation. The presentation of this data is akin to the design of graphical materials. One should always strive for simplicity, clarity, and brevity. Just because a computer can cheaply generate thousands of graphs or tables is no reason to include them in a report.

SUMMARY

One should avoid the temptation to use the enormous power of the computer to produce large quantities of data for inclusion in a report. A premium should be placed on data reduction and interpretation. As with other forms of engineering communication, one should strive for clarity, simplicity, and brevity.

Example   As a routine measure, all of the computer printout at government defense laboratories is classified and marked with red lettering in the margin: "SECRET RESTRICTED DATA." Yet these computers print out millions of pages each month. It has been proposed (somewhat facetiously) that instead of classifying this output, the United States government should pack it up in crates and mail it over to its foreign competitors, reasoning that the workload involved in sifting through the tons and tons of meaningless computer data in search of secrets would quickly overload any foreign intelligence organization.

Exercises

Computers and Communication

1. Prepare the following items on a wordprocessing system:
    (a) A one page paper extolling the virtues of the wordprocessor

    (b) A business letter to the manufacturer of the wordprocessor explaining why you still have not paid for it

    (c) The mathematical equations representing Newton's laws

2. Inquire in your engineering dean's office as to how many wordprocessing systems are presently used in your college of engineering. Compare this to the number of wordprocessors used in a local business enterprise of about the same size.

3. Prepare a three-dimensional illustration of the following items on a computer graphics console:

    (a) A cube   (c) A pyramid

    (b) A sphere   (d) A rectangular building

4. Document one of the programs you prepared in the exercises of Chapter 4 in a form suitable for distribution.

5. It is now common for large computer systems to communicate with each other, transferring data or software, without human intervention. Write a short paper on this development, the ultimate form of "computer communication," briefly describing the mechanics of this communication process and how prevalent it is today.

REFERENCES

Technical Communication

1. Ingrid Brunner, J. C. Mathes, and Dwight W. Stevenson, *The Technician as Writer: Preparing Technical Reports* (Bobbs-Merrill Educational Publishing, Indianapolis, 1980).

2. J. C. Mathes and Dwight W. Stevenson, *Designing Technical Reports: Writing for Audiences in Organizations* (Bobbs-Merrill, Indianapolis, 1976).

3. Kenneth W. Houp and Thomas E. Pearsall, *Reporting Technical Information,* 4th Edition (Collier Macmillan, Encino, California, 1980).

4. Charles T. Brusaw, Gerald J. Alred, and Walter E. Oliu, *Handbook of Technical Writing* (St. Martin's Press, New York, 1976).

5. Thomas E. Pearsall and Donald H. Cunningham, *How to Write for the World of Work* (Holt, Rinehart and Winston, New York, 1978).

Graphics

1. Warren J. Luzadder, *Fundamentals of Engineering Drawing,* 7th Edition (Prentice-Hall, Englewood Cliffs, N.J., 1977).

2. C. H. Jensen, *Engineering Drawing and Design* (McGraw-Hill, New York, 1968).

3. Frederick E. Giesicke, Alva Mitchell, Henry C. Spencer, Ivan L. Hill, and Robert O. Loving, *Engineering Graphics,* 2nd Edition (Macmillan, New York, 1975).

CHAPTER 7

Computer-Aided Engineering: CAD/CAM

The application of computers in engineering has produced some confusing new jargon. Terms such as CAD, CAM, CAD/CAM, and CADD abound in the engineering literature. The employment opportunities section of most newspapers reveal listing after listing for CAD/CAM specialists in the aerospace, automotive, and manufacturing industries. Hundreds of new companies have been formed in recent years to produce and market CAD/CAM systems and services.

What do these magic letters mean? CAD is simply the acronym for *computer-aided design*, a term referring to the use of computers in engineering design activities. The other terms refer to related uses of the computer in engineering: *computer-aided manufacturing* (CAM) and *computer-aided design and drafting* (CADD).

During the early years of development, the term "computer-aided design" was used in a restricted sense to refer to the use of computer graphics in engineering drafting applications. Gradually the computer became an important aid in creating and testing design sketches. The term "computer-aided manufacturing" was also first used in a restricted sense to refer to the use of digital or numerical control of manufacturing machinery such as cutting tools. Today, however, CAD/CAM has come to represent a far more extensive integration of computers into almost every phase of engineering. In fact the computer has now become the instrument that weaves together and extends the traditional tools of the engineer that we have studied in this text, the tools provided by science and mathematics, experiments and testing methods, and communication skills.

An example will make this clearer. Until quite recently the usual procedure for designing and producing a machined part began with the design engineer, who sketched the outlines of the part. This sketch would be passed along to a draftsman for the production of finished drawings. Another engineer would use the drawings to determine geometric properties such as areas, volumes, and weights. A structural engineer would determine stresses and deflections. These data would then be returned to the design engineer for consideration in a redesign of the part, then back to the draftsman for more drawings, and so on, iterating

FIGURE 7.1. An engineer working at a CAD/CAM console. (Courtesy IBM Corporation)

back and forth until a finished design was achieved. The final drawings for the design would then be passed to the manufacturing engineer who would prepare instructions for a machinist. Finally the part would be machined according to the specifications given by these instructions.

The introduction of computer has changed this process quite dramatically. Today CAD/CAM systems have integrated the activities of draftsmen, design engineers, analysis engineers, and manufacturing engineers. Using these systems an engineer can sit down before a computer terminal and sketch the part directly on a CRT (cathode-ray tube) display monitor, using a light pen or electrosensitive graphics tablet. The software and data base provided by the CAD/CAM system can then be used to manipulate this initial design, performing a stress analysis and determining section properties to a very high precision, through an interactive design process between computer and engineer, until a final design is achieved. The CAD/CAM system can then produce finished drawings for this final design. Finally, the CAD/CAM system can prepare instructions (on paper tape or disk) for numerically controlled machine tools and place orders for related parts and production facilities. In essence, the modern CAD/CAM system allows the engineer to sit down at a computer terminal, perform all of the activities of engineering design while interacting with the computer, and then walk over to the numerically controlled machine tool and pick up the finished part.

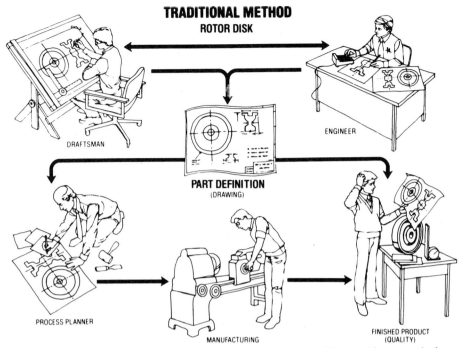

FIGURE 7.2. The traditional engineering activities involved in machine part design and production. (From Engineering* 1979)

FIGURE 7.3 The CAD/CAM approach to machine part design and production. (From Engineering* 1979)

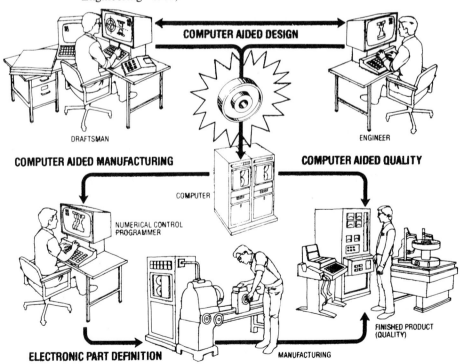

In a very real sense computer-aided design and manufacturing, CAD/CAM, has become the ultimate tool of the engineer, integrating all other aspects of engineering activities, including design, analysis, and communication. In fact, the evolution of such systems is now so advanced, and the role they already play in engineering practice so general, that the term "computer-aided design" is probably outmoded. It would be far more appropriate to refer to these systems *computer-aided engineering* since they span essentially all of the activities of the engineer.

The development of CAD/CAM systems is another example of the profound influence that the computer has had on the engineering profession. It has often been said that the computer age has accomplished for our minds what the industrial revolution did for our muscles. Just as machines have taken over many of the more mundane tasks of labor, thereby allowing us to use our energies for other more sophisticated endeavors, the computer has relieved the engineer of many of the more routine aspects of engineering practice such as drafting, repetitive calculations, and tedious design revisions. The immense data storage and calculating capability of modern computers has provided the engineer with a powerful instrument to rapidly process and analyze the large quantities of information associated with complex engineering systems. It has also facilitated information flow and decision making and provided a new means of communication. The computer has presented engineers with a powerful new lever to use in multiplying their intellectual capabilities.

The key to computer-aided engineering has been *integration:* the synergistic blending together of engineer and computer into a problem solving team so that the strengths of both can be utilized most fully. The computer supplies the raw computational power and the capacity to store and process vast amounts of information. The engineer provides the creativity, judgment, and insight needed to solve complex engineering problems.

The rapid evolution in computer size, capacity, and cost has led to a revolution in the use of CAD/CAM methods in industry. They free engineers from repetitive, time-consuming busywork and allow them to spend more of their time in creative activities. They also promise significant gains in industrial productivity. The potential impact of CAD/CAM methods on engineering practice during the 1980s may well rival the importance of the digital computer to engineering analysis during the 1960s.

SUMMARY

CAD/CAM refers to computer-aided design and manufacturing. CAD/ CAM systems based on interactive computer graphics and engineering analysis programs have revolutionized engineering practice by combining engineering design, analysis, and graphical communication. These computer-aided engineering systems have integrated and extended the traditional tools of engineering provided by science and mathematics, experimental methods, and communication arts. CAD/CAM systems allow the engineer and computer to interact together in a synergistic fashion, thereby combining the abilities of the computer to process rapidly vast amounts of information and the creativity of the engineer to synthesize novel solutions to complex problems.

Exercises

Computer-Aided Engineering

1. Look through the employment opportunities section of a major newspaper and determine the number of available positions for CAD/CAM specialists. What types of firms are searching for these engineers?

2. Determine whether your college of engineering has an operating CAD/CAM system. If it does, briefly describe the system (e.g., capability, number of consoles, degree to which it is used in instructional activities).

8.1. COMPUTER-AIDED DESIGN

Computer-aided design or CAD refers to the application of computers to engineering design. To better understand the role that CAD can play in this activity, recall the steps in the engineering design process:

Identification of need
Information search
Synthesis of possible solutions
Feasibility study
Preliminary design study
Final design
Design documentation

Computers can play an important role in every step of this sequence. The data base maintained by computers can be used to facilitate an information search, to rapidly retrieve past designs or search the technical literature. Although the creativity of the engineer plays the most important role in the synthesis of possible solutions to the design problem, the computer can assist in the expression and examination of these solutions using powerful tools of computer graphics and numerical analysis.

Perhaps the most significant impact of CAD systems has occurred in the more detailed design activities of feasibility studies, preliminary design, and final design. Interactive computer graphics can be used to visualize and rapidly analyze new designs. Many CAD systems are capable of performing sophisticated engineering analyses of preliminary design features, including a determination of geometrical characteristics such as dimensions, surface areas, volumes, weights, moments of inertia; structural and dynamical characteristics; thermal and electrical analysis; and economic analysis. The computer can be used to construct a comprehensive numerical model of the design that the engineer can interrogate and modify in a manner similar to the use of a physical prototype. The computer-based model can actually serve as a prototype to simulate the performance of a given design. Experience gained with these models can be fed back into the design process. In this manner the important processes of iteration and optimization of design options can be accomplished rapidly and inexpensively by the CAD system.

When the engineer has converged on a final design, the computer can then perform the final design analysis and prepare the design documentation. Computer graphics methods can prepare the final design drawings. The computer can also prepare the detailed numerical instructions that will control the machines used to implement the design. It can place orders for parts and schedule production facilities. It can even prepare cost and marketing information.

The single most important factor that has led to the rapid implementation of CAD systems is their use of interactive computing. The engineer is able to communicate instantaneously with the computer system to see the results of design changes, to detect mistakes and implement corrections immediately, and to use the computer to simulate the performance of the design before an expensive prototype is constructed. The computer provides immediate access to a vast data base that can be conveniently manipulated and applied to the design by the CAD system. The key to this interactive capability has been the development of versatile computer graphics and the software for performing engineering analysis of design features. Although the creativity and ability of the engineer still plays the major role in determining the success of any design, CAD/CAM systems have greatly extended the engineer's capabilities and eliminated much of the drudgery and inefficiency from the design process.

SUMMARY

The computer can play an important role in every phase of the engineering design process, including information searches, graphical display, engineering analysis, prototype simulation, final design analysis, and design presentation. The key to CAD systems is interactive computing, allowing the engineer to implement and see the results of design changes immediately.

7.1.1.   Background

Although computers have been applied to engineering analysis since the early days of their development, the integration of computers into the engineering design process in an interactive mode can be traced back to the development of time-sharing computer systems and storage-type cathode-ray-tube (CRT) displays in the 1960s. On-line computer graphics were developed through the cooperative efforts of computer manufacturers and companies in the aerospace and automotive industries. Hardware developments were paralleled by the development of software packages (computer programs) to facilitate the use of computer graphics and integrate this with other aspects of engineering analysis.

When early CAD equipment first became available during the 1960s, it was quite expensive and accompanied only by rather primitive software. Only a few high technology industries for whom speed and precision were essential to product design could afford to implement the early systems. The key to the more recent proliferation of CAD systems has been the rapid evolution in computer hardware coupled with the dramatic decrease in hardware costs. Rapid developments in microelectronics made possible inexpensive graphics display terminals and stand-alone microcomputers. Time-sharing systems brought the power of large, mainframe computers to remote users.

Equally important has been the parallel development of more sophisticated software packages to implement CAD. Simpler, more user-oriented computer languages were developed. Software packages with complete analysis and design modules were developed. But in contrast to declining hardware costs, software costs have increased dramatically and now account for as much as 80% of the expense of a CAD system.

Today the trend is toward general purpose CAD/CAM systems, equipped with the hardware and software to take a design from an early conceptual stage to detailed design drawings and machine tool instructions. For example, a CAD/CAM system for machine part design would allow the engineer to sketch the part at a CAD/CAM console, perform a geometric analysis (including section properties, weights, and moments of inertia), analyze stresses, deflections, dynamics, and temperature distributions, and prepare final design drawings and machine tool instructions. The system might also compile data files for bills of materials, piece part costs or manual machine operation sheets, and part order. For example, the CADAM system developed by Lockheed can handle major portions of the work of draftsmen, design engineers, analysis engineers, and manufacturing engineers.

The implementation of CAD/CAM systems has already revolutionized engineering practice in many industries. It is routinely used for component design and manufacture in many aerospace and automotive activities. CAD/CAM methods have also found extensive use in the electronics and chemical industries. The next decade is likely to see their rapid implementation in general manufacturing and construction activities.

SUMMARY

The development of CAD/CAM systems was stimulated by both computer hardware advances such as graphic displays, as well as by software developments in interactive computing. Initial development and application occurred in the aerospace and automotive industries, although CAD/CAM is now being rapidly implemented in general manufacturing and construction activities.

7.1.2. General Features of CAD/CAM Systems

What are the general components of a CAD/CAM system? The hardware components are similar for the most part to those of other computer systems, with the possible exception of specialized equipment to support sophisticated display graphics applications and numerically controlled machines and robots. CAD/CAM systems can be based either on large, time-sharing computer systems or on dedicated microcomputers. In the latter case, microcomputer CAD/CAM hardware is frequently tied into a much larger data base maintained by a large, mainframe computer.

The most significant developments in CAD/CAM systems have arisen in software, in the development of a variety of computer programs that assist the engineer in performing design and analysis. We can identify four classes of software packages for CAD/CAM systems.

Computer Graphics

Interactive computer graphics has played a major role in assisting the engineer to create and visualize new designs. Computer graphics hardware–software packages typically allow the engineer to interact with two- or three-dimensional color displays of the design and may include additional features such as animation.

Engineering Analysis

The analysis of the design is performed by several computer programs integrated into the CAD/CAM system. These can range from simple geometrical analysis programs to determine dimensions, areas, volumes, and moments of inertia to complex general purpose programs that can perform a wide variety of engineering calculations, including the analysis of stress, dynamics, temperature distributions, fluid flow, and electrical characteristics. The most powerful such analysis packages are based on finite element methods, so-called because they break the design into small components or elements and then solve the mathematical equations that represent the interrelation of these elements. Other more specialized programs may be included in CAD/CAM systems used in particular areas such as chemical process development or electrical circuit design.

Data Processing

CAD/CAM systems must be capable of handling large amounts of data, storing these data until requested by the engineer, and then retrieving and presenting the data in a suitable form. Therefore software has been developed for data handling and process. In some cases, specialized computer languages have evolved that allow easier interaction with the CAD/CAM system than more general computer languages.

Communication

An important component of any CAD/CAM system is the manner in which it facilitates the documentation, communication, and use of design efforts. Software packages have been developed that can generate finished design drawings and instructions for numerically controlled machines. It is becoming more common for engineers to communicate design concepts across a linked network of CAD/CAM consoles using computer graphics.

The present advantages of CAD are most apparent in industrial applications. Although there may initially be some resistance to the role changes caused by the introduction of digital systems, subsequent experience has demonstrated a rapid decrease in the time between design and production. CAD systems can be used to achieve design refinements earlier in the design process, thereby reducing the need for building costly physical prototypes. CAD systems can also eliminate interference between various aspects of the design. Since each designer's console is linked to a single data base, as design changes are entered, these are stored in computer memory and can be immediately taken into account in other aspects of the design. If a hydraulic line is relocated in an aircraft design by structural engineers, control engineers can assess immediately whether this change is consistent with their own design constraints. Through interactive computer graphics and common data bases, CAD systems can display the current status of the project, giving each designer a picture of what fellow engineers have completed or what they are working on.

SUMMARY

The essential features of a CAD/CAM system are a graphics display console supported by a minicomputer or time-sharing computer system, and software to support computer graphics, engineering analysis, data processing, and communication activities.

Example To illustrate the capability of CAD/CAM systems, we have described below a software package for mechanical engineering design:

Computer Graphics Package

> An interactive software package based on CRT screen presentation that provides all of the features of computer-supported drafting (in 2-D or 3-D) including labeling, dimensioning, and geometrical analysis.

Finite Element Analysis Package

> A set of programs capable of static and dynamic structural analysis, heat transfer, aerodynamics, acoustics, electromagnetism, hydroelasticity, and other engineering analysis.

Numerical Control Package

> A software package that produces numerically controlled machine instruction tapes directly from the graphics package.

In Figure 7.4 we show a series of CRT screen simulations of a typical CAD/CAM session using such a system.

Example Figure 7.5 shows the screen display for a CAD program used in marine design to analyze ship hull structural properties. Here the program analyzes a hull "midship," that is, a cross section of the ship's hull. Steel plate dimensions can be chosen and the strength of the midship section can be determined.

Exercises

Computer-Aided Design

1. Briefly describe the role played by computers in each phase of the engineering design process.

2. We have noted the growing use of CAD/CAM methods in the aerospace and automobile industries. In what other industries would you expect to see significant implementation of CAD/CAM systems during the next decade?

3. By using your technical library or writing directly to the appropriate company, obtain and briefly summarize the essential characteristics of a major CAD/CAM system (such as the CADAM system developed by Lockheed).

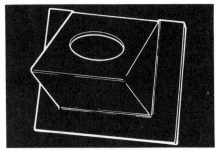

The problem is to describe the geometry of the bracket for structural analysis.

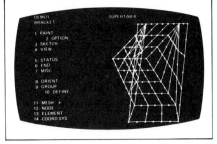

The entire model required 12 minutes to prepare for analysis.

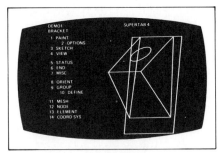

Using the refresh menu, the designer can easily interface with SUPERTAB to input geometry.

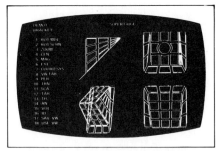

An exploded view of the model permits easy visual verification of all elements.

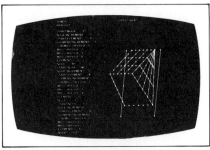

The mesh generation capabilities permit rapid creation of the necessary data.

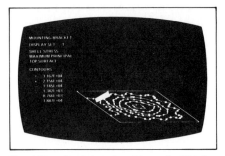

The resulting color stress contour plots permit rapid evaluation of the analysis results.

FIGURE 7.4. A simulation of a CAD/CAM session. (Courtesy of Structural Dynamics Research Corporation. Software Product: SDRC Supertab and Output Display.)

7.2. COMPUTER-AIDED MANUFACTURING AND ROBOTICS

In the same sense that computer-aided design has had a dramatic impact on engineering design and analysis, computer-aided manufacturing or CAM has significantly improved industrial productivity in the areas of manufacturing and assembly. The earliest applications of CAM involved numerically controlled (NC) machine tools introduced in the late 1950s. Control units that responded to instructions coded in digital form on paper tape were used to regulate the speed and

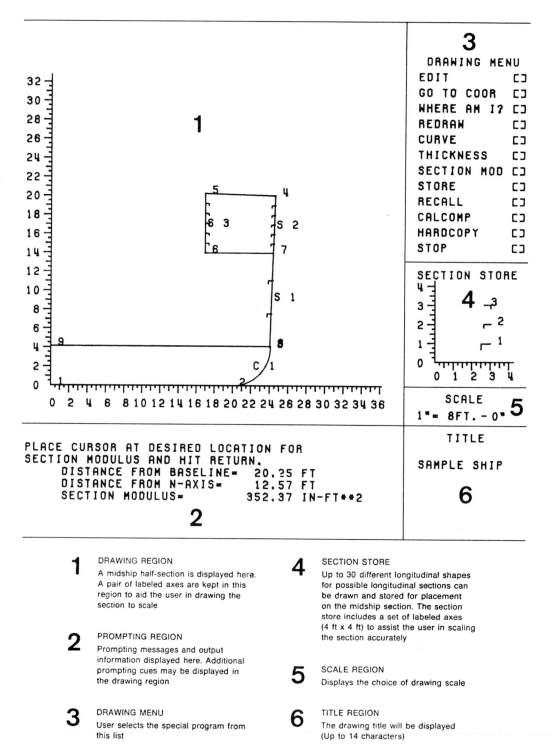

FIGURE 7.5 A CAD software package to design ship hull sections. (Courtesy University of Michigan College of Engineering)

cutting path of tooling machines. Such NC machines could perform complex operations beyond the capability of a human operator and provided dramatic improvements in speed, efficiency, and reliability. Gradually NC machines were developed that could perform many functions by using multiple tools in a programmed sequence.

Eventually, small digital computers took the place of the machine control unit, giving rise to a new term, computer numerical control or CNC. This advance eliminated the paper punch tape, since the machine instructions were stored directly in the computer memory. Since these computers could be reprogrammed, alterations of the finished project could be implemented quite easily. The subsequent implementation of time-sharing computer systems allowed a single large computer to manage several NC machines (so-called direct numerical control or DNC).

More recently, CAM systems have adopted a hierarchical computer organization in which a mainframe computer is used to control a hierarchy of microcomputers, each of which can control a given machine tool. Hierarchical CAM systems have been made possible by the development of the inexpensive microprocessor that allows one to dedicate a tiny computer to control each task of interest. When coupled with artificial intelligence concepts, hierarchical CAM systems have led to a highly flexible and reliable use of computers in all phases of engineering.

An important extension of CAM has involved the development of industrial robots. A robot is generally thought of as a machine capable of imitating human functions. Rapid advances in computer, sensor, and actuator design have led to the development of robots that can replace humans in certain assembly line activities such as welding and painting. Early industrial robots were designed to perform highly specific tasks involving only modest sensing and dexterity requirements. The major effort in robotics today is directed toward the development of flexible and versatile robots, capable of performing a wide range of industrial activities. When combined with microelectronics technology and advances in artificial intelligence, robotics development seems capable of stimulating major increases in industrial productivity.

Today CAM refers to integrated computer networks that oversee all major functions in a manufacturing plant, from initial design sketch to quality control of the final product. When coupled with CAD systems, the computer has led to a "digital unification" of the three major areas of manufacturing: engineering design and analysis, product management and control, and finance and marketing.

The use of such CAD/CAM systems in manufacturing has been a key element of the explosive growth in the industrial productivity of Japan and Europe. In some cases, entire manufacturing and assembly lines have been totally automated with dramatic increases in productivity and quality control. The diffusion of CAM equipment throughout the manufacturing industry in the United States has been much slower. But there seems little doubt that CAD/CAM methods will be implemented quite rapidly by American industry during the 1980s in an effort to match the great strides in industrial productivity achieved by foreign competitors.

FIGURE 7.6. Industrial robots are playing an increasingly important role in heavy manufacturing and production, such as these robots used on an automobile assembly line. (Courtesy Unimation, Inc., from Unimation Robotics)

SUMMARY

Computer-aided manufacturing or CAM has progressed from the early use of numerical controlled machines to the hierarchical systems of today in which mainframe computers manage a large number of dedicated microcomputers that control individual machine tools. CAM systems have been integrated together with computer-aided design or CAD systems to achieve an integrated computer system capable of overseeing all major engineering activities, from initial product design to production and marketing.

Example   The automobile industry has pioneered in the development and implementation of CAD/CAM systems. General Motors began in 1959 to develop interactive computer design and the use of numerical control technology for sculptured body tooling. By the late 1960s they had merged interactive computer graphics with numerical control machining. Today's CAD/CAM systems include both large scale, time-sharing systems with hundreds of computer graphics terminals, as well as stand-alone minicomputer systems that can operate independently or tie into the central base of the large system.

FIGURE 7.7.   CAD/CAM methods are finding increasing application in the aerospace and manufacturing industries. (Courtesy Lockheed California Company. Software Product: CADAM®—registered trademark of the Lockheed California Company)

A major application of CAD/CAM has been to the design and manufacturing of large sheet metal body panels. A design begins with a full-size clay design model of the automobile. Data from the clay model are recorded by a digital scanner and transmitted to a central computer data base. Graphics consoles are used by product designers to recall the data as points and lines, review it for completeness and accuracy, and then represent it as three-dimensional surfaces using computer graphics. The product engineer can add reinforcements, holes, flanges, inner panels, and structural parts directly from the graphics console. The CAD/CAM system produces the instructions for a numerical controlled mill to cut a wood die model. Tool and die groups responsible for building stamping tools then use the wood model as the template for building manufacturing tools.

CAD/CAM also plays an important role in evaluating the structural characteristics of a design. Computer models can be developed to describe the performance of automobile components in various environments (i.e., computer-simulated roads). Computer simulation is used to study passenger visibility or luggage space capacity.

Tooling engineers can use CAD/CAM systems to design welding fixtures and the optimum orientation of welding guns. The computer then automatically puts the weld guns in the proper position. Process engineers use computer monitoring to check part tolerances as they progress through the various machining operations. Die processing engineers use CAD/CAM to determine the die operations required to transform a sheet metal blank into the desired final part shape. Computers are also used to collect data in test laboratories and analyze these data, reducing it to a form suitable for evaluation.

The impact of CAD/CAM on engineering design has been profound. It has eliminated much of the trial and error in the design process, allowing the testing of computer-based models before prototype construction. CAD/CAM has been extended into tool design and manufacturing. The entire body tooling cycle from design concept to production die can be controlled by computer.

Perhaps the most rapid growth of CAD/CAM systems is in manufacturing and assembly. Computer-controlled sensing devices can control production, monitor tolerances, and maintain quality control. Robots are being used for more and more tasks, including inspection and testing, painting, welding, material handling, and assembly and fabrication. Through the use of CAD/CAM methods, the utilization of equipment and material in manufacturing can be optimized.

Exercises

Computer-Aided Manufacturing

1. Why do many numerically controlled machine shops continue to make use of paper tape for NC units rather than magnetic tape or disk?

2. Write a short paper comparing the extent to which CAM and robotics are used in Japanese industry to that in the United States. To what degree would you attribute the marked increase in Japanese productivity in recent years (at a time when United States productivity remained relatively constant or even fell off) to this use of CAM systems?

3. List three applications where CAM could be used effectively.

7.3. COMPUTER GRAPHICS

A major factor in the growth of computer-aided design systems has been the rapid evolution of interactive computer graphics. During the early 1960s software was developed that allowed a computer to display results in graphical form using pen or electrostatic plotters or printers and cathode-ray-tube (CRT) displays. These first applications were rather primitive, and the most common use of computer graphics was in data plotting. In most cases, the time required to plot these graphics by computer and display hardware prevented a truly interactive mode of operation.

However, as computer speeds increased, it soon became possible for the user to sit before a terminal equipped with a video monitor and display immediately the graphical plots of computed results. Software was developed that would allow the computer to draw two-dimensional objects. Some limited capability was achieved for displaying three-dimensional objects in "wire-line" drawing form, that is, as points and lines (rather than solid shapes).

The interaction between the user and the computer was improved with the introduction of light pen and graphics tablet input. These devices allowed the user to input data to the computer system using a light pen to sense a position on a CRT screen or by tracing an electric stylus along an electro-sensitive tablet. This input was then digitized and processed by the computer. In this way the user could actually sketch drawings directly on the display and then use the computer to modify or store these drawings.

Many systems also take advantage of the capability of computers to store a large library of standardized designs or design elements. A preliminary design can be entered by the engineer using light pen of graphics tablet and then compared with a reference standard design. Modifications can then be made based on a comparison of the preliminary design with the stored standards.

Today interactive computer graphics has come to represent a mode of interaction between user and computer through the visual medium of a CRT display screen (Figure 7.8). The user can employ a variety of input devices such as light pen, graphics tablet, or keyboard to produce graphics displays on the terminal screen. Software routines can then act as an analytic assistant. For example, as each element of a sketch is entered into the computer, it can be analyzed by applying precoded rules and constraints—for example, all lines shall be either

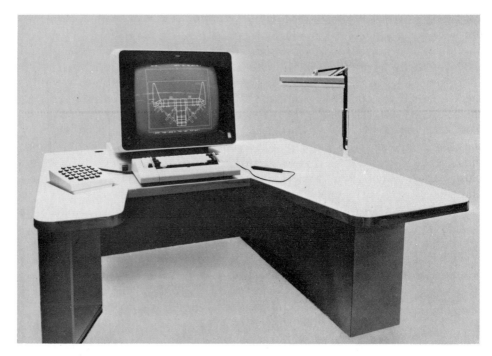

FIGURE 7.8. A modern computer graphics system. (Courtesy Texas Instruments, Inc.)

exactly vertical or horizontal or rough circles shall be made into perfect circles. In this manner the computer can assist in converting a rough freehand sketch into an accurate diagram. In a sense, the graphics software allows the computer to "understand" the sketch, to perform an analysis of it, and to display the results. The detailed dimensions of the sketch can be expanded or rotated. The computer can display alternative views of the object. Or, if the object consists of several parts, the computer can be used to "explode" the drawing and focus on a given component.

The primary emphasis of computer graphics in recent years has been on developing the ability to handle three-dimensional graphics. Ideally one would like to store certain primitive shapes in computer memory, for example, spheres, rectangular blocks, cones, or donut-shaped toroids. The user could then assemble these primitive shapes together by manipulating them on the graphics terminal display screen. Once the appropriate shape has been obtained, the computer can then be asked to provide a geometrical analysis, including quantities such as dimensions, volumes, and weights.

Example Complex three-dimensional shapes can be generated using computer graphics to deform or modify simple shapes such as cubes, cylinders, and spheres.

Examples of the deformation of a cube into various shapes are shown in Figure 7.9.

Interactive computer graphics allows the user to converse directly and immediately with the computer at the most appropriate rate. The responsiveness of the computer allows the engineer to test ideas, concepts, while they are still fresh. Computer graphics facilitates working with the most natural language of design: the language of pictures and images. Engineers can manipulate full three-dimensional images, viewing their effort dynamically and from a perspective that closely mirrors the actual physical situation. These capabilities for visualization and animation are particularly important features of CAD systems.

SUMMARY

The development of CRT displays and time-sharing or microcomputer hardware, coupled with software developments has led to powerful interactive computer graphics systems. The engineer can work interactively with the computer, sketching, analyzing, and modifying design concepts through the use of visual images.

Example   The design of a complex structure such as a power plant requires hundreds of construction drawings and costly scale models. Computer graphics promises to facilitate this design by displaying complex three-dimensional shapes. In Figure 7.10 we show the primitive graphical buildup of a concrete shell for the wall of a nuclear power plant containment structure. In the first diagram, all lines are shown as the shell is "assembled" on the computer, using polyhedrons of various sizes and shapes. A software program can then edit this diagram, searching out interference problems between construction elements and removing hidden lines to achieve the section diagrams shown in Figure 7.10.

Example   Computer graphics can also prove of importance in visualizing architectural design. In Figure 7.11 we have shown a sequence of CRT display images, detailing the design and analysis of an engineering laboratory for the University of Michigan campus.

SAMPLE MODIFICATIONS ON A CUBE

FIGURE 7.9. An example of how elementary shapes can be modified using computer graphics software. (Courtesy University of Michigan College of Engineering)

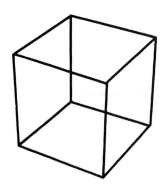

B

OR CREATING
TWO VERTICES
ON A FACE TO
CREATE A
NEW EDGE
AND . . .

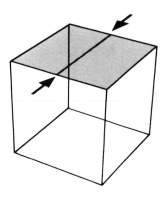

"Split" Face

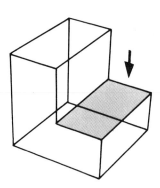

"Push" Right Half
of Split Face Down

C

OR CREATING
A SUBFACE
AND . . .

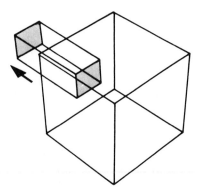

"Pulling" It Out from the Cube

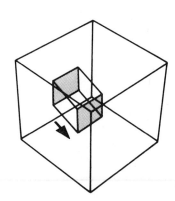

"Pushing" It into the Cube

A

SHAPES CAN BE CHANGED BY....

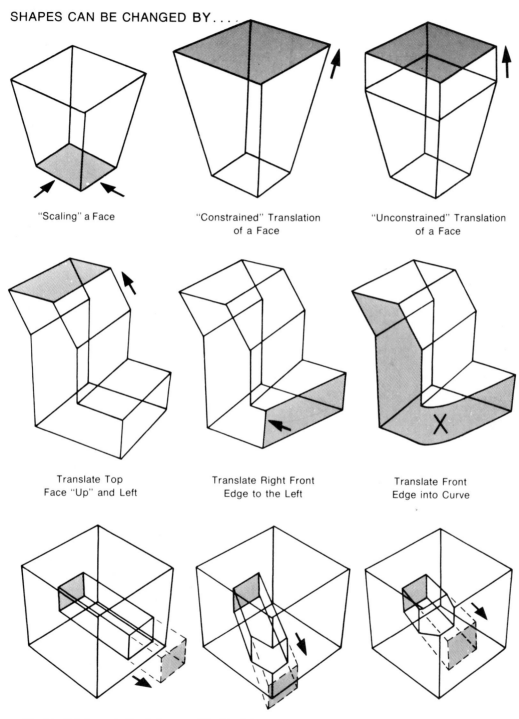

"Scaling" a Face

"Constrained" Translation of a Face

"Unconstrained" Translation of a Face

Translate Top Face "Up" and Left

Translate Right Front Edge to the Left

Translate Front Edge into Curve

"Pushing" It through the Cube

"Pushing" It through an Edge

"Pushing" It through a Vertex

NORTHWEST QUADRANT
OF CONCRETE SHELL
FOR NUCLEAR POWER PLANT

EXTRACTED
SECTION

SECTION
EXTRACTED

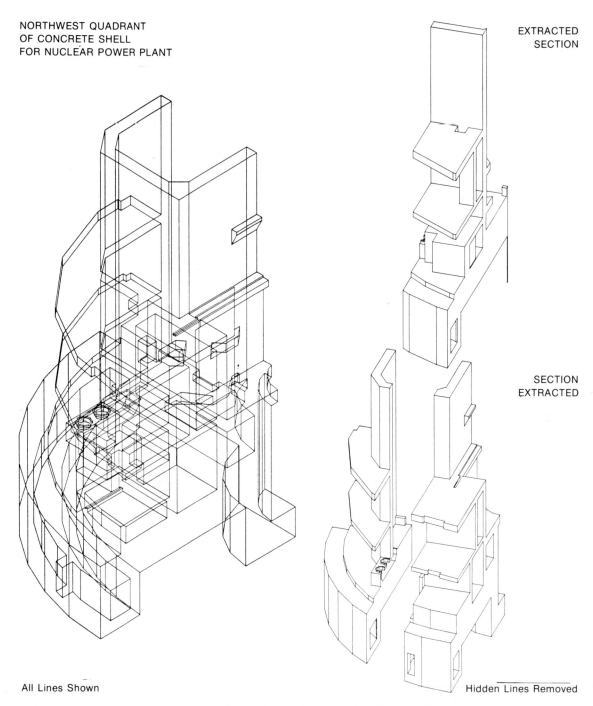

All Lines Shown

Hidden Lines Removed

FIGURE 7.10. Computer graphics can be used to analyze complex shapes such as this nuclear power plant containment wall. (Courtesy University of Michigan College of Engineering)

Exercises

Exercises marked with an asterisk (*) require some degree of programming knowledge.

Computer Graphics

* **1.** Write a computer graphics program that will draw one of the following shapes:
  - (a) A square
  - (b) A triangle
  - (c) A five-pointed star
  - (d) A series of concentric squares of progressively larger (or smaller) size

* **2.** Write a program that will draw a square, then rotate it by a specified number of degrees (entered as input) and redraw it. Continue in this fashion until instructed to stop.

* **3.** Write a graphics program that will permit a user to define and edit (modify) arbitrary polygon shapes. Desirable commands for the program are:
  - (a) ERASE: Initialize the screen by erasing it.
  - (b) SHAPE: Ask the user to input coordinates for each vertex, defaulting to an equilateral polygon centered in the screen.
  - (c) COLOR: Color of polygon lines.
  - (d) SHAPE: Shade in the form of a polygon.
  - (e) MODIFY: Add or delete vertices (and hence sides) of polygon.

* **4.** Write a program that will graph any function over some specified interval.

* **5.** Use the random number generator (e.g., RND in BASIC) to create a "kaleidoscope" of ever-changing color patterns.

* **6.** Write an "etch a sketch" program in which user-controlled ADC input (analog-to-digital converters, or in more common terms, "game paddles") are used to move a tracing point about the screen to draw pictures.

 7. Define the following computer graphics terms:

Pixel	Window
Display buffer	Scroll
Refresh graphics	TURTLEGRAPHICS

 8. Explain how a three-dimensional computer graphics display might be used to represent a four-dimensional object.

* **9.** Write a "bouncing ball" program in which animated graphics is used to display a moving object, for example, a handball game in which the ADC input signal (game paddle) is used to control a paddle that will bounce a ball off a wall (a la ping-pong).

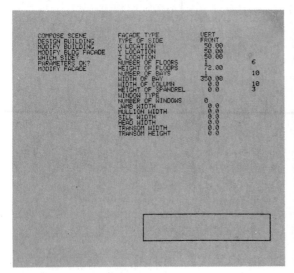

1 Modified parameters for the facade details

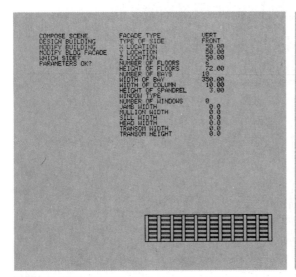

2 Parameters satisfactory for facade

Modify placement

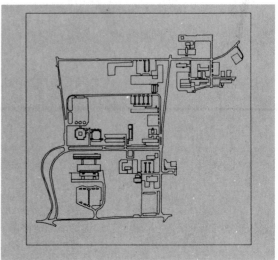

3 Reference map drawn

Indicate position or manipulate map to show area of interest

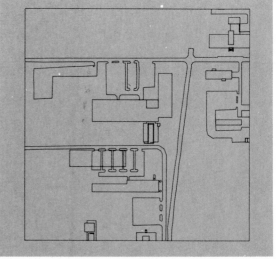

4 Reference map satisfactory

Indicate location of building

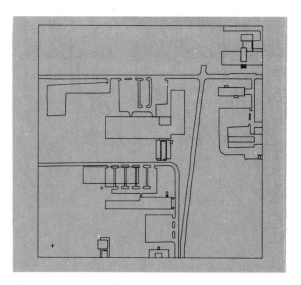

5 End of a sequence of steps

Building modification finished

Program remains open for later modification

Direction and viewpoint indicated for generation of perspective view with hidden lines removed

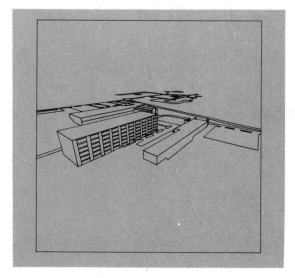

6 Perspective view

FIGURE 7.11. Computer graphics are particularly useful in visualizing new designs. (Courtesy University of Michigan College of Engineering)

10. The use of computer graphics has had a major impact on the design of operator consoles or control panels. Briefly describe the type of graphics display that might prove useful in the following situations:
    (a) Process flow control in a chemical factory
    (b) Pilot instrument panel on the space shuttle
    (c) Nuclear power plant control panel
    (d) Instrument panel for an automobile

11. Explain how computer graphics has been used to produce the special effects in popular science fiction movies (e.g., *Star Wars, Alien, The Black Hole,* or *Startrek*).

7.4 FINITE ELEMENT METHODS

An important aspect of engineering analysis involves taking a complex system, say the wing of an airplane or the piping system for a chemical processing plant, and introducing approximation after approximation until the engineer eventually arrives at a model sufficiently simple for mathematical analysis. This activity sometimes gets carried to an extreme in undergraduate engineering courses. In strength of materials courses, complex beams or machine parts are represented as slender, uniform columns under simple loads. In fluid mechanics courses, the fluids are assumed to flow in uniform pipes or channels. Heat transfer courses usually consider thermal conduction in slabs of infinite width and uniform composition. Electrical circuits are assumed to consist of lumped circuit elements connected by perfectly conducting wires. In course after course, the undergraduate engineer is equipped with an arsenal of analytical tools to handle such simplified, idealized problems.

Hence the new engineering graduate sometimes has a rude awakening when first confronting the complex systems that must be analyzed in engineering practice. Imagine the complexity of performing a stress analysis of an automobile

FIGURE 7.12. Simple models such as those used to describe forces on a beam, fluid flow, heat conduction, and electrical circuits are familiar tools in undergraduate engineering courses.

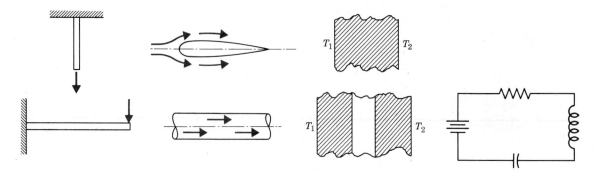

engine or determining the dynamic response of a skyscraper to an earthquake. Of course the initial response is usually to flee for the nearest computer terminal. That is, the engineer is first tempted to write down all of the mathematical equations that describe a complex system and then attempt to solve these equations on a computer.

This task is easier said than done. It takes a considerable amount of effort to cast the relevant equations into a form appropriate for computer solution. If the equations contain derivatives (as they usually do in engineering applications), the engineer must somehow "discretize" the equations by replacing the derivatives with approximate difference formulas. This particular approach is known as the finite difference method. It, and its host discipline, numerical analysis, have provided the basis for the use of computers in engineering analysis for many years. However even finite difference methods are frequently overwhelmed by the complexity of many engineering problems, by the complicated geometrical shapes and the nonuniform composition of engineering systems.

During the 1950s structural engineers developed a quite different approach

FIGURE 7.13. Modern finite element methods can be used to analyze complex systems such as aircraft. (Courtesy Lockheed California Company)

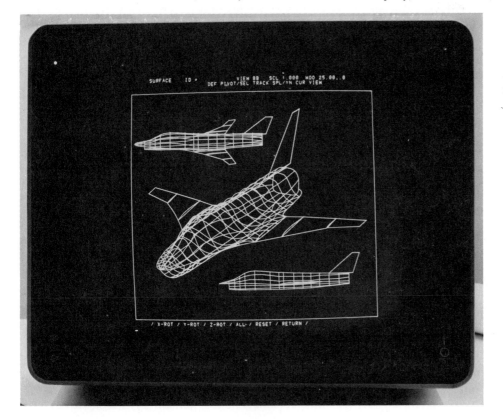

known as the *finite element method*. This approach returns to the simple, idealized models introduced in undergraduate engineering courses. The general method proceeds by noting that while the total system may be very complex, on a fine enough scale, small parts of the system may behave very similarly to idealized models. For example, although the stresses on an automobile are very complex, if we decompose the automobile body into small panels, so-called "finite elements," we can characterize the forces on these panels using simple models. The trick then is to use the computer to reconstruct the analysis of a complex system from idealized models of its component parts, its finite elements.

Finite element methods are ideally suited to computer implementation. Virtually any complex system can be broken into tiny elements. Then each of the elements can be analyzed and reassembled back into the original system. Although finite element methods were first developed and implemented in the area of structural mechanics, their application has since spread to problems in heat transfer, fluid flow, electromagnetism, and many other areas of engineering analysis. You can always spot a finite element analysis by its characteristic art form in which a complex geometrical shape is represented by a grid of small elements (Figure 7.14).

Although finite element methods were first developed as an ad hoc approach to breaking down the complexity of problems in structural mechanics, mathematicians have now provided a powerful mathematical framework justifying why the method works and how it can be modified and improved. Applications of finite element methods have become so widespread that there are several general purpose computer programs in use in engineering with the capability to analyze problems in structural, mechanical, electrical, or chemical engineering. In fact, several finite element computer programs have been integrated into CAD/CAM systems to provide the analysis of engineering design in an interactive mode.

Example   The most popular finite element program for general interactive use (CAD/CAM) is the NASTRAN code. This program, currently used by government, industry, and several universities, was developed under the sponsorship of NASA at a cost of several million dollars. NASTRAN can solve engineering problems in the areas of acoustics, electromagnetism, static and dynamic structural analysis, heat transfer, aerodynamics, hydroelasticity, and other fields of engineering analysis. It is capable of being interfaced with complex computer graphics and data processing codes so that it can be integrated into a total CAD/CAM package.

More specialized computer programs have been developed for use in CAD/CAM applications. For example, programs have been written to analyze the dynamics of general mechanical systems under various constraints. These programs follow a philosophy similar to that of the finite element approach by breaking the system up into a number of nodes, and then solving the equations that describe the dynamics and interactions of these nodes.

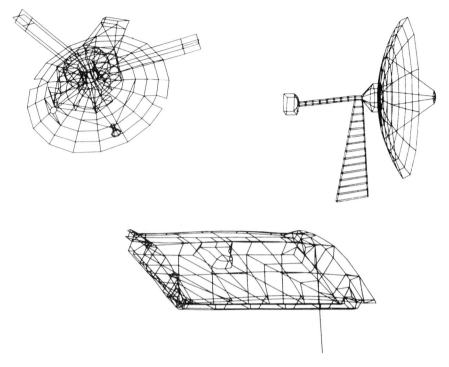

FIGURE 7.14.   One can always identify the characteristic artwork of a finite element analysis. (Courtesy Structural Dynamics Research Corporation)

Example   As an illustration of how the dynamics of a complex mechanical system might be analyzed, Figure 7.15 represents and analyzes a car suspension assembly placed on a road simulation device.

SUMMARY

To analyze the complex systems encountered in practice, engineers have turned to a powerful tool known as the finite element method. This scheme breaks up the system into many small components or elements, each of which can be adequately described by simple models. A computer is then used to reconstruct the analysis of the original system according to the interaction of its component parts. Finite element programs are frequently included as part of a CAD/CAM system to perform structural and thermal analysis or the analysis of fluid or electromagnetic phenomena.

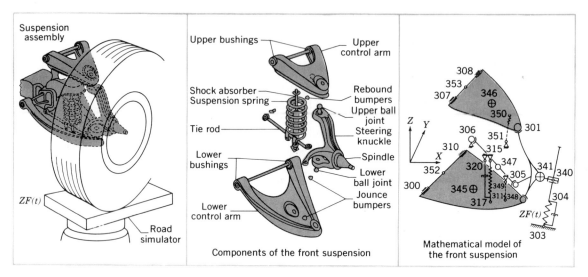

FIGURE 7.15.   Software packages can be developed to model the complex dynamics of systems such as an automobile suspension system. (Courtesy M. Chace, University of Michigan College of Engineering)

7.5   FUTURE DEVELOPMENTS

The past decade has seen a rapid growth in the capability and application of CAD/CAM systems. What was once simply a drafting tool based on computer graphics has now become an important tool in engineering design and analysis. CAD/CAM methods are also being extended into manufacturing and production. There is little doubt that they will play an increasingly important role in all phases of engineering practice. With the rapid strides made in computer hardware development, including the explosive growth in microelectronics, and with new software developments such as simpler programming languages and the use of artificial intelligence concepts to improve engineer-computer interaction, the potential of CAD/CAM as a tool in engineering is almost limitless.

Yet there is some reason for concern. Our past experience with the use of computers as a tool in engineering analysis suggests that there is a danger that the computer will begin to dictate features of the design. That is, the particular computer system hardware or software will structure or constrain the engineer to think along certain lines more compatible with system capabilities. It is important that computer-assisted engineering systems be developed in a manner that imposes the least constraint on the engineer and in a way that allows maximum creativity and imagination.

A second concern arises from the increasing sophistication of these systems. As the methods of computer graphics and engineering analysis become more sophisticated and complex, they also become further removed from the knowledge and experience of the engineer who will use them. Indeed, there is already an alarming tendency for both engineering students and practicing engineers to approach CAD/CAM systems as black boxes, with insufficient knowledge of the limitations of the analytical methods or assumptions built into the software on which they rely. As with any computer application, a healthy skepticism should always be nurtured by the user.

Of course, computer-aided engineering also offers a new tool of remarkable power to the engineer. It provides instant access to vast data bases. It can use unique methods of computer graphics to aid in visualizing and representing complex problems. It allows the engineer to explore different approaches both rapidly and inexpensively. The computer can be used to simulate various aspects of the performance of a design. The interactive nature of computer-aided engineering allows the engineer to correct and modify the design and finally to prepare it in a form most suitable for communication and implementation. Computer-aided engineering has provided engineers with a powerful level to increase their analytical capabilities and their creativity.

Computer-aided engineering may have an even more dramatic impact on the engineering profession. For the past several decades there has been a definite trend toward specialization in engineering. In the manufacturing industry, for example, one rarely encounters simple classifications such as mechanical engineers, electrical engineers, or civil engineers. Rather, we find design engineers, analysis engineers, stress engineers, tool design engineers, process engineers, manufacturing engineers, and product engineers. Interacting with these engineers are drafting and crafts specialists, machine tool operators, and so forth.

However the introduction of CAD/CAM, or in its more general sense, computer-aided engineering, has reversed this trend. These systems have allowed a single engineer to perform the work of draftsman, design engineer, analysis engineer, and manufacturing engineer. In fact, the divisions of responsibility associated with these separate titles are rapidly blurring, as computer-aided engineering systems continue to extend the capability of the engineer.

The age of the engineer as a *specialist* may have reached its zenith and may now be starting to recede. If so, the development and implementation of computer-aided engineering may once again reestablish the importance of the engineer as a *generalist*. In so doing, ironically enough, the computer will lead to a reemphasis of the importance of a broad and liberal education for students intending to enter the profession of engineering.

SUMMARY

CAD/CAM methods will play an increasingly important role in all phases of engineering design and analysis with the rapid evolution of computer hardware and software. CAD/CAM systems must be developed to impose minimum constraints on the creativity of the engineer. Furthermore, it is important that the engineer avoid the tendency to approach CAD/CAM as a black box and, instead, attempt to understand the methods used in these systems. Computer-aided engineering not only provides engineers with a powerful tool to increase their capability, but it also reintegrates the various functions of the engineer, including design, analysis, and production. As computer-aided engineering extends the capability of the engineer, it may reestablish the importance of the engineer as a generalist.

Exercises

Future Developments

1. How might you expect artificial intelligence concepts to impact upon the design of CAD/CAM systems?

2. How would you redesign your college program to stress a general engineering education rather than a specialist major? What would be the advantages and disadvantages of doing so?

PART IV
THE CONSTRAINTS

We have identified engineering as a profession of problem solving, concerned with the application of science and technology to the needs of society. But the activities of the engineer cannot be restricted to technical considerations alone. The problems of society are too complex, too interwoven with social, economic, and political issues to yield merely to the tools of science and mathematics. Engineers should master more than just technical subjects such as physics, chemistry, mechanics, or thermodynamics. They should develop a broader perspective that will allow them to anticipate and analyze the impact of their activities on society. In so doing, they should recognize that their profession will always be subject to the constraints arising from the interaction of engineering with society and its institutions.

Many of these constraints on engineering practice involve economic matters. Engineering activities are frequently dominated by considerations of cost and potential return. In fact, a concern for economic factors might well be identified as one of the characteristics that sets engineering apart from the pure sciences. All engineers, whether involved in research and development, design and production, or management and sales, require some background in the fundamental principles of economics.

As engineers advance in their profession, they may find that human relations take on greater importance in their activities. Engineers are frequently called upon to assume executive roles in management. Furthermore, they should be prepared to interact with other professionals such as accountants, lawyers, business executives, and government administrators. These interactions will usually place constraints on engineering practice.

Engineering affects not only the individual member of society. It also interacts with and affects social organizations and institutions such as government and commerce. As these interactions with societal institutions have increased, new constraints have been placed on engineering practice such as government regulation, environmental impact and public safety legislation, and technology assessment.

In this final part of the text we shall examine the nontechnical constraints that affect the activities of the engineer. We first consider engineering economics. Then we discuss the constraints arising from the interaction of engineering with society, with people and their institutions.

CHAPTER 8

Engineering Economics

The practice of engineering, in addition to employing the principles of science and mathematics, often also requires the application of some basic concepts from the field of economics. Engineers in management roles quickly learn that many of their day-to-day decisions are dominated by considerations of cost and potential return. The need for a background in economic principles is most obvious for engineers employed in commerce and industry. But even engineers less directly concerned with commercial enterprise must be aware of some fundamental principles of economics if they are to make intelligent choices in planning research and development, devising testing programs or carrying out many of the other varied activities that arise in engineering practice.

A concern for economic considerations could be regarded as one of the factors that sets engineering apart from pure science. In the design of a product or the choice of a particular process, considerations of cost are often placed on an equal footing with principles from scientific fields such as thermodynamics or mechanics. History has shown us many examples of ideas that were clever scientific developments but failed to achieve practical applications because of economic factors.

Example   The dream of the alchemist was to change base metal into gold. With the discovery of nuclear reactions, scientists of this century have provided one method of carrying out this long sought-after transmutation. High-powered particle accelerators are used to hurl ions of carbon and neon at a foil target of bismuth, a metal almost as heavy as lead. The incident ions knock off fragments of the bismuth nuclei, leaving behind the lighter element gold.

But the alchemist's dream remains unfulfilled as a practical matter because of economics. The cost of building and operating the massive accelerator needed to produce the ion beam is far greater than the value of the small amount of gold that can be produced. In fact, it costs roughly $10,000 in accelerator operating expenses to produce approximately one million atoms of gold (worth less than one billionth of one cent). Hence, while the nuclear transmutation process has been demonstrated scientifically, economic factors have prevented any real exploitation.

Some principles of economics amount to just plain common sense. If we have to choose between two suppliers of a raw material, with all other factors being equal, we shall choose the supplier that offers us the lowest price. If it will cost more to manufacture an item than we can reasonably expect to sell it for, there is no economic reason to continue its production. However, most economic decisions are not so clear-cut because of complicating factors that must be carefully weighed and analyzed. Very often the responsibility for a preliminary economic analysis of a new project may be assigned to a managing engineer. The engineer may be required to balance various factors based on interest rates, costs, depreciation, and so on in order to recommend a proper course of action. It is the

FIGURE 8.1.   The transmutation of bismuth into gold is limited by economic considerations. (© 1980 by S. Harris. Reprinted with permission of *Science 80*)

*"Unfortunately this lab is funded
only by as much gold as we can make from bismuth."*

purpose of this chapter to outline some of these basic ideas of economics and to show how they often come into play in engineering decision making.

SUMMARY

The principles of economics underlie many decisions made in engineering. Before a scientific development can be exploited, it must be economically feasible. Managing engineers are often responsible for a preliminary economic analysis of proposed new projects.

8.1 RATE OF RETURN

One of the most important concepts of economics is that of the *rate of return* on an investment. This is defined as

$$\text{rate of return} = \frac{\text{annual profit}}{\text{investment}}$$

In the private sector of our economy, prudent investments by individuals or corporations are expected to result in a significant rate of return. As a baseline for comparison, we know that simply investing funds in a savings account will yield a rate of return determined by the interest rate paid by that bank. There is very little risk in this investment; the rate of return is virtually guaranteed.

On the other hand, many investments by private industry in production facilities or other corporate ventures involve a substantial degree of risk. The product might not sell in the volume anticipated, the facilities might develop unexpected maintenance problems, or some aspect of the cost analysis might be in error. In order to compensate for this risk, most corporate investments are made in expectation of a rate of return that is substantially larger than the interest rate from a bank. This assumption of risk by private investment is a necessary element in the current economic structure of the United States.

Example A factory produces 35,000 can openers per month that can be sold at a net profit of $2.00 per unit. If the investment in the production facilities, labor, and required inventory totals $4,000,000 find the rate of return on this investment.

Solution

The annual profit can be calculated as

$$\frac{\text{annual}}{\text{profit}} = (\$2/\text{unit})\,(35{,}000\text{ units/month})\,(12\text{ months/year})$$

$$= \$840{,}000$$

Therefore the rate of return is given by:

$$\text{rate of return} = \frac{\$840{,}000}{\$4{,}000{,}000} = 21\%$$

Complications can arise even in so simple a concept as rate of return. For example, this rate is seldom a constant from year to year. It will usually vary, depending on the circumstances. The rate of return may even be negative during the early years of a project, when startup costs are high. A corporation may still embark on such a venture, however, in the expectation that lower costs in future years will result in an *average* rate of return that will be attractive. In making such projections into the future, it is necessary to recognize that the value of a given monetary investment depends on the time at which it is made.

SUMMARY

The rate of return is defined as the annual profit from an activity expressed as a percentage of the total investment. Investments are made in expectation of a favorable rate of return. The element of risk is usually present in private investment.

Exercises

Rate of Return

1. Assume that an investment of $150 million is required to construct an off-shore oil drilling platform. If this facility is expected to produce one million barrels of oil per year at a production cost of $10 per barrel, find the rate of return on this investment using today's oil prices.

2. Assume that 10% is a fair rate of return on an investment in housing.

If an investor builds a ten-unit apartment house for $1 million, and the annual operating cost for taxes, utilities, depreciation, etc. amount to $50,000 per year, what should the landlord charge for monthly rental rate on each apartment?

8.2   THE TIME VALUE OF MONEY

One of the basic tenets of a capitalistic economy is the assumption that accumulated capital or money has intrinsic worth. As a result, individuals or organizations that wish to use capital accumulated by others must pay for the privilege of doing so. Conversely, those who have accumulated this wealth can expect a reward for allowing their money to be temporarily used by others. Although great philosophical debates have been carried out over the centuries regarding the moral justifications for these assumptions, they underlie the economic systems on which all Western democracies currently operate.

This assumption leads directly to the concept of the *time value of money*. A dollar of accumulated capital no longer has constant worth, but it can be expected to increase in value because of the income it generates. If this income is allowed to accumulate, each dollar of capital on hand today will become two dollars at some future date. In evaluating investment or in carrying out economic analysis of various kinds, we must take into account this change in the dollar's value.

The arguments that follow all assume that the purchasing power of the dollar is constant. As we know all too painfully, the pressures of inflation have substantially reduced the purchasing power of the dollar over recent years. We shall address the effects of inflation in a later section. Here we confine our discussion to the ideal circumstance in which the purchasing power of the dollar remains constant.

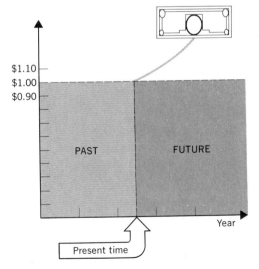

FIGURE 8.2.   The time-dependent value of one dollar received today. The slope of this plot depends on the prevailing interest rate.

SUMMARY

Accumulated capital or money has intrinsic worth. Hence those who have accumulated this capital can expect a reward for allowing its temporary use by others. This leads to the concept of the time value of money, since accumulated capital can be expected to increase in value with time because of the income it can generate.

8.3 INTEREST

8.3.1. Simple Interest

The most direct illustration of the time value of money arises in connection with interest rates. In return for the temporary use of a given sum of money, a borrower must pay to the lender a prearranged fee or *interest*. The amount borrowed is generally called the *principal*. We shall represent it by the symbol P.

The most basic method of computing interest is to assess a fixed percentage of the initial principal per unit time. This is the basis of *simple interest*. The total interest accumulated, I, is calculated from the interest rate i and the number of time units n over which the money is held as follows:

$$I = Pni$$

The unit of time used in quoting interest percentage is almost always taken as one year.

Example A principal of \$10,000 held over two years at an annual interest rate of 12% will yield a total interest charge of \$2400 as shown by the following calculation:

$$I = (\$10{,}000)\,(2)\,(0.12) = \$2400$$

In simple interest calculations, the amount of interest earned per unit time is constant. If we consider the sum of the initial principal and the accumulated interest, this total value is given by $P(1 + ni)$. Its value increases linearly with time. An example is shown in the curve in Figure 8.3.

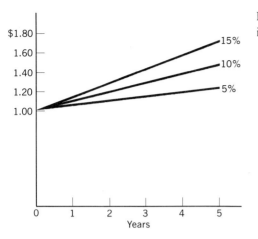

FIGURE 8.3.   Income growth due to various simple interest rates.

8.3.2.   Compound Interest

In the above case of simple interest, the interest payment becomes due and is paid at the end of the prearranged loan period regardless of its length. Until that time, the base principal upon which the interest is calculated is assumed to remain constant.

An alternative method of computing interest is based on the assumption that the loan period is broken up into smaller time intervals. For example, a five-year loan could be assumed to be made up of five individual one-year periods. In calculating *compound interest*, it is assumed that the cumulative interest is paid at the end of each of these individual periods, and this amount is then added to the value of the original principal. The amount of interest accumulated in each individual period is calculated as simple interest over that period, but the principal upon which that interest is calculated increases from period to period. The length of each interest accumulation period must be specified in the terms of the original loan and is known as the *compounding period*. A loan might be compounded annually, semi-annually, monthly, or even weekly.

Example   Assume we have a five-year loan with original principal of $10,000 at an interest rate of 10% compounded annually. At the end of the first year, the interest accumulated is

$$I = Pni = (\$10,000)\,(1)\,(0.10) = \$1000$$

This interest is now assumed to be added to the original principal, bringing it to a total of $11,000. During the second year of the loan, the interest accumulated is now

$$I = Pni = (\$11,000)\,(1)\,(0.10) = \$1100$$

Again this is added to the accumulated principal, bringing the total to $12,100. The amount of interest earned in each successive year is greater than the interest earned in the previous year because of the increasing base upon which it is calculated. The total value of the investment therefore grows at an ever-increasing rate.

In general, let us represent the original principal by P. We shall now let i represent the interest rate per compounding period. For example, in a loan at 10% annual interest compounded semi-annually, i will be one-half of 0.10 or 0.05. During the first period, the interest earned by the principal P is therefore

$$I = Pi$$

The worth of the investment (original principal plus earned interest) at the end of the first period is therefore

$$\text{Cumulative worth} = P + Pi = P(1 + i)$$

This is now the new principal upon which interest is earned during the second period. That interest is given by

$$I = \underbrace{P(1 + i)}\, i$$

New
principal

which leads to a total value at the end of the second period of

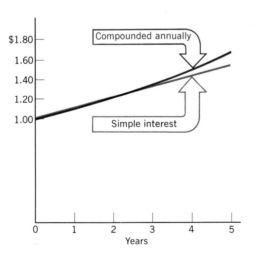

FIGURE 8.4 Investment growth due to a 10% interest rate, calculated both as simple and compound interest.

$$\text{Cumulative worth} = P\,(1 + i) + P\,(1 + i)\,i$$
$$= P\,(1 + i)^2$$

If we carry on similar arguments for further time periods, we find that the value at the end of n periods is

$$\text{Cumulative worth} = P\,(1 + i)^n$$

In Figure 8.5 the cumulative worth of loans made with the same interest rates but different compounding periods are compared. In general, the shorter the compounding period, the faster will the value of the investment grow.

Example   Find the cumulative worth of an initial $5000 investment after 8 years of 10% interest compounded annually.

Solution

We can apply our formula for the cumulative worth after n periods:

$$\text{Cumulative worth} = P\,(1 + i)^n$$

FIGURE 8.5.   Income growth due to 10% compound interest with different compounding periods.

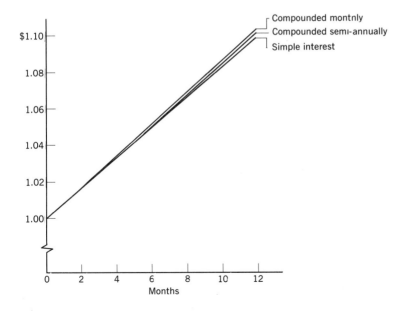

For this problem we set $i = 0.10$ and $n = 8$. Then we find

$$\text{cumulative worth} = \$5{,}000\,(1.00 + 0.10)^8 = \$10{,}718$$

Example   Find the cumulative worth after 5 years of an initial $1000 investment made at an annual interest rate of 9% compounded monthly.

Solution

In this case the compounding period is no longer equal to one year, and we must first calculate the interest per compounding period. Since the annual rate is 9%, the monthly rate is given by

$$\frac{0.09}{12} = 0.0075 \text{ per month}$$

The number of compounding periods is the number of months in 5 years or $5 \times 12 = 60$ months. Therefore we find

$$\text{cumulative worth} = \$1000\,(1.0000 + 0.0075)^{60} = \$1566$$

One might imagine taking this process to the extreme by compounding the interest every day or perhaps even every second. As we divide the time into smaller periods, the interest is compounded more frequently but at a smaller rate per period. Although it is true that the total worth (principal plus interest) of an investment at the end of a fixed investment period is maximized by using the smallest possible compounding period, very small differences will be observed once the compounding period is less than about one day. Nonetheless, it is instructive to imagine subdividing the investment period into an infinite number of periods with infinitesimal length, with the interest compounded after each such interval. This condition is known as *continuous compounding*, and it represents a limit or asymptote beyond which no compounded investment can grow regardless of how fine the compounding period.

In this extreme, imagine that we invest an initial principal P at an annual interest rate r. Let us examine the worth of this investment after a one-year investment period, assuming that one year is subdivided into n compounding periods. The interest over each of these periods is therefore given by r/n. From our earlier formula for compound interest, we find the worth of the investment at the end of one year is given by

$$\begin{matrix}\text{cumulative worth} \\ \text{after one year}\end{matrix} = P\left(1 + \frac{r}{n}\right)^n$$

For continuous compounding, we must now let n approach infinity with the following results:

$$\begin{array}{c} \text{cumulative worth} \\ \text{after one year} \end{array} = \lim_{n \to \infty} P \left(1 + \frac{r}{n} \right)^n = Pe^r$$

It should come as no surprise to see the exponential function arise in this application. The exponential function describes any quantity that grows in direct proportion to the amount already present. The growth rate of an investment when compounded continuously is exactly proportional to the cumulative value of that investment up to that time.

To generalize the expression above to periods other than one year, we need only observe that r represented the interest rate over the full period in the derivation. For a period of T years, the equivalent rate is rT, where r is the annual interest rate. Hence for more general interest periods, we find

$$\begin{array}{c} \text{cumulative worth} \\ \text{after } T \text{ years} \end{array} = Pe^{rT}$$

An example of continuous compounding is given in Figure 8.6

Example   In 1626 the Governor of the Dutch settlement in America, New Amsterdam, bought Manhattan Island from the Indians for beads and trinkets worth $24 (or so the story goes). If that same sum had been invested in mutual

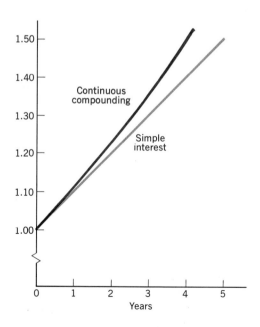

FIGURE 8.6. Cumulative worth resulting from a 10% annual interest rate, calculated both as simple and continuously compounded interest.

funds at 8% annual interest compounded continuously, what would its value have been in 1980?

Solution

We can identify $P = \$24$, $r = 0.08$, and $T = 1980 - 1626 = 354$ in our formula above. Therefore we can calculate

$$\begin{array}{l}\text{cumulative} \\ \text{worth}\end{array} = Pe^{rT} = \$24e^{(0.08)(354)}$$

$$= \$1.99 \times 10^{12} \cong 2 \text{ trillion dollars!}$$

This example shows the power of the exponential function and the surprising accumulation of worth of an investment when extended over long periods of time.

8.3.3.   Effective Annual Interest Rate

It has become commonplace to quote an effective annual rate of simple interest or effective annual yield of an investment to take into account the various possible compounding periods. For example, a principal of $10,000 invested at 10% interest compounded semi-annually will, from our formula, grow to a value of $11,025 at the end of one year. The equivalent simple interest rate of the one-year period is then 10.25%. The same $10,000 principal, if compounded continuously at 10% interest, would accumulate to a value of $11,052 at the end of one year. The equivalent annual interest rate in this latter case would therefore be 10.52%.

SUMMARY

Interest is the fee paid by a borrower to a lender for the temporary use of a sum of money. Simple interest is computed as a fixed percentage of the loan amount or principal and paid at the end of the loan period. In compound interest the loan period is divided into a number of compounding periods. The interest earned during each compounding period is added to the principal for purposes of calculating the interest during the next compounding period. In the limit that the compounding period is made very small, the interest is said to be compounded continuously. For any fixed interest rate, continuous compounding results in maximum rate of growth. Sometimes an effective annual interest rate is used to take into account compounding.

Exercises

Simple Interest

1. What is the total interest earned on a principal of $10,000 invested at 7.5% annual interest over a period of three years?

2. An *endowment* is an investment from which the annual interest is used to fund a specific purpose. If a student scholarship requires $5000 per year, how large an endowment will be required to fund such a scholarship if the investment can be made at a simple annual interest rate of 11%?

3. Find the interest due on a principal of $8000 if held for a period of three years and seven months at a simple interest rate of 8.5%.

Compound Interest

4. A corporation needs to borrow a sum of $100,000 for a period of five years. If the prevailing interest rate is 9% compounded annually, what amount must be repaid at the end of the five years?

5. How much interest is paid on a loan of $5000 at an annual interest rate of 15% compounded weekly over a period of four years?

6. An investment is made at an annual interest rate of 10% compounded weekly over a period of three years. Find the interest rate that would yield the same total interest if compounded on an annual basis.

7. At what interest rate, compounded semi-annually will an invested principal double in value in exactly five years?

8. A father wishes to establish a trust fund that will pay his daughter $10,000 on her twenty-first birthday. If the prevailing interest rate is 9% compounded annually, and his daughter is now observing her twelfth birthday, how much principal must be invested now to meet that objective?

Continuous Compounding

9. Calculate the interest due on a loan of $10,000 after 2.5 years of continuously compounded interest at an annual rate of 11%.

10. Compare the interest earned by $1000 over a one-year period compounded continuously or compounded on a weekly basis.

11. The *doubling time* for an investment is the period of time over which the cumulative worth reaches twice the original investment. For continuous compounding of interest, derive an expression for the doubling time T in terms of the annual interest rate r.

8.4.   PRESENT WORTH

Because money can earn interest, the time at which a payment is made has an influence on the value of that payment. For example, suppose the prevailing interest rate is 10% calculated as simple annual interest. If you receive a payment of $1000 today, that payment has greater value than a similar $1000 payment made a year from now because of the interest that could be earned over the year's time. For the case of our example, the $1000 will have grown to $1100 a year from now and, therefore, its value is 10% greater than the $1000 payment that was delayed a year.

Thus the true value of any payment depends on the time at which that payment is made. If we are to make a fair comparison among several payment schedules, we must pick a fixed point in time at which to calculate the effective worth of each transaction. By choosing the same reference time to evaluate all payments, the time dependence of their value is eliminated, and a fair comparison can then be made between alternative choices.

In the example above, we chose the time of a year from now to compare the worth of a $1000 received today with that of $1000 received in a year's time. In normal economic calculations, it is more conventional to choose instead the present time as the reference point. If we carry out the calculations correctly, the relative value of two different payment options will not be affected by the specific time chosen to make the comparison. Like the zero point or origin on a coordinate axis, the choice is an arbitrary one.

By choosing the present time as a reference point, the *present worth* of a payment is calculated. In the example given above, the present worth of $1000 received today is obviously also $1000. However the worth of $1000 received a year from now is less than $1000. Its present worth is, in fact, $909.09 because the interest earned on this amount over the year's time ($90.91) will increase this present worth to $1000 in a year. Thus we should be equally happy accepting a payment of $909.09 today as opposed to $1000 a year from now; $909.09 is the present worth of the deferred payment.

The present worth both for payments made in the past or the future depends on the rate and type of interest that is assumed to be in effect. The formulas developed earlier for simple and compound interest can be used directly to calculate the present worth of a *past* payment. These formulas fairly account for the gain in worth of past payments up to the present time. The time period used is obviously the length of time between the payment and the present time.

On the other hand, we must discount the value of future payments when determining their present worth. The same interest formulas can be rearranged to give the discounted value of a future payment. For example, recall that we derived the value of a payment P after n periods of compound interest at a rate i per period as

$$\text{cumulative worth} = W = P(1 + i)^n$$

By definition, W is the present worth of P, when P is a past payment. By symmetric logic, P must be the present worth of a payment W made in the future. Thus we find

$$\text{present worth of future payment } W = P = \frac{W}{(1 + i)^n}$$

Similar expressions can be derived from our earlier formulas for simple interest and continuously compounded interest:

$$\text{simple interest:} \quad P = \frac{W}{(1 + ni)}$$

$$\text{continuous compounding}: \quad P = We^{-rT}$$

When carrying out a present worth calculation or other economic analysis, a *cash flow diagram* of the type shown in Figure 8.7 is often of help. Here time is shown on the horizontal axis, with each payment indicated as a vertical arrow. An arrow pointing toward the line represents a payment received, while an arrow pointing away represents a payment made to another party. If the amounts of the payments are entered by the corresponding arrows, one can see at a glance what factors must be included in a present worth calculation.

SUMMARY

The value of a monetary payment depends on the time at which it is made. Early payments have greater value than those made later because of their added ability to earn interest. Comparisons of different payment schemes can be made by calculating their present worth.

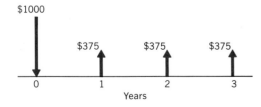

FIGURE 8.7. An example of a cash flow diagram. The case shown is for a $1000 loan to an individual that is paid back by three yearly payments of $375.

Exercises

Present Worth

1. For the following calculations, assume that the interest rate is fixed at 8% per year. In each case find the present worth of the indicated payment schedule.
   (a) A payment of $10,000 made five years ago, with interest compounded annually.
   (b) A payment of $10,000 due in three years with interest compounded continuously.
   (c) A series of $500 annual payments, beginning two years ago, and extending to eight years in the future. Assume interest is compounded annually.

2. A high school student is faced with a decision of whether to attend college for four years or instead go to work directly out of school. Examine the financial implications of this decision by calculating the present worth of both of the following two scenarios. In order to simplify the calculation, we shall ignore the somewhat compensating effects of inflation and future wage increases, and assume that wages remain constant.
   (a) The student takes a job immediately at an annual salary of $19,000. Assume the student's career extends over a period of 45 years. To simplify the calculation, assume that wage payments are received as a lump sum at the end of each year.
   (b) The student first goes to college for four years. In each of these years a tuition payment of $2000 is required but offset in part by $1000 of personal earnings. Following four years in college, the student is able to find a position at $23,000 per year. Assume the student's subsequent career will extend for a period of 41 years.

3. Find the present worth of 12 equal annuity payments of $1500 each to be paid monthly beginning one month from the present date. Assume that 8% interest rate prevails compounded semi-annually.

4. An organization wishes to establish a fund that will provide for five annual payments of $10,000 each, beginning exactly 10 years from now. Ten equal payments will be made into the fund at annual intervals, beginning with the present date. If the prevailing interest rate is assumed to be 9% compounded annually, how large must these 10 annual payments into the fund be if they are to be adequate to cover the desired payments out of the fund?

5. In 1916 the Model T Ford sold for $400. Find the present worth of a payment of that amount made on January 1, 1916, assuming a 7% interest rate compounded continuously.

6. Find the present worth of 12 monthly income tax withholding payments of $400 each, if the first payment was made 14 months ago. Assume an annual interest rate of 10% compounded semiannually.

8.5. INSTALLMENT FINANCING AND SINKING FUNDS

A popular method for financing purchases, both for individuals and corporations, involves deferred payment in the form of a number of equal future payments spaced uniformly in time. The purchaser pays a premium for the privilege of deferring payments, however, because of the principle of the time value of money discussed earlier. The financier of the purchaser, whether it is the seller or a third institution such as a bank, will prescribe the type and rate of interest it demands for this privilege. The size of the payments are then set by the following condition: the sum of the present worth of all future payments must equal the equivalent cash price of the purchase if payment were made at the present time.

Example   A $40,000 piece of equipment is to be financed by an immediate payment of $10,000 with the remaining $30,000 to be covered by three equal payments spaced at one year intervals in the future. Find the size of each of these payments, if the interest rate is to be 12% compounded annually.

Solution

From the terms of the problem, the present worth of the three future payments must sum to $30,000. If we let x represent the size of each payment, we can apply our present worth formula to write the following three present worths, represented as C_1, C_2, and C_3:

$$C_1 = \frac{x}{(1 + 0.12)^1} = 0.893x$$

$$C_2 = \frac{x}{(1 + 0.12)^2} = 0.797x$$

$$C_3 = \frac{x}{(1 + 0.12)^3} = 0.712x$$

Therefore we may write

$$0.893x + 0.797x + 0.712x = \$30,000$$

or solving

$$2.402x = \$30,000$$

$$x = \$12,491$$

A cash flow diagram for this transaction is sketched below:

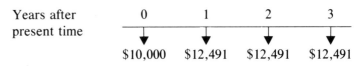

| Years after present time | 0 | 1 | 2 | 3 |
| | $\$10,000$ | $\$12,491$ | $\$12,491$ | $\$12,491$ |

Some industries use an alternative method of financing new or replacement equipment. Funds are set aside for a number of years prior to the time a purchase is planned. The accumulated worth of these funds at the time of purchase will then be sufficient to cover the cash price. This process is known as establishing a *sinking fund* for a specific future purchase.

Again, the principle is the same as in an installment purchase: the present worth of the fund at the time of purchase must equal the cash purchase price. In this case, the interest earned by each payment into the sinking fund is a positive contribution to this present value, and the corporation in effect finances the purchase on its own. This has the advantage of avoiding interest payments to outside agencies, but it also requires that money or capital be accumulated and set aside for this purpose in advance.

Example Find the cash value of a sinking fund established two years ago in which four payments of $10,000 each were made at six-month intervals. Assume that the fund accrues interest at 9% annual rate compounded continuously.

Solution

We shall represent the present worth of the four payments as W_1, W_2, W_3, and W_4. From our formula for continuously compounded interest:

$$W_1 = Pe^{rT} = \$10,000e^{(0.09)(2.0)} = \$11,972$$

Similarly,

$$W_2 = \$10,000e^{(0.09)(1.5)} = \$11,445$$

$$W_3 = \$10,000e^{(0.09)(1.0)} = \$10,942$$

$$W_4 = \$10,000e^{(0.09)(0.5)} = \$10,460$$

The current value of the sinking fund is the sum of these present worths or

$$W_1 + W_2 + W_3 + W_4 = \$44{,}819$$

SUMMARY

Capital equipment is sometimes financed through a series of payments over a period of time rather than in one lump sum. When these payments are deferred to the future, the procedure is known as an installment purchase. When regular payments are accumulated prior to the expenditure, the process is known as establishing a sinking fund.

Exercises

Installment Financing

1. An equipment purchase for $70,000 is to be financed through five equal annual payments, the first of which is to be made six months from the time of purchase. If the prevailing interest rate is 11% compounded annually, find the size of each installment payment.

2. It is anticipated that an item of equipment must be replaced in three years at a cost of $45,000. A sinking fund consisting of four equal payments spaced at one-year intervals and beginning with the present time is to be used to finance this purchase. Assuming a prevailing interest rate of 9% compounded semi-annually, find the required size of the payments to the sinking fund.

3. Five annual payments of $1,000 are made into a sinking fund starting today and repeated for the next four years. Find the total value of this fund at the end of this period assuming an 8% interest rate compounded annually.

4. A purchase of $3,000 is to be financed by monthly payments of $300 beginning one year from the date of purchase. If simple interest applies at an annual rate of 8%, find the number of full payments required in order to reduce the outstanding balance to less than $300.

5. Two purchases of $10,000 each are to be financed using a common set of installment payments. The first purchase was made 2 years ago, the second 1 year ago, and the payments are to begin today. For 24

monthly payments at a semi-annually compounded rate of 7%, find the required size of each payment.

8.6. DEPRECIATION

8.6.1. The Need for Depreciation

Engineering equipment and other physical facilities usually become less valuable with age. In some exceptional cases, such as vintage cars or wines, the true value may actually increase with time. But the normal experience in engineering facilities is a decrease in resale value with use. Some causes of this decline in value are obvious, such as wear and tear, or increased maintenance costs. Even if the facility continues to work as well as it did when it was new, it may become technologically obsolete when compared with newer equipment of the same type.

FIGURE 8.8. Depreciation.

Its resale value will therefore decline. Whatever the cause, the gradual decrease in the value of facilities or equipment is known as *depreciation*.

A recognition of the effects of depreciation is necessary when planning engineering investments. To see why, imagine that a given manufacturing process requires equipment with a cost of $600,000 and a useful life of five years. At the end of life, assume that the salvage value of the equipment is $100,000. Assume also that the labor and materials required to produce the product cost $50,000 per year. At first glance, one might think that the company could make a handsome profit by selling that same product at a price that would yield a yearly revenue of $100,000. Neglecting the time value of money and capital costs (which we shall consider later), the company will have realized $50,000 net income per year, or $250,000 at the end of five years. But at this point the equipment has reached the end of its useful life and must be replaced. This step will require a new investment of $500,000, but the accumulated profits have only amounted to half that figure. The five-year operation has really been a losing venture because of the effects of depreciation of the equipment.

The real problem in this example lies with the bookkeeping technique. The cost of replacing the equipment should have been anticipated by accounting for the depreciation of the original equipment. These depreciation costs must be included when calculating the total cost of producing a manufactured item. These costs are no less real than the labor or materials costs that must be factored into the price to be charged for the finished product.

Depreciation costs are also recognized in some aspects of taxation. Tax laws acknowledge the validity of depreciation costs and allow for their deduction to offset income derived from that equipment before taxes are paid on the resulting income. Therefore careful definition of depreciation policies used in accounting are necessary both in determining a fair market price for a product and in deriving maximum tax benefit from an investment in equipment.

8.6.2.   Methods for Calculating Depreciation

The detailed bookkeeping method used to account for depreciation is obviously of interest to accountants, but engineers must also be aware of the various options. The specific method can influence both the initial price that must be charged for a manufactured item in order to show a profit, and the tax that must be paid by the manufacturer.

The simplest method of accounting is *straight line depreciation* in which the value of a facility is assumed to decrease at a uniform rate from its initial value to its salvage value. A plot of the assumed value versus year is linear and, therefore, the method is called the straight line method in bookkeeping terminology. With this approach, the annual depreciation cost is constant over the entire useful lifetime of the facility.

Other depreciation methods are also widely used. In addition to the straight line method, some of them are recognized by the Internal Revenue Service, and a corporation or individual can generally choose the most favorable option. Most alternative methods recognize that the rate of depreciation is greatest when equipment is new and decreases as the equipment ages. Anyone who has purchased a new car knows that its drop in market value is most severe during the first year of ownership; its value decreases more slowly as the car gets older. The same is true for industrial equipment.

As an example of an accelerated depreciation scheme, the *sum-of-years-digits* method has gained widespread popularity and is recognized as an acceptable method by taxation authorities in many situations. Here the *rate* of depreciation is not held constant as in the straight line method but, rather, is allowed to decrease in a linear fashion from the start of the investment. A situation is then achieved in which the depreciation rate is maximum during the first year and is scheduled to decrease in uniform steps each year until it becomes zero at the end of the equipment's useful life. An example of the type of depreciation is shown in Figure 8.9.

In the example shown in the figure, we assume that a facility with initial cost of $100,000 has a useful life of 8 years, at which time its salvage value is $20,000. The sum-of-years-digits method is implemented as follows: during the first year we will depreciate the value by an amount that is 8 times greater than in the last year, with the *amount depreciated* decreasing uniformly for each intermediate year. The amount depreciated in the first year is thus given by $8/36$ of the total drop in value, with the denominator given by the sum $8 + 7 + 6 + \ldots + 1$. (Sum of years digits over the eight-year period.) In the second year, $7/36$ is depreciated, decreasing to $1/36$ in the eighth year.

SUMMARY

Depreciation represents the gradual decrease in the value of equipment or other facilities. Value is assumed to decrease from the initial cost to the salvage value over the useful lifetime. In any economic analysis these depreciation costs must be included along with all other costs associated with the product or services produced by that facility. Annual depreciation costs depend on the specific method of accounting chosen for each case. These can include a linear or straight line decrease in the value, which results in constant annual depreciation costs, or alternative methods that recognize accelerated depreciation in the early years of the facility lifetime.

Exercises

Depreciation

1. An item of equipment has an initial cost of $100,000, a salvage value of $15,000, and a useful life of eight years. Calculate the depreciation

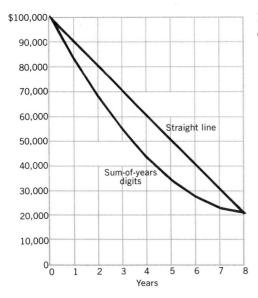

FIGURE 8.9. Straight-line versus sum-of-years-digits depreciation.

costs for each of the first three years of its life under the following assumptions:
    (a)   Straight line or linear depreciation
    (b)   Sum-of-years-digits depreciation

2. A company purchases a piece of equipment for $10,000 with an expected lifetime of five years, after which its salvage value is assumed to be $2000. Make a table of depreciated worth of the equipment at the end of each of the five years using the following schemes:
    (a)   Linear or straight line depreciation
    (b)   Sum-of-years-digit depreciation
For each scheme also show the amount of depreciated loss assumed to take place in each of the calendar years.

3. A manufacturing facility with an initial cost of $120,000 produces 80 units per day. If its useful lifetime is 10 years and its estimated salvage value is $5,000, find the depreciation cost per manufactured unit assuming linear depreciation.

4. Find the ratio of depreciation in the first year to that in the last year, using the sum-of-years-digits methods and a lifetime of 12 years.

8.7. CATEGORIES OF COST

In carrying out a fair economic analysis, the engineer must be familiar with the different types of costs that are associated with a given product or service. The different sources of costs can be sorted into several broad categories.

8.7.1. Fixed Cost

This category includes any costs that do not depend on the number of units produced or the level of activity in a given venture. For example, once a production facility has been constructed, the depreciation costs mentioned in the previous section are committed for the life of the facility. Other types of fixed costs include taxes, security, insurance, some types of regular maintenance, and an important category known as *capital costs*.

By capital costs we mean the inherent expense of owning a facility that required a capital investment to acquire. This important concept should not be confused with depreciation, since capital costs are present whether or not the value of the property is assumed to decrease as it ages. Rather, capital costs are a consequence of the time value of money. In building a plant that costs a million dollars, a company has made the decision to forgo the interest that the same million dollars could accrue if otherwise invested. Thus the million dollar plant really costs the company an annual charge equal to the deferred interest. The revenue obtained by the company from operation of this plant must be adequate to cover these costs as well as all others associated with the operation of the plant, or else the decision to build the plant would have led to an unwise investment for the company. When large and expensive facilities are needed for production, the capital costs of these facilities may dominate the remainder of the fixed costs involved in the operation of that facility.

The annual charge for the capital investment is derived from the initial investment by using one of the methods for calculating interest described earlier. The annual charge, of course, depends on the prevailing rate of interest assumed to be in effect at the time. It does not matter whether the facility is self-financed by the company from its own assets (as in the example above), or whether money is borrowed from another institution to make the initial investment. In the first case, the capital cost represents the lost interest that could have been earned by making the investment elsewhere. In the second case, the capital cost is directly given by the interest paid to the lender. In either case, the annual cost is identical if the same rate of interest is assumed.

8.7.2. Variable Costs

In any manufacturing operation or other activity, there is generally a component of costs that depends on the number of units produced or level of the activity. These are categorized as *variable costs* in contrast to the fixed costs mentioned above. For example, the total cost of materials that go into a manufactured product will increase as more units are produced. Labor costs go up as more production workers are needed. Many maintenance and utility costs will also increase as the level of activity goes up.

Variable costs are seldom constant per unit produced. Production beyond a certain level may require adding workers at overtime wage rates, raising the labor cost component per unit. On the other hand, increasing output may allow the purchase of raw materials in larger volume and therefore at more favorable prices. An analysis of variable costs is therefore rather complex and often relies on past experience with similar operations as an essential guide.

8.7.3. Incremental Costs

The increase in variable cost per unit of output required to raise the output rate from one level to another is known as *incremental cost*. For example, the combined fixed and variable costs involved in producing 1000 units per year of a product may be $10,000 or $10 per unit. An analysis might be made of increasing production to 1200 units per year. The fixed cost will not change and, thus, it may be possible to carry out this increase in production for a new annual total cost of $11,000. Thus the incremental cost of adding 200 units to the annual production is $1000 per 200 units, or $5 a unit. Because of the effect of fixed costs, it is usually true that incremental costs go down as production volume increases.

SUMMARY

Costs of an engineering operation can generally be divided into fixed costs and variable costs. Fixed costs do not depend on the number of units produced or the level of the activity, whereas variable costs do change with these indices. An important part of fixed costs is capital cost, which reflects the fixed investment made in equipment and facilities. Variable costs include labor, supplies, and other expenses that depend on the volume of output. Incremental cost is the increase in variable cost per unit of output that is required to raise production from one level to another.

Exercises

Cost

1. For the following activities, make lists of fixed and variable costs that must be reflected in the total cost of production:
    (a) Operating a lemonade stand
    (b) Manufacturing integrated circuits for use in hand-held calculators
    (c) Producing raisin bagels at the local bagel factory

(d)   Providing engineering consultation services to a banker contemplating investment in a high technology area

(e)   Writing computer programs on a consulting basis to local industry

In each case, comment on the degree to which incremental costs change with the volume of the activity.

2.   Large central station electric generating plants of 1000 megawatt electric generating capacity now cost in the vicinity of two billion dollars to construct. Assuming an 8% simple interest rate, what is the corresponding capital cost in units of cents per kilowatt hour? Compare this figure with the prevailing electric utility rates in your area.

3.   A college professor is contemplating buying his own cap and gown for use at graduation ceremonies twice a year. He presently rents this costume at $15 for each use. If a new gown is purchased, it can be expected to last for 15 years, at which point its value is reduced to zero. Assuming a 10% simple interest rate and linear depreciation, find the maximum purchase price that would justify buying the costume. (At this price, the capital costs of owning the gown plus the depreciation charges will just equal the rental now paid.)

4.   For a manufacturing facility in your city, make a list of all important cost categories and divide the list into fixed and variable cost categories.

5.   A modern jet airliner costs approximately 10 million dollars when purchased new. In a particular application that aircraft will provide 800,000 passenger-miles of service per day. At 10% simple interest, what are the corresponding capital costs per passenger-mile?

6.   A man paid $10 an hour performs a job that can be done equally well by a computer. At an annual interest rate of 12% simple interest, at what computer purchase price will the capital costs be equal to this figure? With a useful life of 5 years and no salvage value, what is the additional hourly rate corresponding to depreciation? Assume 8 hour operation per day, 5 days per week.

8.8.   INFLATION

All of our discussion to this point has assumed that the purchasing power of a dollar does not change with time. As we are all aware from our daily lives, the inexorable rise in prices known as *inflation* leads to a continual decrease in the real value of the dollar. The causes of inflation are many and complex, and it has become a common malady of the economies of virtually all the developed countries of the world. In the United States every year since 1940 has shown a measurable degree of inflation as indicated by various federal price indexes. Since 1970 the rate of inflation has been particularly severe, and it is no longer unusual to see

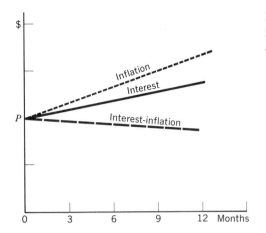

double-digit inflation (greater than 10%) in any given year. Prudent economic planning in today's climate must take into account the effects of this inflation.

Some obvious effects of inflation are felt in predicting future variable costs of an operation. Inflation increases the cost of raw materials. Inflation also drives up labor costs as salary and wage increases are needed to keep pace with increasing prices. These increasing costs for a manufacturing company cause it to raise prices, resulting in something of a vicious spiral. Governments have found it difficult to eliminate or even control inflation rates. Therefore it is only prudent that economic planning be done under the assumption that inflation will always be with us.

Inflation has a considerable effect on the economic implications of interest rates discussed earlier in this chapter. For example, assume that we borrow $1000 at 10% simple interest for repayment in one year's time. The repaid principal and interest will amount to $1100. However, if there has been a general 5% inflation over that same year, each of the repaid dollars will only be worth approximately 95¢ when measured against the goods that one dollar would have purchased when the loan was made. Thus half of the accrued interest is really eaten away by the effects of inflation, and the effective amount earned is only $500. By allowing debts to be paid back with currency of less real value, the effect of inflation is to favor the borrower and to penalize the lender. When measured in true purchasing power, the effective interest earned by a lender (or paid by a borrower) is the difference between nominal interest rates and the rate of inflation.

If it is assumed that a constant rate of inflation prevails, then the principles of present worth accounting outlined earlier in this chapter still apply. In other words, if two different schedules of loans and/or payments are intercompared in the absence of inflation, they shall bear the same relationship to each other independent of any assumed inflation rate. If one schedule of payments is shown to be more favorable than another without inflationary effects, then that preference will not be changed by any constant inflation factor.

SUMMARY

Inflation has become an important factor in economic analysis and decision making. The declining purchasing power of the dollar must be taken into account when evaluating the true future value of an investment or in projecting future costs. Comparisons based on present worth accounting remain valid in the presence of fixed rate inflation.

Exercises

Inflation

1. Inflation percentages are quoted yearly and therefore correspond to annual compounding. Find the percentage decrease in purchasing power corresponding to 10 years of 7% inflation.

2. A 3-year investment is made at an interest rate of 11% compounded semi-annually. Inflation rates for those 3 years are 8%, 7%, and 10%. By what factor has the real purchasing power of that investment increased over the 3 years?

REFERENCES

1. Eugene L. Grant, W. Ireson, and Richard S. Leavenworth, *Principles of Engineering Economy*, 7th Edition (Wiley, New York, 1981).

2. John A. White, *Principles of Engineering Economic Analysis* (Wiley, New York, 1977).

3. W. J. Fabrycky and G. J. Thuesen, *Engineering Economy*, 5th Edition (Prentice-Hall, Englewood Cliffs, N.J., 1977).

4. Anthony J. Tarquin and Leland T. Blank, *Engineering Economy* (McGraw-Hill, New York, 1975).

CHAPTER 9

Interaction With People

As engineers advance in their profession, they usually find that human relations play an important role, even replacing some of their technical activities. Many engineers will find themselves in technical management. They might be asked to lead a small group of engineers or supervise a design project. Some engineers will move out of technical activities altogether into executive management positions that are more concerned with personnel or budget matters than engineering analysis and design. Their interaction with other professionals such as accountants, business executives, and lawyers will become increasingly important.

The more management responsibility engineers assume, the greater will be their need for knowledge of business and management methods and a general understanding of the law. A strong case can be made for the inclusion of a course on modern management methods in the formal education of an engineer, and many undergraduate curricula provide for this opportunity. Similarly, some exposure to economics and accounting methods is desirable. In addition, although engineers are not expected to master the complex and intricate (and confusing) details of our legal code, they should be able to recognize those occasions when legal advice is necessary and to interact intelligently with lawyers.

9.1. ENGINEERS AND MANAGEMENT

About five years into their careers most engineers will have to decide if they should continue on in a technical specialty or enter technical management. It has been estimated that as many as two-thirds of all engineers will spend a good part of their careers as managers.

This trend should not be surprising. Even recent engineering graduates find themselves faced with decisions concerning the allocation of human and material resources. They must consider the economic implications of their technical decisions. They are also frequently faced with human relations activities, whether interacting with other engineers in their organization, or supervising the work of support personnel such as draftsmen, technicians, and craftsmen. Since so many of the engineer's activities involve "management," and since the reward system in most organizations tends to favor those in executive positions, it is not surprising that many engineers move rapidly into formal management roles.

9.1.1.   What is Management?

Just what is management? A typical picture is that of the cigar-chomping executive, seated behind an enormous desk in a plush office, surrounded by subordinates anxious to carry out every request. This stereotypical image suggests that *authority* over people is an important part of the management function. But the executive must also assume *responsibility*, that is, the obligation to accomplish certain goals. Of course, managers may play an important role in determining these goals and the methods used to pursue them. In management activities responsibility and authority go hand-in-hand.

Perhaps the most famous definition of management was provided by an early pioneer of modern management methods, Federick W. Taylor: "Management is knowing exactly what you want men to do, and then seeing that they do it in the best and cheapest way." It is important to recognize two aspects of Taylor's

FIGURE 9.1.   The most common perception of management fails to take into account the dual aspects of authority and responsibility.

definition. The first is concerned with defining objectives and planning the use of resources in an efficient manner. The second aspect is concerned with achieving satisfactory performance from employees. Management not only involves careful planning and organization, but leadership as well. In a very real sense, management is the function of getting things done through people. It involves both economic and behavioral aspects.

The major stimulus for the development of modern management methods was the recognition that management activities are simply another form of problem solving and therefore subject to scientific methods. Whether seeking to minimize the production costs of an item or to increase the productivity of subordinates, a manager can apply the same problem solving approach so common in other aspects of engineering. As in the scientific method, modern management methods involve reasoning from facts rather than conjecture, moving from cause to effect based upon observation or experience.

In fact, the similarity between modern management methods and engineering analysis is quite striking. For this reason most engineers find the transition to managerial activities to be a rather natural step. They have already developed the discipline and approach necessary for engineering problem solving. To apply these to management problems is a straightforward extension (although it may involve the acquisition of new ''tools'' such as finance and psychology). Just as in engineering, the complexity of management problems will require a creativity and imagination in the application of scientific principles that can challenge and stimulate engineers.

Although management roles vary greatly in authority and responsibility, ranging from the leadership of a small technical group to the direction of a large corporation with thousands of employees, we can identify several general activities involved at all levels of the management heirarchy. Certainly one important component is *planning*, of establishing objectives and planning the work of others to achieve these objectives. A second activity is *organization*, arranging the planned work and the structure of the group of employees to function as efficiently and effectively as possible. The third activity is *leadership*, the activity of motivating others to carry out the planned work as organized and then controlling the performance of their work. We shall consider each of these activities in turn.

SUMMARY

Management involves planning objectives, organizing resources, and leading others to achieve these objectives in the most efficient and effective manner. Management activities involve both authority and responsibility. Since management is essentially an activity of problem solving, the similarity between modern management methods and engineering analysis is quite striking.

9.1.2. Planning and Decision Making

The most basic of management functions is planning. In this step, broad objectives are defined and goals set for various times into the future. In many organizations, detailed planning is carried out on a year-by-year basis, often to coincide with annual budgets. Equally important, however, is long-range planning that looks beyond the year ahead in an attempt to anticipate major trends or goals as much as five or ten years into the future.

An essential ingredient in planning is decision making. Managers make many different types of decisions in establishing their objectives, including those concerned with planning, organization, motivation, and control. They, in turn, will be influenced by decisions made by others. While their duties may, in fact, be defined in terms of other functions such as quality control or research and development, the ability of managers to make the proper decision is an important factor in the success of the organization.

It is important to recognize that decision making refers not just to the final act of choosing among alternatives. It involves the entire process leading up to and including this choice. Indeed, decision making follows the same format as engineering problem solving: One must find an occasion for making a decision (define the problem), then determine possible courses of action (synthesize solutions), and finally choose among various courses of action (solution evaluation). Just as in technical problem solving, the managerial task of decision making is frequently complicated by a degree of ill-definition and the existence of multiple (and sometimes conflicting) objectives.

Many aspects of managerial decision making lend themselves to quantitative analysis. For example, one can apply the engineering methods of systems analysis to the organization involved (which is certainly a system, albeit large and complex). This organization is comprised of subsystems of a social, economic, and technological nature. It also functions itself as a subsystem within the larger environment that includes competitive organizations as interrelated subsystems. A factory can be analyzed as a complex system, receiving raw material as input and manufacturing finished products as output. It is comprised of numerous subsystems such as assembly lines, maintenance groups, and quality control and inspection units. The factory itself is a subsystem of a larger corporate structure.

The discipline that deals with the analysis of such organizational systems in a quantitative manner is known as *decision theory*. Mathematical methods have been developed to assist in determining the course of action that will optimize the system behavior (for example, minimize the costs of production). Although not all aspects of the decision situation lend themselves to such a quantitative approach, it is still advantageous to carry the quantitative analysis as far as possible before adding the subjective elements.

The manager can employ various instruments to accomplish the objectives established by planning and decision making. These include policies, rules, procedures, and budgets. In a sense these instruments act as constraints that guide the path leading to the objective. A *policy* usually takes the form of a general statement or understanding that serves to channel activity toward the objective. A *policy* is a plan, in that it establishes, in advance, some ground rules to assist in making later decisions. In contrast, a *rule* is more narrow and rigid, more specific, than a policy. Rules are designed to define in advance what alternative must be selected or what decision must be made in various situations. A *procedure* is a rule that usually incorporates a time element and requires sequential activity. It can be a means for implementing a policy. Finally, a *budget* is a projection (and a plan) that defines the anticipated costs of attaining an objective, either in economic or human terms. The budget may be used as a control device; however it begins as a forecast and therefore involves planning.

This discussion of planning and decision making is highly idealized, particularly since it has ignored the fact that most managers are under serious time constraints that may prevent the detailed analysis and thought required for careful planning. Managers may be forced to react to unanticipated and immediate problems. For example, a breakdown in a key piece of equipment or the threat of a wildcat strike will be more pressing than reviewing a plan for next year's budget. This element of time should be taken into account when determining planning procedures.

SUMMARY

Planning involves determining goals and the selecting among alternatives, that is, decision making. Decision making is very similar to engineering problem solving. One can apply quantitative methods of systems analysis to assist in the decision process. The manager can use policies, rules, procedures, and budgets to channel the activities toward the desired objectives.

9.1.3 Organization

An essential aspect of management involves organization—that is, an administrative structure that can accomplish an objective. Organization means assigning both authority and responsibility together with accountability to members of a group for the accomplishment of various tasks. In a general sense, organizing is the process of developing the most effective and efficient relationship between work, workplace, and people. Here we use "relationship" to include the structure of the system used to divide up and assign tasks and the procedures that coordinate the authority and responsibility for these tasks.

Organization requires careful planning. When a large amount of standardized work is to be done, great economies can be realized by breaking the complex task into subtasks and assigning these to groups of specialists. We need only think of the assembly line approach to manufacturing to realize the success of subdivision and specialization. However, as the division of work becomes finer and finer, so does the need for coordination among the groups of specialists. Each specialized group becomes more dependent on other groups for support or services.

Various criteria can be used to develop an organizational structure. One can determine the division of labor based on, for example, product or process similarities, technological function, and authority. Of particular importance is the division of work into managerial and nonmanagerial activities, usually referred to as "line" and "staff" activities. *Line* authority implies primary responsibility for the conduct of a specific division of work, say design, production, or marketing. For example, a production manager in charge of overseeing the day-to-day operation of a manufacturing facility would be a member of line management. *Staff* authority implies secondary responsibility for work conduct. The supervisor of quality control, a staff group, is only indirectly responsible for production activity. Staff groups aid line management in discharging its responsibilities.

All management functions, planning, organizing, and leading, are performed within the line organization. These management activities are shared vertically. Portions of the total planning function, organization, and so on are assigned to each manager in the vertical line. In contrast, the role of staff units are to aid line management, to provide it with information and services.

An organizational structure should be flexible. It should be capable of responding to changes in the environment or objectives within the organization. An excellent example is provided by a structure known as *project* management that is frequently adopted for large scale construction or production projects such as the development and assembly of military aircraft. Here a unit headed by a project manager is given responsibility and authority to make key decisions in managing the project. This authority is ad hoc, and it lasts only until the project is completed. The project manager's responsibility and authority cuts across the traditional functional organization units such as production control, quality control, and cost control. Such an organization structure possesses the flexibility to adapt to changing requirements of new projects.

A closely related organizational structure is known as *matrix* management. Here, the traditional line or pyramid structure of authority is replaced by a grid (a "matrix"). Two types of managers are assigned authority: program managers and functional managers. Program managers concentrate on a given program or project; functional managers are responsible for personnel and facilities management (hiring, firing, and personnel and facilities development). Program managers can negotiate across the matrix structure for manpower and resources, although it is still the job of top level managers to resolve conflicts. Although the delegation of authority and harnessing of talent provided by a matrix management organization can prove quite valuable in certain activities, it does require a high degree of cooperation and teamwork within the organization.

SUMMARY

Organization is the development of an administrative structure for accomplishing an objective. It means assigning authority and responsibility to members of an organization for various subtasks. As the subdivision of the tasks becomes finer, the dependence among specialized groups increases. Line and staff management refer to primary and secondary responsibility for work conduct, respectively. Organization structures should be flexible, capable of responding to changes in the organization's environment or objectives. This sometimes requires alternative organization systems such as project management or matrix management.

9.1.4. Leadership

The responsibility of managers extends beyond planning and organization. They are required to take actions to implement policy decisions reached by others

FIGURE 9.2. Leadership is an important aspect of management.

higher up in the organization. Directives reaching them are typically broad, leaving them the task of converting overall objectives and policies into operational programs of action. A major aspect of this management activity is *leadership*, or the process of influencing the behavior of others.

Leadership involves persuading others to work toward an objective. To do this, a manager should develop the ability to view activities from the perspective of various individuals within the organization. In the early days of management science physiological needs were thought to provide the basic drive among employees. However industry has passed beyond the era of paternalism, during which many leaders assumed that the key to obtaining human cooperation was to offer security and other economic satisfactions. Managers now recognize human motivation as a dynamic force, with multiple kinds of drives, shifting in importance as satisfaction levels change and aspirations evolve. Individuals are seen as joining an organization to achieve personal goals. They follow orders or otherwise contribute their effort and talent only if such activities satisfy their needs and abilities and are compatible with their sense of propriety. They may or may not be especially concerned with the objectives management has established for the organization. This aspect of human behavior has come to be recognized as a critical component of successful management.

Successful managers, through their own behavior, can also provide leadership by example. The influence that a dedicated and enthusiastic leader can have in instilling the same spirit throughout an organization should not be underestimated. When dealing with subordinates, managers cannot expect a sustained level of effort that exceeds their own.

SUMMARY

> *Leadership involves influencing the behavior of others. An important element of leadership is behavioral, that is, persuading or motivating individuals to contribute their efforts toward achieving the objectives of the organization.*

9.1.5. Preparation for Management Responsibilities

Even though most engineers will eventually work as managers, the typical engineering education currently does not provide much material on modern management methods. However this deficiency is being corrected as many programs are adding formal courses in technical management to the engineering curriculum.

There has also been recent growth in specialized programs in engineering management at both the baccalaureate and graduate level. Typically these programs include quantitative courses in management theory, engineering economics, statistics, decision theory, and operations research. They also include behavioral courses such as personnel management, psychology, and human relations. Engineering management programs differ from business administration programs (e.g., M.B.A. programs) in that they tend to be more quantitative, stressing aspects of technical management.

SUMMARY

There is a need in the undergraduate engineering curriculum for formal courses in management methods. Degree programs have been introduced in technical management at the baccalaureate and graduate level.

Exercises

Management

1. List the principal aspects of authority and responsibility for each of the following management positions:
   (a) Design group leader
   (b) Manager of quality control
   (c) President of a large corporation
   (d) Chairman of the board of a large corporation
   (e) Football coach

2. Sometimes a manager will be ineffective because of an imbalance between authority and responsibility. Suggest three examples of such a situation.

   It was noted that the general problem solving methods used in engineering can also be applied to problems in management. Demonstrate this by briefly describing how our procedure from Chapter 2 (problem definition, solution synthesis, verification and evaluation, presentation) might be applied to each of the following tasks:

3. Determine how to introduce robotics into an assembly line (recognizing the constraints posed by union labor contracts).

4. Determine an optimum balance between merit and cost of living factors in determining salary increases of professional staff during a period of high inflation.

5. Develop a marketing strategy for a new microcomputer system that will control every electrical outlet (and thereby any applicances plugged into it) in a typical home.

6. An assembly line worker (who also happens to be a union shop steward) has been stopped at the plant gate and found to have several plant tools in the trunk of his car. Determine what action should be taken.

7. Develop an affirmative action program to increase the percentage of women engineers in your company to 50%, recognizing that only 20% of graduating engineers are female.

Planning and Decision Making

8. Give an example of a policy, a rule, a procedure, and a budget for each of the following situations:
   (a) A power plant construction project
   (b) An automotive assembly plant
   (c) Your college of engineering
   (d) Your personal life

9. Identify each of the following as a policy, rule, procedure, or budget:
   (a) Individuals should not remain in top line management positions past the age of 65.
   (b) A manual containing instructions for nuclear power plant operators in the event of an accident.
   (c) The Ten Commandments.
   (d) New plans for bringing into balance expenditures with receipts over the next calendar year.

10. Briefly define the following terms that arise in modern technical management:
    (a) Cost benefit analysis
    (b) Environmental Impact Statement
    (c) PERT (Program Analysis and Review Technique)
    (d) Critical path scheduling
    (e) Linear programming

The computer has had a significant impact on modern management methods. Write simple computer programs (on either microcomputer or time-sharing mainframe computer systems) to accomplish each of the following management tasks:

11. *Salary Determination:* Develop a mathematical model to assist a manager in determining salary increases for his or her employees. For example, you might represent the salary increment as a linear function of various factors such as productivity, seniority, potential, cost of living, and so on:

$$\Delta\$ = c_1 f_1 + c_2 f_2 + \cdots + c_n f_n$$

The coefficients $c_1, c_2, \cdots$ could then be adjusted by the manager to stress whatever factors were felt to be of most importance in determining salaries.

12. *Inventory Control:* Develop a simple inventory control program. The program should keep track of the numbers of various items (gadgets, widgets, or wombats) as they are removed from or added to storage. Design the program to be interactive, in the sense that it will interrogate the user for the appropriate information.

13. *Capital Cost Accounting:* Develop a program that will calculate the capital cost of a project as a function of the labor, materials, and site costs; years to completion; interest rates; and so on. Remember to use the present worth accounting methods described in Chapter 8.

Organization

14. Using an organization diagram, design a suitable administrative structure for each of the following:
    - (a) A technical group of 10 engineers charged with the responsibility for the design of an all-terrain vehicle.
    - (b) A corporation of moderate size (1000 employees) that manufactures and markets personal computers.
    - (c) A college of engineering.
    - (d) A hamburger stand employing 20 people.

15. Identify each of the following as line or staff positions:
    - (a) Design group leader
    - (b) Director of plant security
    - (c) Director of plant maintenance
    - (d) Manager of advanced products research

16. Identify each of the engineering functions in Section 1.3 as corresponding to a line or a staff activity.

Leadership

17. Identify the principal approach to leadership of each of the following (e.g., authority, persuasion, respect, intimidation, . . .):
    - (a) Your class instructor
    - (b) A Marine drill sergeant
    - (c) A college dean
    - (d) The President of the United States

18. Factors frequently regarded as important in motivating employees include: salary and wages, job security, job esteem or prestige, possibility for advancement, physical environment, mental challenge, pressure, fringe benefits, and perquisites. Identify and rank the factors you feel are of most importance in each of the following occupations:

  (a) Design engineer
  (b) Craftsman
  (c) Assembly line worker
  (d) College professor

19. You are a manager. Discuss how you might develop a specific plan of implementation for each of the following broad directives.
  (a) Improve productivity in an assembly line by 10%
  (b) Develop an economic wind turbine generator.
  (c) Develop a more effective recruiting program for new engineers.
  (d) Meet newly imposed limits on stack emissions for a cement plant.

Preparation for Management:

20. Determine the courses in your college's curriculum that would most properly prepare you for a career in engineering management.

21. What factors in your career planning do you feel are of most importance in helping you to decide between a graduate degree in engineering (e.g., M.S., M.Eng., or Ph.D.) and a graduate degree in business administration (M.B.A.)?

9.2. ENGINEERS AND THE LAW

The activities of engineers are frequently influenced by legal considerations. For example, the preparation of contract documents such as drawings and specifications requires knowledge of contract law. Research and development engineers should be acquainted with patent law. Consulting engineers may be asked to testify as expert witnesses in product liability suits or license hearings. In fact, all engineers should be aware of their legal liability in engineering activities.

It is not our intent here to present a comprehensive survey of the many ways in which the law influences the engineering profession. Rather we shall illustrate several aspects of this interaction to show the importance of legal constraints on engineering practice.

9.2.1. Background

When we use the term "the law," we refer to the rules of civil conduct, of right and wrong, under which civilized individuals and communities live and maintain their relationships with one another. The body of law governing human activities includes not only the rules enacted by legislative bodies, but also the refinement and elaboration of these rules by interpretation in the courts.

To be more precise, a *statute* (or statute law) is a rule of conduct enacted by a legislative authority. For example, the National Environmental Policy Act is a statute enacted in 1969 by the United States Congress; it requires (among many other things) that an Environmental Impact Report be filed with any application for a federal license or permit. In contrast, *common law* refers to doctrines that are not founded upon statute but rather have their origin in court decisions. The particular interpretation of the National Environmental Policy Act by the United States Supreme Court in 1971 (in the case of the Calvert Cliffs Coordinating Committee vs. the Atomic Energy Commission) greatly broadened and extended the application of the Act by requiring that the Environmental Impact Report address all aspects of environmental impact, including social and aesthetic considerations.

The common law established through judicial processes is an ever-changing body of rules, continually subject to modification and reinterpretation. As circumstances change and the need arises, legislative bodies may enact statutes that result in modifying or eliminating entirely some particular aspect or rule of the common law. For example, for many years there has been a general principle of common law known as "caveat emptor," or "let the buyer beware." Continued abuses in the manufacture and sale of food and drugs eventually led to the passage of the Pure Food and Drug Act. This statute was designed to safeguard the public health and welfare; in effect it replaced the caveat emptor principle in the food and drug area.

The judicial courts are mechanisms for settling legal disputes. The judicial structure in the United States consists of courts at the federal, state, and local level. The *federal courts* deal with civil and criminal cases that involve federal laws or the constitution. They also handle disputes between citizens or corporations of different states. Sitting atop the federal court system is the United States Supreme Court. Virtually all of the Supreme Court's activities are reviews of lower court decisions. Below the Supreme Court in the federal court hierarchy are the Circuit Courts of Appeals. These intermediate level courts, sitting in 11 cities across the country, hear cases appealed from the more than 100 federal district courts that exercise broad jurisdiction over federal matters. There are also special courts created by Congress as the need arises, such as the Court of Claims, the Court of Customs and Patent Appeals, and the Tax Court of the United States.

Complementing the federal court system are local judicial systems such as the state and municipal courts. These have concurrent jurisdiction with federal courts in many instances. But they also deal with disputes not within the jurisdiction of federal courts. For example, most *state* court systems have a Supreme Court that will hear appeals from lower state courts. They may also have District Courts of Appeals similar to the federal court system. Superior or Circuit Courts located in each county of the state handle trials of general jurisdiction including both civil and criminal cases, felonies, divorce, and contract disputes. At the lowest level are Municipal or Justice Courts, with jurisdiction over misdemeanors and civil cases.

One can distinguish among several types of law. The most familiar classification is criminal law versus civil law. Criminal law includes rules that define and prohibit various crimes; it also makes provision for punishment. Civil law refers to rules governing disputes between parties (individuals or organizations). Contract disputes, product liability, and license intervention all are examples of civil law areas. There are still other types of law such as tax law, patent law, maritime law, and various forms of international law.

SUMMARY

The law refers to rules of civil conduct that govern the behavior and relationship among individuals. The body of law includes not only rules enacted by legislative bodies (statute law), but also the refinement and elaboration of these rules by interpretation in the courts (common law). Courts at the federal, state, and local level interpret and apply the law. Most aspects of engineering law such as contract disputes and product liability can be classified as civil law.

9.2.2. Contracts

One field of law of particular importance to practicing engineers is contract law. A civil engineer might bid successfully on a contract to survey and plat a group of residential building lots. Or a large aerospace corporation might sign a contract with the federal government to design and build a spacecraft for a manned mission to Mars. Although contract law is extensive and complex, we shall illustrate several of its more general aspects that have proven to be important in engineering activities.

In an abstract sense, a *contract* is simply the result of the concurrence of agreement and obligation. It is an understanding made between two or more persons or organizations enforceable by law. In this action rights are acquired by one side, requiring acts or obligations from the other. An engineering firm agrees to provide the detailed plans for a power plant for an agreed-upon sum.

For an agreement to result in a contract several ingredients are needed. First there must be an offer, followed by a corresponding acceptance. Furthermore the promises that stem from the offer and acceptance must be bound by law, so as to invest this agreement with the character of obligation.

There are many different kinds of contracts. In an *express contract*, the terms are declared by the parties in words, either orally or in writing, at the time the agreement is made. An *implied contract* requires inference or deduction from facts and circumstances showing a mutual intention to contract. There are also *unilateral contracts* when only one of the contracting parties makes a promise in exchange for an act or an executed consideration, and *bilateral contracts* involv-

FIGURE 9.3.   Engineers should become familiar with contract law.

ing mutual promises with each contracting party playing the dual role of promisor
and promisee. For a contract to be binding, the agreement must be entered into by
competent parties who express a definite assent in the form required by law.

Precisely defined terms characterize legal concepts such as contracts (much
as the engineer utilizes mathematical symbolism to characterize technical concepts). Terms of most relevance to engineering contracts include the following: an
owner, which refers to an individual or organization for which something is to be
built or furnished under contract; the *engineer*, which refers to the engineer or
architect (or organization) who acts in behalf of the owner in the transaction; and
the *contractor*, who is the party (either individual or organization) who undertakes
for a stated price to provide the services for the owner. The term *engineering* is
used by lawyers in a broad sense to include the work of all those engaged in the

provision of engineering services, ranging from architects in construction projects to scientists in research and development projects.

One particular form of contract quite common in engineering is the *construction contract*. Documents are usually prepared in support of such contracts: design drawings, which portray the structure to be built; specifications, which describe the qualities of materials and workmanship to be provided by the contractor; and various clauses setting forth specific features that apply to the work and obligations of the parties involved.

Contracts vary as to the manner of payment. In *lump sum contracts* the contractor agrees to perform the activity for a stated sum. For this compensation all obligations specified by the contract must be fulfilled even though their eventual cost may exceed the stipulated payment. A *unit price contract* is one in which payment for the work is made on the basis of an agreed-upon price for units of work actually performed or materials furnished.

When a premium is placed on construction time, the contractor may enter a *cost-plus-percentage contract* to perform the work on the basis of its actual cost plus a percentage of the cost, the owner paying all of the bills. A variation on this theme is the cost-plus-fixed-fee contract in which the owner pays all costs of the project plus a fixed fee as a compensation for the contractor. Some organizations make a specialty of both designing and building a project. In other words, they will sign a contract with an owner to make all the preliminary studies and the final design and then go ahead and build the structures involved. This is sometimes called a *turnkey* type of contract. Such a contract may be made on the basis of the cost-plus-fixed-fee or percentage.

To illustrate more clearly the various activities involved in contract preparation, let us briefly review how a construction contract might be arranged. The first stage involves the preparation of drawings that detail the design of the project. These drawings are prepared by the engineer and architect (in legal language, the design professionals). Such drawings are usually referred to as design drawings when they are prepared for bidding purposes; after the contract is let, they become known as the contract drawings. The preparation of design drawings can become quite complex since they must accurately document all details of the construction project. Such drawings play an important role in the contract because the contractor is required to provide and perform exactly what the drawings specify.

The architect-engineer will also provide detailed specifications concerning the particular items and materials to be used in the project along with specifications related to the quality of the workmanship required of the contractor. Here workmanship is intended to denote the contractor's operations in the shop or field rather than the materials used in the performance of the contract itself.

The complete contract drawings, specifications, and all other papers and data are put in a form suitable to enable contractors to submit bids on the project. Included in this material is usually the engineer's estimate of the cost of the project. All contractors must be provided with the same proposal form requirements if each bid is to be entered on the same basis.

The next phase involves advertising for the bids and furnishing plans and specifications to prospective contractors. The contractors then submit proposals or bids (normally confidential "sealed bids") in which the contractor agrees to furnish all materials and perform all work required by the contract documents prepared by the owner for a stated price. Furthermore, the proposal or bid is framed to constitute an undertaking by the bidder to sign the contract if the bid is accepted. Notice that a proper bid and its acceptance together contain the essential elements of a contract: (1) There is an agreement. (2) The agreement concerns specific subject matter. (3) The agreement is made between competent parties. (4) The parties promise to do certain lawful things. (5) There is proper consideration.

The engineer, under supervision of the owner, receives the proposals or bids from the contractors, compares these proposals, and determines the successful bidder. The contract is then signed by both the owner and the contractor. The final stage involves proceeding with construction, that is, converting the plans into reality. A knowledge of the law of contracts and the general conduct of business is quite important during this last phase.

SUMMARY

A contract is an agreement in which rights are acquired by one side to acts or obligations from the other. Contracts may be unilateral or bilateral, and various means of payment are possible, including lump sum, unit price, cost-plus-percentage, and turnkey. The most common form of contract is that used for a construction project, in which an engineer prepares drawings and specifications, contractors then submit bids to perform the specified work, the engineer determines the successful bidder, and then the engineer supervises the performance of the contract obligations.

9.2.3. Professional Liability

The activities of an engineer can affect countless people. Therefore it is not surprising that engineers may be held liable under law for their efforts or mistakes. In antiquity engineers were a combination of designer and builder. They were subject to rather severe liability. For example, the code of Hammurabi in ancient Samaria required that "If a builder builds a house for a man and does not make its construction firm, and the house which he has built collapses and causes the death of the owner, then the builder shall himself be put to death." Similar liabilities applied to lesser forms of damage or injury.

Today, although the punishment is certainly less severe, engineers bear a strong liability for their work. By accepting the responsibility for a task, engineers imply that they possess reasonable competence and the diligence necessary to perform the required activity. They do not guarantee perfection—or even satisfactory end results. Neither do they warrant that miscalculations will not occur. However they are expected to use all the skills at their command and to be acquainted with the state-of-the-art of their profession at the time. They are expected to display skill and knowledge at the "professional level," beyond that expected of nonskilled persons.

Engineers or architects may be held liable if the plans they draw prove grossly defective. They are also potentially responsible for damage stemming from their failure to supervise the construction work, if their supervision is required, or for handling their duties in a negligent manner. They may also be liable if they fraudulently or carelessly underestimate the cost of a project, thereby causing additional expense to the owner.

In most states the law requires a professional engineer or architect to obtain a certificate of registration as a prerequisite to practicing in that state. This certification is needed if the engineer is to carry on professional activities either in an individual capacity or as the supervisor of an engineering project within an organization (see Section 1.5).

SUMMARY

Engineers may be held liable under law for their efforts. They are expected to possess reasonable competence and display the skill and knowledge characteristic of their profession. Most states require professional engineers to obtain a certificate of registration as a prerequisite for certain types of practice.

9.2.4.   Expert Witnesses

Engineers are occasionly called upon to testify in court as expert witnesses. By definition, an *expert witness* is one whose background, training, and professional experience provides a superior knowledge in a particular field of endeavor that can serve as foundation for meaningful conclusions and opinions. Engineers may be asked for testimony or statements of opinion in a variety of civil court matters, including product liability suits or patent disputes. They may also be asked to testify at commission hearings, arbitration proceedings, legislative committee hearings, conferences pertaining to contract disputes, hearings to establish property evaluations, rates, or services, hearings before zoning boards, and building code matters.

FIGURE 9.4. Engineers may be held liable under law for their activities.

It is essential that the engineer who is to serve as an expert witness become familiar with the technical aspects of the case. This may involve close cooperation with attorneys for both sides. There should be a definite understanding about what the expert witnesses will be asked, what information they are in a position to provide, and what the limitations of their knowledge and experience may be. Careful preparation for such testimony is essential since an engineer's conduct in court may not only determine the outcome of a case, but it can also influence the engineer's reputation and professional future.

SUMMARY

Engineers are occasionally called upon for testimony as expert witnesses in court or other legal proceedings. Careful preparation for such testimony is essential.

9.2.5. Patents

Patents play an important role in engineering. A *patent* is a right granted by the government to a person(s) for a limited period of time to exclude everyone else from making, selling, and using the process or device that has been invented and patented. In effect a patent is a contract between the inventor and the government. The inventor is obliged to describe the invention fully through the medium of the issued patent; the government undertakes to protect the right created by the patent. Of course, a patent on a particular article or process will be granted only to the ''first'' inventor.

The life of a patent in the United States is 17 years from the date of issue. The government gives the inventor the right to exclusive use of the patented device for this period. However this privilege is conditioned upon the complete disclosure of all details of the invention, so that, after expiration of the patent, anyone with the technical knowledge and financial resources will be in a position to make, sell, or use the device. In this sense, the word patent means to ''open'' or ''disclose.''

A patent does not necessarily mean that the inventor has the right to manufacture the article involved, since other licenses or permits may be required for this. A patent merely grants to the holder the privilege of temporarily preventing others from using the process or making the article in question. It provides inventors with protection to encourage them to develop their inventions without concern that their ideas will be exploited by others.

The preparation of an application for a patent is a complex task. An engineer who intends to apply for a patent should secure the services of an experienced agent or lawyer who is registered with the United States Patent Office. The first stage of the application is to determine whether the proposed invention is indeed novel. The Patent Office contains extensive libraries detailing all previous patents for the use of all who wish to learn what has been developed. (We have already noted in Chapter 2 that these patent descriptions provide excellent resources for engineering design.

Engineers involved in research and development (and therefore invention) are advised to keep a careful record of their activities. Preferably, this record should be kept in a bound notebook so that others would find it difficult to claim

FIGURE 9.5.   Engineers should be concerned with patenting novel aspects of their work.

that the inventor has inserted information at a date later than is actually the case, or that pages were deleted (as might be possible with a looseleaf notebook). It is also desirable for inventors to date and sign each page as they record their notes. Furthermore at intervals they might have these notes read, dated, and notarized.

After the tedious process of patent application has been successfully completed and the patent has been awarded, inventors may assign their patent or the rights pertaining to it, or they may issue another person a grant or permit to make use of any of the rights covered under the patent. A license may or may not be treated as exclusive, and the agreement between parties should clarify this point. The licensee will then provide payment to the patent holder in the form of royalties.

SUMMARY

A patent is a right granted by the government to an inventor to exclude others from using or making the invention. The inventor is obliged to describe the invention fully in the patent; the government undertakes to protect the right created by the patent. The application for a patent is a complex task requiring the assistance of experts. Research and development engineers should always keep careful records of their activities to support possible patent applications.

Exercises

The Law

1. Give three examples of statute laws and common laws.

2. Describe the court system in your state.

3. Suggest circumstances in which an engineer might be prosecuted under criminal statutes for improper activities related to professional practice.

Contracts

4. Contrast the contract procedure used in public projects with that in private projects.

5. Summarize briefly the advantages and disadvantages of
   (a) Lump sum or fixed cost contracts
   (b) Cost-plus-percentage contracts
   (c) Cost-plus-fixed-fee contracts
   (d) Turnkey contracts

6. Write a brief contract to be used in requesting bids for each of the following tasks:
   (a) Providing a computer program to balance your checkbook.
   (b) Cleaning your room (assuming that this is a possible task).
   (c) Conducting a research and development program to improve the design of the corkscrew.
   (d) Developing a new automobile engine design that will be fueled with alcohol.

7. Prepare a bid for each of the following tasks:
   (a) Analyzing the energy use and recommending energy conservation measures for your college.
   (b) Redesigning your college's football field in SI units (including laying out and marking the new field).
   (c) Washing your instructor's car.
   (d) Modifying all Interstate highway signs in your state to implement the changeover to SI units.

8. Prepare a bid (cost estimate) for the following engineering task:

   Develop a computer program to perform a statistical analysis and least squares fit correlation of the data obtained in a chemistry laboratory. The estimated effort is 100 hours. The project must be completed in one month's time.

   Be certain to include the cost of computer time, report preparation, a contingency factor and, of course, a suitable profit margin.

9. Define the following terms frequently used in contract preparations:
   (a) Boiler plate clauses
   (b) RFP (Request for Proposal)
   (c) Contingency factors
   (d) Direct and indirect costs
   (e) TPR (Technical Proposal Requirements)

Patents

10. Why does a patent grant the right to exclude others from making, using, or selling the invention rather than the right to make, use, or sell it?

11. As an engineer working for a company, you sign a patent agreement assigning your rights to patents on any products developed during your employment over to the company. During your off-hours, you suddenly hit upon a idea that possesses a considerable potential. Under what circumstances (if any) would you be justified in seeking an independent patent on the idea?

12. Why are records of invention development necessary in patent applications?

REFERENCES

Engineering Management

1. Benjamin S. Blanchard, *Engineering Organization and Management* (Prentice-Hall, Englewood Cliffs, N.J., 1976).

2. Nicholas P. Chironis, ed., *Management Guide for Engineers and Technical Administrators* (McGraw-Hill, New York, 1969).

3. Val Cronstedt, *Engineering Management and Administration* (McGraw-Hill, New York, 1961).

4. Victor G. Hajik, *Management of Engineering Projects*, Second Edition (McGraw-Hill, New York, 1977).

5. Herbert Popper, ed., *Modern Technical Management Techniques* (McGraw-Hill, New York, 1971).

Legal Aspects of Engineering

1. Clarence W. Dunham, *Contracts, Specifications, and Law for Engineers* (McGraw-Hill, New York, 1979).

2. Richard C. Vaughn, *Legal Aspects of Engineering*, 2nd Edition (Kendall-Hunt Publishing Co., Dubuque, Iowa, 1977).

CHAPTER 10

Interaction with Society

Throughout history the tools, machines, structures, and processes of technology that engineers have created have profoundly affected the course of civilization. The impact of the invention of the printing press, steam power, and gun powder are familiar topics in history courses. The impact of technology on society in modern times is perhaps even more dramatic and accelerated. We need only think of modern inventions such as the computer, telecommunications, and atomic energy to realize the extent to which engineering and technology have affected each of us.

Engineering has affected not only individual members of society. It has had also a strong influence on social organizations and institutions and especially on government. Governments, whether monarchy, democracy, or oligarchy, have always been influential in determining the direction of technological activities and in sponsoring these activities. Certainly the size of many engineering activities makes this a necessity—for example, the pyramids of Egypt or the aqueducts of Rome. History has also demonstrated that engineering has been used as a tool to realize the ambitions of rulers and governments, as an instrument of both peace and war.

An important interaction in recent times has been between engineering and members of society, the "public." The degree to which public attitudes toward science and technology influence the development and implementation of engineering achievements is a more recent phenomenon deserving of serious attention.

As the scale of engineering activities increases, so too does the magnitude of their impact on society. New technologies require a careful evaluation before implementation. No longer can we rely simply on technical feasibility as sufficient justification for engineering accomplishment. Social desirability must also be considered. Will recombinant DNA and genetic engineering create new life forms that cannot be controlled? Will the radioactive waste produced by our nuclear power plants contaminate our environment for generations to come? Will the continued combustion of fossil fuels such as oil and coal disturb the carbon dioxide balance in the atmosphere and trigger catastrophic climatic changes? The increased complexity of engineering activities makes it extremely difficult to analyze the potential impact of engineering projects. We have come to recognize that technology presents us with mixed blessings. It has an enormous potential for good; yet, if implemented in a careless fashion, it can also cause great harm.

10.1. ENGINEERING AND GOVERNMENT

Governments have always had a significant influence as sponsors of engineering activities. Since society's resources are limited, it is understandable that government should feel obliged to determine priorities and make choices among various technical options.

Recently, the role of government has broadened to include regulation of all types of engineering activity, whether supported by public or private funds. Federal regulations now governing areas such as product safety, environmental impact, and construction codes exemplify the ever-increasing regulatory role of government.

Today, government has not only a regulatory role, but has assumed the responsibility for determining whether certain technologies should even be developed. The public debates over the development of the supersonic transport and the fast breeder reactor are evidence of the increased involvement of government in technology assessment.

Assessment and regulation of technology are particularly complex tasks for public officials. If these officials are to participate wisely in decisions concerning technology, they must be acquainted with the scientific and technical foundations of the engineering activities involved. The resolution of complex issues involving energy production, pollution abatement, public transportation, and national defense illustrate the difficulty in making essentially political decisions about the implementation of future technologies.

10.1.1. Government Involvement in Technology

To cope with the increasing complexity and array of options involving the implementation of technology, government has spawned numerous federal departments, commissions, boards, and administrations such as the National Aeronautics and Space Administration (NASA), the Department of Energy (DOE), the Environmental Protection Agency (EPA), and the Civil Aeronautics Board (CAB). Each of these agencies employs hundreds (in some cases, thousands) of engineers to assist them in identifying, developing, and regulating an array of

FIGURE 10.1. Science and engineering have become increasingly regulated by the federal government.

technologies. These institutions have become powerful forces in our society, both technically and politically.

The role of elected public officials in determining policies involving technology has also become quite complex. Not only must they understand the technical aspects of their decisions, but in reaching policy decisions they must also deal with powerful groups such as private industry, consumer advocate or environmental groups, and government bureauracy.

Government has become increasingly involved in both the development of new technology and the implementation of existing technology. The large resources needed for the development of a new technology (research and development) almost require government participation. Indeed, as we have mentioned, government has come to play the dominant role in the decisions about whether a new technology should be developed at all.

Government's role has also spread to include questions on the implementation of presently available technology. Government now largely determines where new highways are routed or factories are built, how products such as automobiles, airplanes, or electric toasters are designed, and even the foods we purchase at the grocery store.

Unfortunately, the complexity of many of these decisions and the increasing involvement of government has isolated the general public from technical decision making. More decisions are being made by a relatively small handful of people within key federal and state offices. We have developed a new elite of persons in critical government positions who have acquired unusual influence as a result of their access to technical and scientific knowledge that the general public no longer comprehends. This new technocratic elite even tends to draw away from elected public officials who are frequently ill-equipped to deal with questions of a highly technical nature.

SUMMARY

Government has spawned new agencies to cope with the increasing complexity of technology. It has assumed a growing role in both the development of new technology and the implementation of existing technology.

10.1.2.   Government Regulation

Although technology has given us the means to improve our lives, it has also provided us with new means to abuse society and the environment. The power plant that generates needed electricity can also produce smoke that may pollute the atmosphere and harm both public health and the environment. The automobile has given us a new mobility, but it has also polluted the air we breathe, paved over the environment in which we live, and threatens to exhaust our resources of energy fuels.

Therefore it is apparent that some restraint on our use of technology is required. We cannot blindly implement new technologies. We must first evaluate their potential impact on us and our environment and, once implemented, technologies must be carefully controlled.

The type and degree of control, however, are controversial. The growing awareness of the damage that we can cause to our society and our environment through technology has stimulated public resistance. Powerful new groups such as those in the consumer advocate and environmental movements have rejected laisse-faire or self-restraint. They demand strong government control and regulation of technology. Indeed, the general public now appears to be of the attitude that the opportunities for abuse provided by technology are too numerous and too tempting to rely on self-restraint. There appears to be a strong conviction in many quarters that business interests cannot be trusted to refrain voluntarily from using technology to exploit natural resources or the public. The feeling is that given the opportunity, we shall exploit each other and nature. Furthermore, economic disincentives (e.g., billing industry through taxes for the true environmental and human costs of their activities) are usually considered to be impractical. Hence the public is demanding more and more that government assume the primary responsibility for managing, controlling, and regulating both new and existing technology.

Of course government regulation of technology is nothing new, just as government regulation of all human activities has always been a part of many cultures. What is different is the mounting pressure on government to legislate against *all* technology. Whether it be the control of chemicals in our food, pollutants in the air we breathe, the efficiency of our appliances, the location of our factories, highways, or housing developments, government regulation is playing an increasingly dominant role at all levels of government federal, state, and local.

Several examples illustrate this growing involvement of the government in the regulation of technology. We noted earlier the passage of the National Environmental Policy Act of 1969 to regulate the impact of technology on the environment. This Act led to the establishment of the Environmental Protection Agency that formulated federal regulations on all aspects of environmental impact. It required federal agencies and contractors to include a comprehensive environmental impact statement with every recommendation or proposal for legislation or other major federal action. Subsequent court decisions have broadened the application of this legislation to require a thorough analysis of the environmental costs and benefits of any project involving federal licenses, permits, or financing. Such reports must include all significant primary and secondary environment effects, aesthetic considerations, and a thorough evaluation of alternatives to the proposed project (including the elimination of the project altogether). Other environmental legislation has included the Clean Air Acts of 1967 and 1970, the Noise Control Act of 1972, the Federal Water Pollution Control Act, the Federal Environmental Pesticide Control Act of 1972, and so on.

Federal legislation has also had great importance in regulating safety. For example the Occupational Safety and Health Act (OSHA) of 1970 has established occupational health and safety regulations plus an ongoing procedure for adopting additional regulations as the need is perceived. Other safety legislation established the Consumer Product Safety Commission, the Nuclear Regulatory Commission, the Pure Food and Drug Commission, among many other agencies concerned with public safety.

Although some degree of government constraint on technology is certainly needed, there are serious concerns about the timing and the degree of such regulation. Frequently the impact of a particular technology is realized only after its implementation, and it can be difficult at that point to develop adequate regulations and standards to control these effects. The legal and legislative bureauracy surrounding the formulation and administration of regulations can be overwhelming. For example, there is frequently a high degree of uncertainty about the form that government regulations may take. To the extent that such regulations may be applied retroactively, they can greatly complicate the planning for or implementation of technology. Regulation usually means additional costs for the producer. In marginal cases, an activity may become unprofitable because of operational restrictions or administrative expense.

Unlimited regulation can strangle technology and prevent its implementation. The unfortunate tendency of the public to become estranged from science and technology, coupled with their growing suspicion of technical enterprises, can lead to demands for stringent overregulation that effectively blocks needed technology.

Engineers too often view the constraints imposed by government regulation with impatience and irritation. They often regard regulators as adversaries. Certainly the bureaucracy that encumbers many government agencies gives rise to this attitude. Yet while the precise details and enforcement of a given regulation may seem unnecessarily burdensome, frequently the motivation behind the regulation is quite valid. Usually it is far more constructive for the engineer to attempt to work *with* rather than *against* regulatory agencies in an effort to arrive at constraints on the technology that adequately protect the public and the environment without unnecessarily impeding its implementation.

SUMMARY

Technology requires restraint. The government has assumed the primary responsibility for managing, controlling, and regulating both new and existing technology. The timing and degree of regulation can be very important. Excessive regulation can strangle a technology, preventing its implementation. Yet engineers should learn to respect regulations and attempt to work with rather than against regulatory agencies.

10.1.3. Technology Assessment

As government becomes increasingly involved in evaluating technology, in making decisions concerning its development and implementation, in its regulation, it becomes increasingly important for public officials to approach technical decisions in an informed and intelligent manner. In the late 1960s the United States Congress coined the term *technology assessment* to refer to a mechanism that would enable it to gain a legislative capability for policy determination in applied science and technology and to anticipate social decisions involving technical matters. More specifically, Congress proposed the following definition:

> *Technology assessment is the process of taking a purposeful look at the consequences of technological change. It includes the primary cost/benefit balance of short term localized market place economics, but particularly goes beyond these to identify affected parties and unanticipated impacts in as broad and long range a fashion as possible. It is neutral and objective, seeking to enrich the information for management decisions. Both "good" and "bad" side effects are investigated since a missed opportunity for benefit may be detrimental to society just as is an unexpected hazard.*

Formal mechanisms have been established for technology assessment. Many committees and commissions and advisory groups have been set up to perform such assessments (including, naturally enough, Congress's own Office of Technology Assessment). There have also been attempts to formalize technology assessment by applying abstract methods of systems analysis (such as decision theory).

Although there has been disagreement over the success of various approaches to technology assessment, there is general agreement that the goals are commendable. Technology assessment aims at providing a total conceptual framework, complete in both scope and time, for decisions about appropriate utilization of technology for social purposes. However there have been difficulties in its implementation. Society is ill-equipped to handle conflicting interests. It is difficult to attach a quantitative value to goals such as a cleaner environment, public risk, and the preservation of future choices. A quantitative approach to risk and environmental impact assessment is still in its infancy.

Nevertheless, to the extent that technology assessment evaluates alternative means to achieve the same end by comparing social and economic costs, it has the potential for providing useful information for policy decisions.

SUMMARY

Government has attempted to establish formal methods for technology assessment. Although it is still rather difficult to attach a quantitative measure to public risk and environmental impact, such methods do have the potential to provide useful information for policy decisions.

Exercises

Interaction with Society

1. Rank in order of importance 10 engineering accomplishments that you feel have had the most significant impact on the course of civilization.

2. Rank in order of importance (i.e., probable impact on society) five engineering accomplishments likely to occur during the next 20 years. Give reasons for each of your choices.

Engineering and Government

3. List the various ways that you are affected by government agencies (e.g., FDA, EPA, DOE) on a typical day.

4. Using library resources, draw a graph of the number of federal agencies involved in the development and regulation of technology versus year from 1950 to the present. Also plot the total number of

employees of these agencies versus year. (Only crude estimates are necessary.)

5. List all of the federal agencies that you can identify which are involved with each of the following areas:
   (a) Energy
   (b) Environment
   (c) Transportation
   (d) Communication

6. Choose a technical issue that has come before a legislative committee during recent years and obtain the transcripts of the testimony. Then perform a critical analysis of the degree of understanding of technical issues exhibited by the participating legislators in their questions and statements.

7. Would the following technologies be more appropriately developed by government or private industry? Why?
   (a) A synthetic organism to biodegrade oil spills
   (b) The fast breeder reactor
   (c) A process to produce synthetic fuel from coal
   (d) Controlled thermonuclear fusion

8. Using a biographical source such as *Who's Who* or *American Men and Women of Science*, attempt to determine the technical qualifications of 10 different spokespersons for technology that are frequently quoted in the news media (including government officials, elected public representatives, and advocates or lobbyists).

9. You are staff assistant to Congressman Horatio Hornblower. Prepare a one-page briefing and suggested position statement on each of the following issues:
   (a) Carbon dioxide buildup in the atmosphere
   (b) Plutonium reprocessing
   (c) Mandatory use of automobile air bag restraints
   (d) Banning general aviation aircraft from the vicinity of major metropolitan airports

Technology Assessment

10. Perform a brief technology assessment of the benefits versus risks (human, social, and environmental) of the following:
    (a) A massive "Apollo scale" program to rebuild American industry based on automation and robotics technology
    (b) The fast breeder reactor
    (c) A massive synthetic fuels program, converting coal into liquid and gaseous fuels
    (d) Electric-powered automobiles
    (e) Aquaculture (use of the oceans to grow plants and aquatic life

to meet a significant fraction of the world's food and fiber needs)

(f) Production of synthetic lifeforms (recombinant DNA and genetic engineering)

(g) A national data bank, which would contain and coordinate the collection of data on individuals by various government agencies

11. The assessment and comparison of various energy options can be conveniently classified into four criteria: (1) resource base, (2) environmental impact, (3) public risk, and (4) economics. Prepare a table analyzing each of the following energy options in terms of these criteria (perhaps on a scale of 1 to 10):

(a) Petroleum
(b) Natural gas
(c) Coal
(d) Hydroelectric
(e) Nuclear fission (conventional)
(f) Nuclear fission (breeder reactors)
(g) Geothermal
(h) Wind power
(i) Solar thermal
(j) Solar electric (photovoltaic cell)
(k) Biomass conversion (e.g., gasohol)
(l) Thermonuclear fusion
(m) Oil shale and tar sands
(n) Synthetic fuels from coal
(o) Conservation
(p) Indecision

12. One of the most difficult areas in technology assessment involves a quantitative assessment of risk—that is, an estimate of the probability of human injury or fatality arising from a given technology. While the use of such methods to estimate risk on an absolute scale is a subject of considerable controversy, there is an increasing use of quantitative methods to make relative risk comparisons among alternative technologies. Using the resources available in your technical library, compare the relative risk of five different technologies.

10.2. ENGINEERING AND THE PUBLIC

Has the relationship between the engineer and the public changed in recent years? Certainly the pace of technological innovation has quickened. The time lag between discovery and implementation has grown shorter. It is also true that there have been changes in the public's attitudes toward science and technology.

The public's knowledge and understanding have not kept pace with the rapid progress of science and technology. This is due in part to the quickening rate and complexity of technological advances. The esoteric language employed by scientists and engineers to describe new concepts and discoveries is also partly to blame. The complexity of modern technology frequently requires a high degree of specialized knowledge that cannot be easily communicated to the public (or even to other engineers, for that matter).

Therefore it is understandable that many of us feel estranged from technology. Although most people have enjoyed the fruits of technological innovation such as modern transportation and telecommunication, few of us really understand these technologies. There is a growing feeling among laypersons of technological disenfranchisement, that their knowledge is insufficient to exert control over the very technology that is affecting their live (in both positive and negative ways). This has led to a sense of suspicion and distrust, a loss of confidence in the ability of science and technology to deal with the needs and problems faced by modern society.

This public attitude has led to several interesting reactions. Today we hear demands for social responsibility on the part of scientists and engineers. No longer is technology hailed and accepted as an irresistible force bringing unquestioned technical advantages and intrinsically containing seeds of economic and social progress. Indeed, the idea that economic growth is necessarily dependent on technological innovation and progress is being called into question. The public disaffection with technology has given rise to a call for a shift away from high technology and toward technologies more appropriate for our society, more compatible with the understanding and control of the individual.

The most extreme reaction has been a growth in those elements of society showing an overt hostility toward technology. Today one finds technology indicted for past failures such as the Love Canal, Three Mile Island, and unsafe and fuel-inefficient automobiles. Technology has been blamed for creating weapons of mass destruction, with disrupting ecological and environmental equilibrium, and with depleting our natural resources. When carried to the extreme, this attitude views technology as essentially foreign to humanity. In this view technology has done too much too fast, leaving us without appropriate facilities to adapt to the change.

In a sense our society is divided into two camps: those who fear modern technology and distrust its intrusion into their lives, and those who advocate technology as the solution to many of society's problems. Aspects of this division can be traced back to our basic approach to education. We attempt to identify at an early stage students with scientific ability and separate them out to receive special attention in their education. Other students with stronger interests in the humanities and arts are encouraged to pursue studies in these directions, with little scientific content. Hence the paths begin to diverge very early in our educational system. This difference is reinforced by our present approach to teaching topics such as literature, history, and art without any reference to the scientific and technological factors that have influenced our culture. As a result, the liberal

FIGURE 10.2. Many members of society face modern technology with suspicion.

arts student is frequently graduated into a highly technical world without any significant acquaintance with technology. Our system of formal education has not kept up with the technological age in which we live.

At the same time engineers are frequently allowed to pursue their technical studies without adequate exposure to the humanities. Certainly a liberal education is as important to a scientist or engineer as an understanding and appreciation of science and technology is to the liberal arts major.

Until these education reforms have been achieved, many people in our society will become ever more alienated and estranged from technology, even as they become ever more dependent upon it. Perhaps it is this frustration, this constant reminder of their inadequate understanding of the technology around them, that has caused them to distrust and even reject technological innovation.

More and more we are hearing proposals that we return to "the simple life," that we discard high technology and return to technologies more appropriate for our society. The path away from high, centralized technology was blazed by E. F. Schumacher. It combines "appropriate technology" based on small-scale, decentralized processes and social approaches to minimize environmental impact, public risk, and cost. Perhaps an outstanding example of the appropriate technology route is the "soft energy" technology that relies on decentralized processes involving renewable energy sources such as solar power and biomass conversion. The proponents of the "small is beautiful" approach do not advocate returning to the primitive technology of bygone eras, but rather utilizing advanced, decentralized, small-scale technology more in harmony with humanity and nature. In this way perhaps we can draw society closer together by reducing the present gulf between the technologist and the layperson, thereby allowing greater participation in the decisions concerning a simpler technology.

Have we not already become too dependent on high technology to make the switch? Can a world teeming with a population of 4 billion that doubles every 20 years ever return to the simple life, that is to the land? There may be a better alternative. We should take more care in choosing which technologies we develop and implement. We should only encourage those technologies that are manifestly compatible with our humanity and our environment. Here again we encounter a flaw, since history has shown us that it is exceedingly difficult to predict just where new technologies are going to burst forth. Perhaps by curtailing research on recombinant DNA we shall also prevent research leading to a cure for cancer; by restricting nuclear research we might well discard the development of a limitless source of energy. We simply may not have the wisdom or foresight to allow us to pick and choose among developing technologies.

Of course, a third option is to continue along our present path, to allow science and technology to develop in the haphazard fashion they have always followed. This is also an alternative with difficulties, for while an unrestrained technology may well proceed most rapidly to provide society with solutions to its problems, it could also lead to a rapid exhaustion of our natural resources and perhaps the destruction of our natural environment. Certainly as technology becomes more complex and as the pace of technological innovation becomes more rapid, we must take more care to assess the potential impact of our endeavors.

We, as engineers, must accept a professional, and indeed, a moral and ethical responsibility for the impact of our efforts. In today's world it is not sufficient to merely pursue technology for its own sake, without regard for the individuals or the institutions it may affect. Rather we must always keep foremost in mind our professional obligation to serve society and to assume the responsibility for our activities.

SUMMARY

The public's knowledge and understanding have not kept pace with the rapid progress of science and technology. This loss of control has led to a suspicion and distrust of technology. Today we hear proposals to redirect our society away from high technology toward technologies more appropriate for humanity and nature. Perhaps we should carefully choose among those technologies we develop and implement. Engineers have a professional responsibility for assessing the potential impact of their endeavors.

REFERENCES

1. Edward Krick, *An Introduction to Engineering: Concepts, Methods, and Issues* (Wiley, New York, 1976).

2. James Burke, *Connections* (Little, Brown and Company, Boston, 1978).

APPENDIX A

Suggested Guidelines For Use
With The Fundamental Canons of Ethics*

1. Engineers shall hold paramount the safety, health, and welfare of the public in the performance of their professional duties.

   a. Engineers shall recognize that the lives, safety, health, and welfare of the general public are dependent upon engineering judgments, decisions, and practices incorporated into structures, machines, products, processes, and devices.

   b. Engineers shall not approve nor seal plans and/or specifications that are not of a design safe to the public health and welfare and in conformity with accepted engineering standards.

   c. Should the Engineers' professional judgment be overruled under circumstances where the safety, health, and welfare of the public are endangered, the Engineers shall inform their clients or employers of the possible consequences and notify other proper authority of the situation, as may be appropriate.

   (c.1) Engineers shall do whatever possible to provide published standards, test codes, and quality control procedures that will enable the public to understand the degree of safety or life expectancy associated with the use of the design, products, and systems for which they are responsible.

   (c.2) Engineers will conduct reviews of the safety and reliability of the design, products, or systems for which they are responsible before giving their approval to the plans for the design.

*Source: Developed by the Ethics Committee of the Engineers' Council for Professional Development and adopted by the American Association of Engineering Societies, 1980.

513

(c.3) Should Engineers observe conditions which they believe will endanger public safety or health, they shall inform the proper authority of the situation.

d. Should Engineers have knowledge or reason to believe that another person or form may be in violation of any of the provisions of these Guidelines, they shall present such information to the proper authority in writing and shall cooperate with the proper authority in furnishing such further information or assistance as may be required.

(d.1) They shall advise proper authority if an adequate review of the safety and reliability of the products or systems has not been made or when the design imposes hazards to the public through its use.

(d.2) They shall withhold approval of products or systems when changes or modifications are made which would affect adversely its performance insofar as safety and reliability are concerned.

e. Engineers should seek opportunities to be of constructive service in civic affairs and work for the advancement of the safety, health, and well-being of their communities.

f. Engineers should be committed to improving the environment to enhance the quality of life.

2. Engineers shall perform services only in areas of their competence.

a. Engineers shall undertake to perform engineering assignments only when qualified by education or experience in the specific technical field of engineering involved.

b. Engineers may accept an assignment requiring education or experience outside of their own fields of competence, but only to the extent that their services are restricted to those phases of the project in which they are qualified. All other phases of such projects shall be performed by qualified associates, consultants, or employees.

c. Engineers shall not affix their signatures and/or seals to any engineering plan or document dealing with subject matter in which they lack competence by virtue of education or experience, nor to any such plan or document not prepared under their direction supervisory control.

3. Engineers shall issue public statements only in an objective and truthful manner.

a. Engineers shall endeavor to extend public knowledge, and to prevent misunderstandings of the achievements of engineering.

b. Engineers shall be completely objective and truthful in all professional reports, statements, or testimony. They shall include all relevant and pertinent information in such reports, statements, or testimony.

c. Engineers, when serving as expert or technical witnesses before any court, commission, or other tribunal, shall express an engineering opinion only when it is founded upon adequate knowledge of the facts in issue, upon a background of technical competence in the subject matter, and upon honest conviction of the accuracy and propriety of their testimony.

d. Engineers shall issue no statements, criticisms, nor arguments on engineering matters which are inspired or paid for by an interested party, or parties, unless they have prefaced their comments by explicitly identifying themselves, by disclosing the identities of the party or parties on whose behalf they are speaking, and by revealing the existence of any pecuniary interest they may have in the instant matters.

e. Engineers shall be dignified and modest in explaining their work and merit, and will avoid any act tending to promote their own interests at the expense of the integrity, honor, and dignity of the profession.

4. Engineers shall act in professional matters for each employer or client as faithful agents or trustees, and shall avoid conflicts of interest.

a. Engineers shall avoid all known conflicts of interest with their employers or clients and shall promptly inform their employers or clients of any business association, interests, or circumstances which could influence their judgment or the quality of their services.

b. Engineers shall not knowingly undertake any assignments which would knowingly create a potential conflict of interest between themselves and their clients or their employers.

c. Engineers shall not accept compensation, financial or otherwise, from more than one party for services on the same project, not for services pertaining to the same project, unless the circumstances are fully disclosed to, and agreed to, by all interested parties.

d. Engineers shall not solicit nor accept financial or other valuable considerations, including free engineering designs, from material or equipment suppliers for specifying their products.

e. Engineers shall not solicit nor accept gratuities, directly or indirectly, from contractors, their agents, or other parties dealing with their clients or employers in connection with work for which they are responsible.

f.   When in public service as members, advisors, or employees of a governmental body or department, Engineers shall not participate in considerations or actions with respect to services provided by them or their organization in private or product engineering practice.

g.   Engineers shall not solicit nor accept an engineering contract from a governmental body on which a principal officer or employee of their organization serves as a member.

h.   When, as a result of their studies, Engineers believe a project will not be successful, they shall so advise their employer or client.

i.   Engineers shall treat information coming to them in the course of their assignments as confidential, and shall not use such information as a means of making personal profit if such action is adverse to the interests of their clients, their employers, or the public.

    (i.1)   They will not disclose confidential information concerning the business affairs or technical processes of any present or former employer or client or bidder under evaluation, without his consent.

    (i.2)   They shall not reveal confidential information nor findings of any commission or board of which they are members.

    (i.3)   When they use designs supplied to them by clients, these designs shall not be duplicated by the Engineers for others without express permission.

    (i.4)   While in the employ of others, Engineers will not enter promotional efforts or negotiations for work or make arrangements for other employment as principals or to practice in connection with specific projects for which they have gained particular and specialized knowledge without the consent of all interested parties.

j.   The Engineer shall act with fairness and justice to all parties when administering a construction (or other) contract.

k.   Before undertaking work for others in which Engineers may make improvements, plans, designs, inventions, or other records which may justify copyrights or patents, they shall enter into a positive agreement regarding ownership.

l.   Engineers shall admit and accept their own errors when proven wrong and refrain from distorting or altering the facts to justify their decisions.

m.   Engineers shall not accept professional employment outside of their regular work or interest without the knowledge of their employers.

n. Engineers shall not attempt to attract an employee from another employer by false or misleading representations.

o. Engineers shall not review the work of other Engineers except with the knowledge of such Engineers, or unless the assignments or contractual agreements for the work have been terminated.

    (o.1) Engineers in governmental, industrial, or educational employment are entitled to review and evaluate the work of other engineers when so required by their duties.

    (o.2) Engineers in sales or industrial employment are entitled to make engineering comparisons of their products with products of other suppliers.

    (o.3) Engineers in sales employment shall not offer nor give engineering consultation or designs or advice other than specifically applying to equipment, materials, or systems being sold or offered for sale by them.

5. Engineers shall build their professional reputation on the merit of their services and shall not compete unfairly with others.

a. Engineers shall not pay nor offer to pay, either directly or indirectly, any commission, political contribution, or a gift, or other consideration in order to secure work, exclusive of securing salaried positions through employment agencies.

b. Engineers should negotiate contracts for professional services fairly and only on the basis of demonstrated competence and qualifications for the type of professional service required.

c. Engineers should negotiate a method and rate of compensation commensurate with the agreed upon scope of services. A meeting of the minds of the parties to the contract is essential to mutual confidence. The public interest requires that the cost of engineering services be fair and reasonable, but not the controlling consideration in selection of individuals or firms to provide these services.

    (c.1) These principles shall be applied by Engineers in obtaining the services of other professionals.

d. Engineers shall not attempt to supplant other Engineers in a particular employment after becoming aware that definite steps have been taken toward the others' employment or after they have been employed.

(d.1) They shall not solicit employment from clients who already have Engineers under contract for the same work.

(d.2) They shall not accept employment from clients who already have Engineers for the same work not yet completed or not yet paid for unless the performance or payment requirements in the contract are being litigated or the contracted Engineers' services have been terminated in writing by either party.

(d.3) In case of termination or litigation, the prospective Engineers before accepting the assignment shall advise the Engineers being terminated or involved in litigation.

e. Engineers shall not request, propose, nor accept professional commissions on a contingent basis under circumstances under which their professional judgments may be compromised, or when a contingency provision is used as a device for promoting or securing a professional commission.

f. Engineers shall not falsify nor permit misrepresentation of their, or their associates' academic or professional qualifications. They shall not misrepresent nor exaggerate their degree of responsibility in or for the subject matter of prior assignments. Brochures or other presentations incident to the solicitation of employment shall not misrepresent pertinent facts concerning employers, employees, associates, joint ventures, or their past accomplishments with the intent and purpose of enhancing their qualifications and work.

g. Engineers may advertise professional services only as a means of identification and limited to the following:

(g.1) Professional cards and listings in recognized and dignified publications, provided they are consistent in size and are in a section of the publication regularly devoted to such professional cards and listings. The information displayed must be restricted to firm name, address, telephone number, appropriate symbol, names of principal participants, and the fields of practice in which the firm is qualified.

(g.2) Signs on equipment, offices, and at the site of projects for which they render services, limited to firm name, address, telephone number, and type of services, as appropriate.

(g.3) Brochures, business cards, letterheads, and other factual representations of experience, facilities, personnel and capacity to render service, providing the same are not misleading relative to the extent of participation in the projects cited and are not indiscriminately distributed.

(g.4) Listings in the classified section of telephone directories, limited to name, address, telephone number and specialties in which the firm is qualified without resorting to special or bold type.

h. Engineers may use display advertising in recognized dignified business and professional publications, providing it is factual, and relates only to engineering, is free from ostentation, contains no laudatory expressions or implication, is not misleading with respect to the Engineers' extent of participation in the services or projects described.

i. Engineers may prepare articles for the lay or technical press which are factual, dignified, and free from ostentations or laudatory implications. Such articles shall not imply other than their direct participation in the work described unless credit is given to others for their share of the work.

j. Engineers may extend permission for their names to be used in commercial advertisements, such as may be published by manufacturers, contractors, material suppliers, etc., only by means of a modest dignified notation acknowledging their participation and the scope thereof in the project or product described. Such permission shall not include public endorsement of proprietary products.

k. Engineers may advertise for recruitment of personnel in appropriate publications or by special distribution. The information presented must be displayed in a dignified manner, restricted to firm name, address, telephone number, appropriate symbol, names of principal participants, the fields of practice in which the firm is qualified, and factual descriptions of positions available, qualifications required, and benefits available.

l. Engineers shall not enter competitions for designs for the purpose of obtaining commissions for specific projects, unless provision is made for reasonable compensation for all designs submitted.

m. Engineers shall not maliciously or falsely, directly or indirectly, injure the professional reputation, prospects, practice, or employment of another engineer, nor shall they indiscriminately criticize another's work.

n. Engineers shall not undertake nor agree to perform any engineering service on a free basis, except professional services which are advisory in nature for civic, charitable, religious, or non-profit organizations. When serving as members of such organizations, engineers are entitled to utilize their personal engineering knowledge in the service of these organizations.

o. Engineers shall not use equipment, supplies, laboratory nor office facilities of their employers to carry on outside private practice without consent.

p. In case of tax-free or tax-aided facilities, engineers should not use student services at less than rates of other employers of comparable competence, including fringe benefits.

6. Engineers shall act in such a manner as to uphold and enhance the honor, integrity, and dignity of the profession.

a. Engineers shall not knowingly associate with nor permit the use of their names nor firm names in business ventures by any person or firm which they know, or have reason to believe, are engaging in business or professional practices of a fraudulent or dishonest nature.

b. Engineers shall not use association with non-engineers, corporations, nor partnerships as ''cloaks'' for unethical acts.

c. Engineers should encourage engineering employees to attend and present papers at professional and technical society meetings.

d. Engineers should support the professional and technical societies of their discipline.

e. Engineers shall give proper credit for engineering work to those to whom credit is due, and recognize the proprietary interests of others. Whenever possible, they shall name the person or persons who may be responsible for designs, inventions, writings, or other accomplishments.

f. Engineers shall endeavor to extend the public knowledge of engineering, and shall not participate in the dissemination of untrue, unfair, or exaggerated statements regarding engineering.

g. Engineers shall uphold the principle of appropriate and adequate compensation for those engaged in engineering work.

h. Engineers should assign professional engineers duties of a nature which will utilize their full training and experience insofar as possible, and delegate lesser functions to subprofessionals or technicians.

i. Engineers shall provide prospective engineering employers with complete information on working conditions and their proposed status of employment, and after employment shall keep them informed of any changes.

APPENDIX B

List of Unit Conversion Factors*

TO CONVERT FROM	TO	MULTIPLY BY
	ACCELERATION	
foot/second²	meter/second² (m/s²)	3.048 000*E−01
free fall, standard	meter/second² (m/s²)	9.806 650*E+00
inch/second²	meter/second² (m/s²)	2.540 000*E−02
	AREA	
acre	meter² (m²)	4.046 856 E+03
barn	meter² (m²)	1.000 000*E−28
circular mil	meter² (m²)	5.067 075 E−10
foot²	meter² (m²)	9.290 304*E−02
inch²	meter² (m²)	6.451 600*E−04
mile² (U.S. statute)	meter² (m²)	2.589 988 E+06
section	meter² (m²)	2.589 988 E+06
township	meter² (m²)	9.323 957 E+07
yard²	meter² (m²)	8.361 274 E−01
	BENDING MOMENT OR TORQUE	
dyne-centimeter	newton-meter (N · m)	1.000 000*E−07
kilogram-force-meter	newton-meter (N · m)	9.806 650*E+00

*Metric Practice Guide, "American Society for Testing and Materials, E 380-72 (approved by American National Standards Institute ANSI Z210.1-1973).
Note: Asterisks denote exact equivalence.

TO CONVERT FROM	TO	MULTIPLY BY
ounce-force-inch	newton-meter (N · m)	7.061 552 E−03
pound-force-inch	newton-meter (N · m)	1.129 848 E−01
pound-force-foot	newton-meter (N · m)	1.355 818 E+00

(BENDING MOMENT OR TORQUE)/LENGTH

pound-force-foot/inch	newton-meter (N · m/m)	5.337 866 E+01
pound-force-inch/inch	newton-meter/meter (N · m/m)	4.448 222 E+00

ELECTRICITY AND MAGNETISM

abampere	ampere (A)	1.000 000*E+01
abcoulomb	coulomb (C)	1.000 000*E+01
abfarad	farad (F)	1.000 000*E+09
abhenry	henry (H)	1.000 000*E−09
abmho	siemens (S)	1.000 000*E+09
abohm	ohm (Ω)	1.000 000*E−09
abvolt	volt (V)	1.000 000*E−08
ampere, international U.S. ($A_{INT\text{-}US}$)	ampere (A)	9.998 43 E−01
ampere, U.S. legal 1948 ($A_{US\text{-}48}$)	ampere (A)	1.000 008 E+00
ampere-hour	coulomb (C)	3.600 000*E+03
coulomb, international U.S. ($C_{INT\text{-}US}$)	coulomb (C)	9.998 43 E−01
coulomb, U.S. legal 1948 ($C_{US\text{-}48}$)	coulomb (C)	1.000 008 E+00
EMU of capacitance†	farad (F)	1.000 000*E+09
EMU of current	ampere (A)	1.000 000*E+01
EMU of electric potential	volt (V)	1.000 000*E−08
EMU of inductance	henry (H)	1.000 000*E−09
EMU of resistance	ohm (Ω)	1.000 000*E−09
ESU of capacitance	farad (F)	1.112 650 E−12
ESU of current	ampere (A)	3.335 6 E−10
ESU of electric potential	volt (V)	2.997 9 E+02
ESU of inductance	henry (H)	8.987 554 E+11
ESU of resistance	ohm (Ω)	8.987 554 E+11
farad, international U.S. ($F_{INT\text{-}US}$)	farad (F)	9.995 05 E−01
faraday (based on carbon-12)	coulomb (C)	9.648 70 E+04
faraday (chemical)	coulomb (C)	9.649 57 E+04
faraday (physical)	coulomb (C)	9.652 19 E+04
gamma	tesla (T)	1.000 000*E−09
gauss	tesla (T)	1.000 000*E−04

†ESU means electrostatic cgs unit. EMU means electromagnetic cgs unit.

TO CONVERT FROM	TO	MULTIPLY BY
gilbert	ampere-turn	7.957 747 E−01
henry, international U.S. (H_{INT-US})	henry (H)	1.000 495 E+00
maxwell	weber (Wb)	1.000 000*E−08
oersted	ampere/meter (A/m)	7.957 747 E+01
ohm, international U.S. (Ω_{INT-US})	ohm (Ω)	1.000 495 E+00
ohm-centimeter	ohm-meter ($\Omega \cdot$ m)	1.000 000*E−02
statampere	ampere (A)	3.335 640 E−10
statcoulomb	coulomb (C)	3.335 640 E−10
statfarad	farad (F)	1.112 650 E−12
stathenry	henry (H)	8.987 554 E+11
statmho	siemens (S)	1.112 650 E−12
statohm	ohm (Ω)	8.987 554 E+11
statvolt	volt (V)	2.997 925 E+02
unit pole	weber (Wb)	1.256 637 E−07
volt, international U.S. (V_{INT-US})	volt (V)	1.000 338 E+00
volt, U.S. legal 1948 (V_{US-48})	volt (V)	1.000 008 E+00

ENERGY (INCLUDES WORK)

British thermal unit (International Table)	joule (J)	1.055 056 E+03
British thermal unit (mean)	joule (J)	1.055 87 E+03
British thermal unit (thermochemical)	joule (J)	1.054 350 E+03
British thermal unit (39 F)	joule (J)	1.059 67 E+03
British thermal unit (60 F)	joule (J)	1.054 68 E+03
calorie (International Table)	joule (J)	4.186 800*E+00
calorie (mean)	joule (J)	4.190 02 E+00
calorie (thermochemical)	joule (J)	4.184 000*E+00
calorie (15 C)	joule (J)	4.185 80 E+00
calorie (20 C)	joule (J)	4.181 90 E+00
calorie (kg, International Table)	joule (J)	4.186 800*E+03
calorie (kg, mean)	joule (J)	4.190 02 E+03
calorie (kg, thermochemical)	joule (J)	4.184 000*E+03
electron volt	joule (J)	1.602 10 E−19
erg	joule (J)	1.000 000*E−07
foot-pound-force	joule (J)	1.355 818 E+00
foot-poundal	joule (J)	4.214 011 E−02
joule, international U.S. (J_{INT-US})	joule (J)	1.000 182 E+00
joule, U.S. legal 1948 (J_{US-48})	joule (J)	1.000 017 E+00
kilocalorie (International Table)	joule (J)	4.186 800*E+03
kilocalorie (mean)	joule (J)	4.190 02 E+03
kilocalorie (thermochemical)	joule (J)	4.184 000*E+03

TO CONVERT FROM	TO	MULTIPLY BY
kilowatt-hour	joule (J)	3.600 000*E+06
kilowatt-hour, international U.S. (kWh$_{\text{INT-US}}$)	joule (J)	3.600 655 E+06
kilowatt-hour, U.S. legal 1948 (kWh$_{\text{US-48}}$)	joule (J)	3.600 061 E+06
ton (nuclear equivalent of TNT)	joule (J)	4.20 E+09
watt-hour	joule (J)	3.600 000*E+03
watt-second	joule (J)	1.000 000*E−00

ENERGY/AREA TIME

TO CONVERT FROM	TO	MULTIPLY BY
Btu (thermochemical)/foot^2-second	watt/meter^2 (W/m^2)	1.134 893 E+04
Btu (thermochemical)/foot^2-minute	watt/meter^2 (W/m^2)	1.891 489 E+02
Btu (thermochemical)/foot^2-hour	watt/meter^2 (W/m^2)	3.152 481 E+00
Btu (thermochemical)/inch^2-second	watt/meter^2 (W/m^2)	1.634 246 E+06
calorie (thermochemical)/ centimeter^2-minute	watt/meter^2 (W/m^2)	6.973 333 E+02
erg/centimeter^2-second	watt/meter^2 (W/m^2)	1.000 000*E−03
watt-centimeter^2	watt/meter^2 (W/m^2)	1.000 000*E+04

FORCE

TO CONVERT FROM	TO	MULTIPLY BY
dyne	newton (N)	2.000 000*E−05
kilogram-force	newton (N)	9.806 650*E+00
kilopound-force	newton (N)	9.806 650*E+00
kip	newton (N)	4.448 222 E+03
ounce-force (avoirdupois)	newton (N)	2.780 139 E−01
pound-force (lbf avoirdupois)	newton (N)	4.448 222 E+00
poundal	newton (N)	1.382 550 E−01

FORCE/LENGTH

TO CONVERT FROM	TO	MULTIPLY BY
pound-force/inch	newton/meter (N/m)	1.751 268 E+02
pound-force/foot	newton/meter (N/m)	1.459 390 E+01

HEAT

TO CONVERT FROM	TO	MULTIPLY BY
Btu (thermochemical) · in/s · ft^2 · °F (k, thermal conductivity)	watt/meter-kelvin (W/m · K)	5.188 732 E+02
Btu (International Table) · in/s · ft^2 · °F (k, thermal conductivity)	watt/meter-kelvin (W/m · K)	5.192 204 E+02
Btu (thermochemical) · in/h · ft^2 · °F (k, thermal conductivity)	watt/meter-kelvin (W/m · K)	1.441 314 E−01

TO CONVERT FROM	TO	MULTIPLY BY
Btu (International Table) · in/h · ft$_2$ · °F	watt/meter-kelvin	
(k, thermal conductivity)	(W/m · K)	1.442 279 E−01
Btu (International Table)/ft²	joule/meter² (J/m²)	1.135 653 E+04
Btu (thermochemical)/ft²	joule/meter² (J/m²)	1.134 893 E+04
Btu (International Table)/h · ft² · °F	watt/meter²-kelvin	
(C, thermal conductance)	(W/m² · K)	5.678 263 E+00
Btu (thermochemical)/h · ft² · °F	watt/meter²-kelvin	
(C, thermal conductance)	(W/m² · K)	5.624 466 E+00
Btu (International Table)/1bm	joule/kilogram (J/kg)	2.326 000*E+03
Btu (International Table)/lbm · °F		
Btu (thermochemical)/lbm · °F	joule/kilogram-kelvin	
(c, heat capacity)	(J/kg · K)	4.184 000 E+03
Btu (International Table)/s · ft² · °F	watt/meter²-kelvin	2.044 175 E+04
	(W/m² · K)	
Btu (thermochemical)/s · ft² · °F	watt/meter²-kelvin	2.042 808 E+04
	(W/m² · K)	
cal (thermochemical)/cm²	joule/meter² (J/m²)	4.184 000*E+04
cal (thermochemical)/cm² · s	watt/meter² (W/m²)	4.184 000*E+04
cal (thermochemical)/cm · s · °C	watt/meter-kelvin	4.184 000*E+02
	(W/m · K)	
cal (International Table)/g	joule/kilogram (J/kg)	4.186 800*E+03
cal (International Table)/g · °C	joule/kilogram-kelvin	4.186 800*E+03
	(J/kg · K)	
cal (thermochemical)/g	joule/kilogram (J/kg)	4.184 000*E+03
cal (thermochemical)/g · °C	joule/kilogram-kelvin	4.184 000*E+03
	(J/kg · K)	
clo	kelvin-meter²/watt	2.003 712 E−01
	(K · m²/W)	
°F · h · ft²/Btu (thermochemical)	kelvin-meter²/watt	
(R, thermal resistance)	(K · m²/W)	1.762 280 E−01
°F · h · ft²/Btu (International Table)	kelvin-meter²/watt	
(R, thermal resistance)	(K · m²/W)	1.761 102 E−01
ft²/h (thermal diffusivity)	meter²/second (m²/s)	2.580 640*E−05

LENGTH

angstrom	meter (m)	1.000 000*E−10
astronomical unit	meter (m)	1.495 98　E+11
caliber (inch)	meter (m)	2.540 000*E−02
fathom	meter (m)	1.828 800*E+00
fermi (femtometer)	meter (m)	1.000 000*E−15
foot	meter (m)	3.048 000*E−01

TO CONVERT FROM	TO	MULTIPLY BY
foot (U.S. survey)†	meter (m)	3.048 006 E−01
inch	meter (m)	2.540 000*E−02
league (international nautical)	meter (m)	5.556 000*E+03
league (statute)	meter (m)	4.828 032*E+03
league (U.K. nautical)	meter (m)	5.559 552*E+03
light year	meter (m)	9.460 55 E+15
microinch	meter (m)	2.540 000*E−08
micron	meter (m)	1.000 000*E−06
mil	meter (m)	2.540 000*E−05
mile (international nautical)	meter (m)	1.852 000*E+03
mile (U.K. nautical)	meter (m)	1.853 184*E+03
mile (U.S. nautical)	meter (m)	1.852 000*E+03
mile (U.S. statute)	meter (m)	1.609 344*E+03
parsec	meter (m)	3.083 74 E+16
pica (printer's)	meter (m)	4.217 518 E−03
point (printer's)	meter (m)	3.514 598*E−04
rod	meter (m)	5.029 200*E+00
statute mile (U.S.)	meter (m)	1.609 344*E+03
yard	meter (m)	9.144 000*E 01

LIGHT

footcandle	lumen/meter² (lm/m²)	1.076 391 E+01
footcandle	lux (lx)	1.076 391 E+01
footlambert	candela/meter² (cd/m²)	3.426 259 E+00
lux	lumen/meter² (lm/m²)	1.000 000*E+00

MASS

carat (metric)	kilogram (kg)	2.000 000*E−04
grain	kilogram (kg)	6.479 891*E−05
gram	kilogram (kg)	1.000 000*E−03
hundredweight (long)	kilogram (kg)	5.080 235 E+01
hundredweight (short)	kilogram (kg)	4.535 924 E+01
kilogram-force-second²/meter (mass)	kilogram (kg)	9.806 650*E+00
kilogram-mass	kilogram (kg)	1.000 000*E+00
ounce-mass (avoirdupois)	kilogram (kg)	2.834 952 E−02
ounce-mass (troy or apothecary)	kilogram (kg)	3.110 348 E−02
pennyweight	kilogram (kg)	1.555 174 E−03

†The exact conversion factor is 1200/3937.

TO CONVERT FROM	TO	MULTIPLY BY
pound-mass (lbm avoirdupois)	kilogram (kg)	4.535 924 E−01
pound-mass (troy or apothecary)	kilogram (kg)	3.732 417 E−01
slug	kilogram (kg)	1.459 390 E+01
ton (assay)	kilogram (kg)	2.916 667 E−02
ton (long, 2240 lbm)	kilogram (kg)	1.016 047 E+03
ton (metric)	kilogram (kg)	1.000 000*E+03
ton (short, 2000 lbm)	kilogram (kg)	9.071 847 E+02
tonne	kilogram (kg)	1.000 000*E+03

MASS/AREA

ounce-mass/yard^2	kilogram/meter^2 (kg/m^2)	3.390 575 E−02
pound-mass/foot^2	kilogram/meter^2 (kg/m^2)	4.882 428 E+00

MASS/TIME (INCLUDES FLOW)

perm (0 C)	kilogram/pascal-second-meter^2 (kg/Pa · s · m^2)	5.721 35 E−11
perm (23 C)	kilogram/pascal-second-meter^2 (kg/Pa · s · m^2)	5.745 25 E−11
perm-inch (0 C)	kilogram/pascal-second-meter (kg/Pa · s · m)	1.453 22 E−12
perm-inch (23 C)	kilogram/pascal-second-meter (kg/Pa · s · m)	1.459 29 E−12
pound-mass/second	kilogram/second (kg/s)	4.535 924 E−01
pound-mass/minute	kilogram/second (kg/s)	7.559 873 E−03
ton (short, mass)/hour	kilogram/second (kg/s)	2.519 958 E−01

MASS/VOLUME (INCLUDES DENSITY AND MASS CAPACITY)

grain (lbm avoirdupois/7000)/gallon (U.S. liquid)	kilogram/meter^3 (kg/m^3)	1.711 806 E−02
gram/centimeter^3	kilogram/meter^3 (kg/m^3)	1.000 000*E+03
ounce (avoirdupois)/gallon (U.K. liquid)	kilogram/meter^3 (kg/m^3)	6.236 027 E+00
ounce (avoirdupois)/gallon (U.S. liquid)	kilogram/meter^3 (kg/m^3)	7.489 152 E+00
ounce (avoirdupois) (mass)/inch^3	kilogram/meter^3 (kg/m^3)	1.729 994 E+03
pound-mass/foot^3	kilogram/meter^3 (kg/m^3)	1.601 846 E+01
pound-mass/inch^3	kilogram/meter^3 (kg/m^3)	2.767 990 E+04
pound-mass/gallon (U.K. liquid)	kilogram/meter^3 (kg/m^3)	9.977 644 E+01
pound-mass/gallon (U.S. liquid)	kilogram/meter^3 (kg/m^3)	1.198 264 E+02
slug/foot^3	kilogram/meter^3 (kg/m^3)	5.153 788 E+02
ton (long, mass)/yard^3	kilogram/meter^3 (kg/m^3)	1.328 939 E+03

TO CONVERT FROM	TO	MULTIPLY BY
POWER		
Btu (International Table)/hour	watt (W)	2.930 711 E−01
Btu (thermochemical)/second	watt (W)	1.054 350 E+03
Btu (thermochemical)/minute	watt (W)	1.757 250 E+01
Btu (thermochemical)/hour	watt (W)	2.928 751 E−01
calorie (thermochemical)/second	watt (W)	4.184 000*E+00
calorie (thermochemical)/minute	watt (W)	6.973 333 E−02
erg/second	watt (W)	1.000 000*E−07
foot-pound-force/hour	watt (W)	3.766 161 E−04
foot-pound-force/minute	watt (W)	2.259 697 E−02
foot-pound-force/second	watt (W)	1.355 818 E+00
horsepower (550 ft · lbf/s)	watt (W)	7.456 999 E+02
horsepower (boiler)	watt (W)	9.809 50 E+03
horsepower (electric)	watt (W)	7.460 000*E+02
horsepower (metric)	watt (W)	7.354 99 E+02
horsepower (water)	watt (W)	7.460 43 E+02
horsepower (U.K.)	watt (W)	7.457 0 E+02
kilocalorie (thermochemical)/minute	watt (W)	6.973 333 E+01
kilocalorie (thermochemical)/second	watt (W)	4.184 000*E+03
watt, international U.S. (W_{INT-US})	watt (W)	1.000 182 E+00
watt, U.S. legal 1948 (W_{US-48})	watt (W)	1.000 017 E+00
PRESSURE OR STRESS (FORCE/AREA)		
atmosphere (normal = 760 torr)	pascal (Pa)	1.013 25 E+05
atmosphere (technical = 1 kgf/cm²)	pascal (Pa)	9.806 650*E+04
bar	pascal (Pa)	1.000 000*E+05
centimeter of mercury (0 C)	pascal (Pa)	1.333 22 E+03
centimeter of water (4 C)	pascal (Pa)	9.806 38 E+01
decibar	pascal (Pa)	1.000 000*E+04
dyne/centimeter²	pascal (Pa)	1.000 000*E−01
foot of water (39.2 F)	pascal (Pa)	2.988 98 E+03
gram-force/centimeter²	pascal (Pa)	9.806 650*E+01
inch of mercury (32 F)	pascal (Pa)	3.386 389 E+03
inch of mercury (60 F)	pascal (Pa)	3.376 85 E+03
inch of water (39.2 F)	pascal (Pa)	2.490 82 E+02
inch of water (60 F)	pascal (Pa)	2.488 4 E+02
kilogram-force/centimeter²	pascal (Pa)	9.806 650*E+04
kilogram-force/meter²	pascal (Pa)	9.806 650*E+00
kilogram-force/millimeter²	pascal (Pa)	9.806 650*E+06

TO CONVERT FROM	TO	MULTIPLY BY
kip/inch² (ksi)	pascal (Pa)	6.894 757 E+06
millibar	pascal (Pa)	1.000 000*E+02
millimeter of mercury (0 C)	pascal (Pa)	1.333 224 E+02
poundal/foot²	pascal (Pa)	1.488 164 E+00
pound-force/foot²	pascal (Pa)	4.788 026 E+01
pound-force/inch² (psi)	pascal (Pa)	6.894 757 E+03
psi	pascal (Pa)	6.894 757 E+03
torr (mm Hg, 0 C)	pascal (Pa)	1.333 22 E+02

TEMPERATURE

° Celsius	kelvin (K)	$t_K = t_C + 273.15$
° Fahrenheit	kelvin (K)	$t_K = (t_F + 459.67)/1.8$
° Rankine	kelvin (K)	$t_K = t_R/1.8$
° Fahrenheit	° Celsius	$t_C = (t_F - 32)/1.8$
Kelvin	° Celsius	$t_C = t_K - 273.15$

TIME

day (mean solar)	second (s)	8.640 000 E+04
day (sidereal)	second (s)	8.616 409 E+04
hour (mean solar)	second (s)	3.600 000 E+03
hour (sidereal)	second (s)	3.590 170 E+03
minute (mean solar)	second (s)	6.000 000 E+01
minute (sidereal)	second (s)	5.983 617 E+01
month (mean calendar)	second (s)	2.628 000 E+06
second (sidereal)	second (s)	9.972 696 E−01
year (calendar)	second (s)	3.153 600 E+07
year (sidereal)	second (s)	3.155 815 E+07
year (tropical)	second (s)	3.155 693 E+07

VELOCITY (INCLUDES SPEED)

foot/hour	meter/second (m/s)	8.466 667 E−05
foot/minute	meter/second (m/s)	5.080 000*E−03
foot/second	meter/second (m/s)	3.048 000*E−01
inch/second	meter/second (m/s)	2.540 000*E−02
kilometer/hour	meter/second (m/s)	2.777 778 E−01
knot (international)	meter/second (m/s)	5.144 444 E−01
mile/hour (U.S. statute)	meter/second (m/s)	4.470 400*E−01
mile/minute (U.S. statute)	meter/second (m/s)	2.682 240*E+01
mile/second (U.S. statute)	meter/second (m/s)	1.609 344*E+03
mile/hour (U.S. statute)	kilometer/hour†	1.609 344*E+00

TO CONVERT FROM	TO	MULTIPLY BY
VISCOSITY		
centipoise	pascal-second (Pa · s)	1.000 000*E−03
centistoke	meter²/second (m²/s)	1.000 000*E−06
foot²/second	meter²/second (m²/s)	9.290 304*E−02
poise	pascal-second (Pa · s)	1.000 000*E−01
poundal-second/foot²	pascal-second (Pa · s)	1.488 164 E+00
pound-mass/foot-second	pascal-second (Pa · s)	1.488 164 E+00
pound-force-second/foot²	pascal-second (Pa · s)	4.788 026 E+01
rhe	meter²/newton-second (m²/N · s)	1.000 000*E+01
slug/foot-second	pascal-second (Pa · s)	4.788 026 E+01
stoke	meter²/second (m²/s)	1.000 000*E−04
VOLUME (INCLUDES CAPACITY)		
acre-foot	meter³ (m³)	1.233 482 E+03
barrel (oil, 42 gal)	meter³ (m³)	1.589 873 E−01
board foot	meter³ (m³)	2.359 737 E−03
bushel (U.S.)	meter³ (m³)	3.523 907 E−02
cup	meter³ (m³)	2.365 882 E−04
fluid ounce (U.S.)	meter³ (m³)	2.957 353 E−05
foot³	meter³ (m³)	2.831 685 E−02
gallon (Canadian liquid)	meter³ (m³)	4.546 090 E−03
gallon (U.K. liquid)	meter³ (m³)	4.546 092 E−03
gallon (U.S. dry)	meter³ (m³)	4.404 884 E−03
gallon (U.S. liquid)	meter³ (m³)	3.785 412 E−03
gill (U.K.)	meter³ (m³)	1.420 654 E−04
gill (U.S.)	meter³ (m³)	1.182 941 E−04
inch³‡	meter³ (m³)	1.638 706 E−05
liter	meter³ (m³)	1.000 000*E−03
ounce (U.K. fluid)	meter³ (m³)	2.841 307 E−05
ounce (U.S. fluid)	meter³ (m³)	2.957 353 E−05
peck (U.S.)	meter³ (m³)	8.809 768 E−03
pint (U.S. dry)	meter³ (m³)	5.506 105 E−04
pint (U.S. liquid)	meter³ (m³)	4.731 765 E−04
quart (U.S. dry)	meter³ (m³)	1.101 221 E−03

†Although speedometers may read km/h, the correct SI unit is m/s.
‡The exact conversion factor is 1.638 706 4*E−05.

TO CONVERT FROM	TO	MULTIPLY BY
quart (U.S. liquid)	meter³ (m³)	9.463 529 E−04
stere	meter³ (m³)	1.000 000*E+00
tablespoon	meter³ (m³)	1.478 676 E−05
teaspoon	meter³ (m³)	4.928 922 E−06
ton (register)	meter³ (m³)	2.831 685 E+00
yard³	meter³ (m³)	7.645 549 E−01

VOLUME TIME (INCLUDES FLOW)

foot³/minute	meter³/second (m³/s)	4.719 474 E−04
foot³/second	meter³/second (m³/s)	2.831 685 E−02
inch³/minute	meter³/second (m³/s)	2.731 177 E−07
yard³/minute	meter³/second (m³/s)	1.274 258 E−02
gallon (U.S. liquid)/day	meter³/second (m³/s)	4.381 264 E−08
gallon (U.S. liquid)/minute	meter³/second (m³/s)	6.309 020 E−05

APPENDIX C

Physical Tables*

Atomic Weight and Melting Points of the Elements

NAME	SYMBOL	ATOMIC NUMBER	ATOMIC WEIGHT	MELTING POINT, °C	BOILING POINT, °C
Actinum	Ac	89	~227	1,050	3,200
Aluminum	Al	13	26.98	660	2,467
Americium	Am	95	~243	994	2,607
Antimony	Sb	51	121.75	630.74	1,750
Argon	Ar	18	39.94	−189.2	−185.7
Arsenic (gray)	As	33	74.92	817(28 atm)	613
Astatine	At	85	~210	302	337
Barium	Ba	56	137.33	725	1,640
Berkelium	Bk	97	~247	–	–
Beryllium	Be	4	9.01	1,278	2,970
Bismuth	Bi	83	208.98	271	1,560
Boron	B	5	10.81	2,300	2,550
Bromine	Br	35	79.90	−7.2	58.78
Cadmium	Cd	48	112.41	320.9	765
Calcium	Ca	20	40.08	839	1,484
Californium	Cf	98	251	–	–
Carbon	C	6	12.01	~3,550	4,827
Cerium	Ce	58	140.12	798	3,257
Cesium	Cs	55	132.90	28.40	678.4
Chlorine	Cl	17	35.45	−100	−34.6
Chromium	Cr	24	51.99	1,857	2,672
Cobalt	Co	27	58.93	1,495	2,870
Copper	Cu	29	63.54	1,083	2,567
Curium	Cm	96	~247	1,340	
Dysprosium	Dy	66	162.50	1,409	2,335
Einsteinium	Es	99	~254	–	–

* *CRC Handbook of Chemistry and Physics*, ed. Robert C. West (CRC Press, 1977).

Atomic Weight and Melting Points of the Elements *(continued)*

NAME	SYMBOL	ATOMIC NUMBER	ATOMIC WEIGHT	MELTING POINT, °C	BOILING POINT, °C
Erbium	Er	68	167.26	1,522	2,510
Europium	Eu	63	151.96	822	1,597
Fermium	Fm	100	~257	–	–
Fluorine	F	9	18.99	−219	−188
Francium	Fr	87	~223	~27	~677
Gadolinium	Gd	64	157.25	1,311	3,233
Gallium	Ga	31	69.72	29.78	2,403
Germanium	Ge	32	72.59	937	2,830
Gold	Au	79	196.96	1,064	2,807
Hafnium	Hf	72	178.49	2,227	4,602
Helium	He	2	4.00	−272.2	−268
Holmium	Ho	67	164.93	1,470	2,720
Hydrogen	H	1	1.0079	−259	−252
Indium	In	49	114.82	156	2,080
Iodine	I	53	126.90	113	184
Iridium	Ir	77	192.22	2,410	4,130
Iron	Fe	26	55.84	1,535	2,750
Krypton	Kr	36	83.80	−156	−152
Lanthanum	La	57	138.90	920	3,454
Lawrencium	Lr	103	~260	–	–
Lead	Pb	82	207.2	327	1,740
Lithium	Li	3	6.94	180	1,347
Lutetium	Lu	71	174.97	1,656	3,315
Magnesium	Mg	12	24.30	648	1,090
Manganese	Mn	25	54.93	1,244	1,962
Mendelevium	Md	101	~258	–	–
Mercury	Hg	80	200.59	−38	356
Molybdenum	Mo	42	95.94	2,617	4,612
Neodymium	Nd	60	144.24	1,010	3,127
Neon	Ne	10	20.17	−248	−246
Neptunium	Np	93	237.048	640	3,902
Nickel	Ni	28	58.71	1,453	2,732
Niobium (Columbium)	Nb	41	92.90	2,468	4,742
Nitrogen	N	7	14.00	−209	−195
Nobelium	No	102	~259	–	–
Osmium	Os	76	190.2	3,045	5,027
Oxygen	O	8	15.99	−218	−182
Palladium	Pd	46	106.4	1,552	3,140

Atomic Weight and Melting Points of the Elements *(continued)*

NAME	SYMBOL	ATOMIC NUMBER	ATOMIC WEIGHT	MELTING POINT, °C	BOILING POINT, °C
Phosphorus	P	15	30.97	44.1	280
Platinum	Pt	78	195.09	1,772	3,827
Plutonium	Pu	94	~244	641	3,232
Polonium	Po	84	~210	254	962
Potassium	K	19	39.09	63	774
Praeseodymium	Pr	59	140.90	931	3,212
Promethium	Pm	61	~145	~1,080	2,460
Protactinium	Pa	91	231.03	<1,600	–
Radium	Ra	88	226.02	700	1,140
Radon	Rn	86	~222	−71	−61
Rhenium	Re	75	186.2	3,180	5,627
Rhodium	Rh	45	102.90	1,966	3,727
Rubidium	Rb	37	85.46	38.89	688
Ruthenium	Ru	44	101.07	2,310	3,900
Samarium	Sm	62	150.4	1,072	1,778
Scandium	Sc	21	44.95	1,539	2,832
Selenium	Se	34	78.96	217	684.9
Silicon	Si	14	28.08	1,410	2,355
Silver	Ag	47	107.86	961	2,212
Sodium	Na	11	22.98	97	882
Strontium	Sr	38	87.62	769	1,384
Sulfur	S	16	32.06	112	444
Tantalum	Ta	73	180.94	2,996	5,425
Technetium	Tc	43	98.90	2,172	4,877
Tellurium	Te	52	127.60	449	989
Terbium	Tb	65	158.92	1,360	3,041
Thallium	Tl	81	204.37	303.5	1,457
Thorium	Th	90	232.03	1,750	~4,790
Thulium	Tm	69	168.93	1,545	1,727
Tin	Sn	50	118.69	231	2,270
Titanium	Ti	22	47.90	1,660	3,287
Tungsten	W	74	183.85	3,410	5,660
Uranium	U	92	238.02	1,132	3,818
Vanadium	V	23	50.94	1,890	3,380
Xenon	Xe	54	131.30	−111	−107
Ytterbium	Yb	70	173.04	824	1,193
Yttrium	Y	39	88.90	1,523	3,337
Zinc	Zn	30	65.38	419	907
Zirconium	Zr	40	91.22	1,852	4,377

Specific Gravity and Density of Common Substances

SUBSTANCE	SPECIFIC GRAVITY	AVERAGE DENSITY, kg/m^3
Metals, Alloys, Ores		
Aluminum, cast-hammered	2.55–2.80	2,643
Brass, cast-rolled	8.4–8.7	8,553
Copper, cast-rolled	8.8–8.95	8,906
Copper ore, pyrites	4.1–4.3	4,197
Gold, cast-hammered	19.25–19.35	19,300
Iridium	21.78–22.42	22,160
Iron, cast, pig	7.2	7,207
Iron, wrought	7.6–7.9	7,658
Iron ore, limonite	3.6–4.0	3,796
Iron ore, magnetite	4.9–5.2	5,046
Lead	11.34	11,370
Lead ore, galena	7.3–7.6	7,449
Manganese	7.42	7,608
Manganese ore	3.7–4.6	4,149
Mercury	13,546	13,570
Monel metal, rolled	8.97	8,688
Nickel	8.9	8,602
Platinum, cast-hammered	21.5	21,300
Silver, cast-hammered	10.4–10.6	10,510
Steel, tool	7.70–7.73	7,703
Tin, cast-hammered	7.2–7.5	7,352
Tin ore, cassiterite	6.4–7.0	6,695
Tungsten	19.22	18,820
Uranium	18.7	18,740
Zinc, cast-rolled	6.9–7.2	7,049
Zinc, ore, blende	3.9–4.2	4,052
Various Solids		
Cereals, wheat, bulk	0.77	769
Cork	0.22–0.26	240
Cotton, flax, hemp	1.47–1.50	1,491
Fats	0.90–0.97	925
Flour, loose	0.40–0.50	448
Glass, common	2.40–2.80	2,595
Hay and straw, bales	0.32	320
Leather	0.86–1.02	945
Paper	0.70–1.15	929
Rubber goods	1.0–2.0	1,506
Salt, granulated, piled	0.77	769
Wool	1.32	1,315

SUBSTANCE	SPECIFIC GRAVITY	AVERAGE DENSITY, kg/m³
Timber, Air-Dry		
Cedar, white, red	0.35	352
Hickory	0.74–0.80	769
Mahogany	0.56–0.85	705
Oak	0.87	866
Pine, white	0.43	433
Redwood, California	0.42	417
Teak, African	0.99	994
Walnut, black	0.59	593
Various Liquids		
Alcohol	0.789	802
Acid, nitric, 91%	1.50	1,506
Acid, sulfuric, 87%	1.80	1,795
Lye, soda, 66%	1.70	1,699
Oils, vegetable	0.91–0.94	930
Oils, mineral, lubricants	0.88–0.94	914
Turpentine	0.861–0.867	866
Various Liquids		
Water, 4°C, max density	1.0	999.97
Water, 100°C	0.9584	958.10
Water, ice	0.88–0.92	897
Water, snow, fresh fallen	0.125	128
Water, seawater	1.02–1.03	1,025
Masonry		
Granite, syenite, gneiss	2.4–2.7	2,549
Limestone	2.1–2.8	2,450
Marble	2.4–2.8	2,597
Sandstone	2.0–2.6	2,290
Medium brick	1.6–2.0	1,794
Cement, stone, sand	2.2–2.4	2,309
Cement, cinder, etc.	1.5–1.7	1,602
Ashes, cinders	0.64–0.72	640–721
Earth, etc., Excavated		
Clay, dry	1.0	1,009
Clay, damp, plastic	1.76	1,761
Clay and gravel, dry	1.6	1,602
Earth, dry, loose	1.2	1,217
Earth, dry, packed	1.5	1,521
Earth, moist, packed	1.6	1,538
Sand, gravel, dry, loose	1.4–1.7	1,441–1,681
Sand, gravel, wet	1.89–2.16	2,019

SUBSTANCE	SPECIFIC GRAVITY	AVERAGE DENSITY, kg/m³
Minerals		
Asbestos	2.1–2.8	2,451
Basalt	2.7–3.2	2,950
Bauxite	2.55	2.549
Borax	1.7–1.8	1,746
Chalk	1.8–2.8	2,291
Granite	2.6–2.7	2,644
Gypsum, alabaster	2.3–2.8	2,549
Limestone	2.1–2.86	2,484
Marble	2.6–2.86	2,725
Pumice, natural	0.37–0.90	641
Quartz, flint	2.5–2.8	2,645
Sandstone	2.0–2.6	2,291
Shale, slate	2.6–2.9	2,758
Bituminous Substances		
Asphaltum	1.1–1.5	1,298
Coal, anthracite	1.4–1.8	1,554
Coal, bituminous	1.2–1.5	1,346
Coal, lignite	1.1–1.4	1,250
Coal, peat, turf, dry	0.65–0.85	753
Coal, charcoal, pine	0.28–0.44	369
Coal, charcoal, oak	0.47–0.57	481
Coal, coke	1.0–1.4	1,201
Graphite	1.64–2.7	2,163
Paraffin	0.87–0.91	898
Petroleum	0.87	856
Petroleum, refined (kerosene)	0.78–0.82	801
Petroleum, gasoline	0.70–0.75	721
Pitch	1.07–1.15	1,105
Tar, bituminous	1.20	1,201
Coal and Coke, Piled		
Coal, anthracite	0.75–0.93	753–930
Coal, bituminous, lignite	0.64–0.87	641–866
Coal, peat, turf	0.32–0.42	320–417
Coal, charcoal	0.16–0.23	160–224
Coal, coke	0.37–0.51	369–513

Selected Physical constants

Avogadro's number $= 6.022\ 57 \times 10^{23}/\text{mol}$
Density of dry air at 0°C, 1 atm $= 1.293\ \text{kg/m}^3$
Density of water at 3.98°C $= 9.999\ 973 \times 10^2\ \text{kg/m}^3$
Equatorial radius of the earth $= 6378.39\ \text{km}$
Gravitational acceleration (standard) at sea level $= 9.806\ 65\ \text{m/s}^2$
Heat of fusion of water, 0°C $= 3.3375 \times 10^5\ \text{J/kg}$
Heat of vaporization of water, 100°C $= 2.2591 \times 10^6\ \text{J/kg}$
Mass of hydrogen atom $= 1.673\ 39 \times 10^{-27}\ \text{kg}$
Molar gas constant $= 8.3144\ \text{J/(mol} \cdot \text{K)}$
Planck's constant $= 6.625\ 54 \times 10^{-34}\ \text{J/Hz}$
Velocity of light in a vacuum $= 2.9979 \times 10^8\ \text{m/s}$
Velocity of sound in dry air at 0°C $= 331.36\ \text{m/s}$

APPENDIX D

Selected Topics in Mathematics

D.1 TRIGONOMETRY

Geometric Definition of the Trigometric Functions (see Figure D.1)

	θ	45°	30°	60°	0°	90°
$\sin\theta = \dfrac{\text{opposite side}}{\text{hypotenuse}} =$	$\dfrac{y}{r}$	$\dfrac{1}{\sqrt{2}}$	$\dfrac{1}{2}$	$\dfrac{\sqrt{3}}{2}$	0	1
$\cos\theta = \dfrac{\text{adjacent side}}{\text{hypotenuse}} =$	$\dfrac{x}{r}$	$\dfrac{1}{\sqrt{2}}$	$\dfrac{\sqrt{3}}{2}$	$\dfrac{1}{2}$	1	0
$\tan\theta = \dfrac{\text{opposite side}}{\text{adjacent side}} =$	$\dfrac{y}{x}$	1	$\dfrac{1}{\sqrt{3}}$	$\sqrt{3}$	0	∞
$\cot\theta = \dfrac{\text{adjacent side}}{\text{opposite side}} =$	$\dfrac{x}{y}$	1	$\sqrt{3}$	$\dfrac{1}{\sqrt{3}}$	∞	0
$\sec\theta = \dfrac{\text{hypotenuse}}{\text{adjacent side}} =$	$\dfrac{r}{x}$	$\sqrt{2}$	$\dfrac{2}{\sqrt{3}}$	2	1	∞
$\csc\theta = \dfrac{\text{hypotenuse}}{\text{opposite side}} =$	$\dfrac{r}{y}$	$\sqrt{2}$	2	$\dfrac{2}{\sqrt{3}}$	∞	1

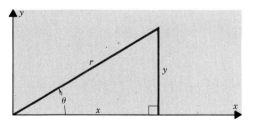

FIGURE D.1

D.2. RADIAN MEASURE

1 radian corresponds to 57.29578 degrees.
θ (radians) = θ (degrees) × $(\pi/180°)$

ANGLE	RADIANS
0°	0.0
1°	0.0175
10°	0.1754
45°	$\dfrac{\pi}{4} = 0.7854$
90°	$\dfrac{\pi}{2} = 1.5708$
180°	$\pi = 3.1416$
270°	$\dfrac{3}{2}\pi = 4.7123$
360°	$2\pi = 6.2832$

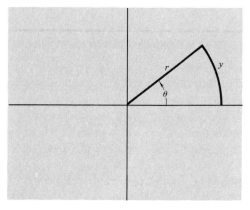

FIGURE D.2

D.3. CALCULUS

A brief table of derivatives

1. $\dfrac{d}{dx}(c) = 0$

2. $\dfrac{d}{dx}(x) = 1$

3. $\dfrac{d}{dx}(cx) = c$

4. $\dfrac{d}{dx}(x^n) = nx^{n-1}$

5. $\dfrac{d}{dx}(x^{1/2}) = \dfrac{1}{2}x^{-1/2}$

6. $\dfrac{d}{dx}(e^x) = e^x$

7. $\dfrac{d}{dx}(\ln x) = \dfrac{1}{x}$

8. $\dfrac{d}{dx}(\sin x) = \cos x$

9. $\dfrac{d}{dx}(\cos x) = -\sin x$

10. $\dfrac{d}{dx}(\tan x) = \sec^2 x$

11. $\dfrac{d}{dx}(\sin^{-1} x) = (1 - x^2)^{-1/2}$

12. $\dfrac{d}{dx}(\cos^{-1} x) = -(1 - x^2)^{-1/2}$

13. $\dfrac{d}{dx}(\tan^{-1} x) = (1 + x^2)^{-1/2}$

14. $\dfrac{d}{dx}(f + g) = \dfrac{df}{dx} + \dfrac{dg}{dx}$

15. $\dfrac{d}{dx}(fg) = \dfrac{df}{dx}g + f\dfrac{dg}{dx}$

16. $\dfrac{d}{dx}\left(\dfrac{1}{f}\right) = -\dfrac{1}{f^2}\dfrac{df}{dx}$

17. $\dfrac{d}{dx}F(f) = \dfrac{dF}{df}\dfrac{df}{dx}$

18. $\dfrac{df}{dx} = \left(\dfrac{dx}{df}\right)^{-1}$

A Brief Table of Integrals

1. $\displaystyle\int a\, dx = ax$

2. $\displaystyle\int a \cdot f(x)dx = a\int f(x)dx$

3. $\displaystyle\int \phi(y)\, dx = \int \dfrac{\phi(y)}{y'}\, dy,$ where $y' = \dfrac{dy}{dx}$

4. $\displaystyle\int (u + v)\, dx = \int u\, dx + \int v\, dx,$ where u and v are any functions of x

5. $\displaystyle\int u\, dv = u\int dv - \int v\, du = uv - \int v\, du$

6. $\int u \dfrac{dv}{dx} dx = uv - \int v \dfrac{du}{dx} dx$

7. $\int x^n \, dx = \dfrac{x^{n+1}}{n+1}$,   except $n = -1$

8. $\int \dfrac{f'(x) \, dx}{f(x)} = \log f(x)$, $(df(x) = f'(x) \, dx)$

9. $\int \dfrac{dx}{x} = \log x$

10. $\int \dfrac{f'(x) \, dx}{2\sqrt{f(x)}} = \sqrt{f(x)}$, $(df(x) = f'(x) \, dx)$

11. $\int e^x \, dx = e^x$

12. $\int e^{ax} \, dx = e^{ax}/a$

13. $\int b^{ax} \, dx = \dfrac{b^{ax}}{a \log b}$, $(b > 0)$

14. $\int \ln x \, dx = x \ln x - x$

15. $\int a^x \quad a \, dx = a^x$, $(a > 0)$

16. $\int \dfrac{dx}{a^2 + x^2} = \dfrac{1}{a} \tan^{-1} \dfrac{x}{a}$

17. $\int (\sin ax) \, dx = -\dfrac{1}{a} \cos ax$

18. $\int (\cos ax) \, dx = \dfrac{1}{a} \sin ax$

19. $\int (\tan ax) \, dx = -\dfrac{1}{a} \log \cos ax = \dfrac{1}{a} \log \sec ax$

20. $\displaystyle\int_0^{\pi/2} \sin^2 x \, dx = \int_0^{\pi/2} \cos^2 x \, dx = \dfrac{\pi}{4}.$

21. $\displaystyle\int_0^{\pi} \sin^2 x \, dx = \int_0^{\pi} \cos^2 x \, dx = \dfrac{\pi}{2}.$

22. $\displaystyle\int_0^{\infty} e^{-ax} \, dx = \dfrac{1}{a},$

23. $\displaystyle\int_0^{\infty} x \, e^{-ax} \, dx = \dfrac{1}{a^2},$

24. $\displaystyle\int_0^{\infty} x^2 \, e^{-ax} \, dx = \dfrac{2}{a^3},$

25. $\displaystyle\int_0^{\infty} e^{-r^2 x^2} \, dx = \dfrac{\sqrt{\pi}}{2r},$

26. $\displaystyle\int_0^{\infty} x \, e^{-r^2 x^2} \, dx = \dfrac{1}{2r^2}.$

27. $\displaystyle\int_0^{\infty} x^2 \, e^{-r^2 x^2} \, dx = \dfrac{\sqrt{\pi}}{4r^3},$

D.4. GRAPHS OF SIMPLE FUNCTIONS

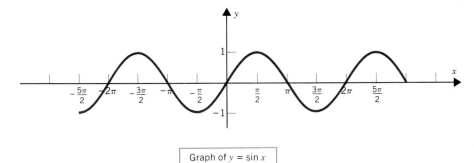

Graph of $y = \sin x$

FIGURE D.3

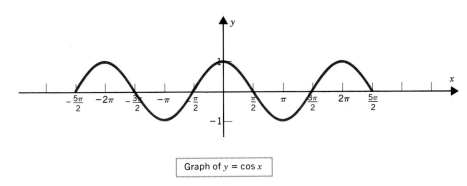

Graph of $y = \cos x$

FIGURE D.4

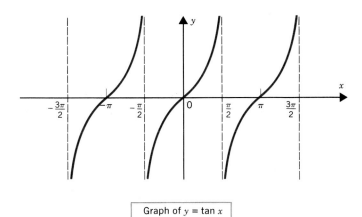

Graph of $y = \tan x$

FIGURE D.5

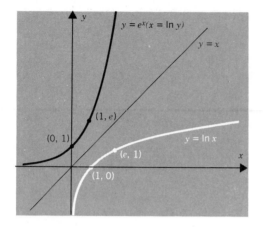

FIGURE D.6

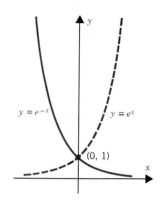

FIGURE D.7

APPENDIX E

Answers To Selected Exercises

Chapter 1

1.0-4. Human beings, wood, animals, wind, water wheel, coal, oil, gas, geothermal, nuclear fission, solar (thermal, biomass, electric), nuclear fusion

1.1-4. (a) scientist or engineer, (b) engineer, (c) engineering technician, (d) craftsmen, (e) engineering technician, (f) engineer, (g) engineering technologist, (h) scientist or engineer

1.2-6. (a) input/output devices (keyboards, card readers, disk drives), component heat transfer, packaging (c) site evaluation and selection, plant construction, structural analysis

1.3-4. Research, academic, development, design, test, construction and production, operations, sales, management

1.4-5. Electrical engineering—communication: analytic geometry, calculus and physics, differential equations, partial differential equations, electromagnetic theory, antenna design

1.6-3. Agriculture, hunting and fishing, mining, urbanization, combustion of fossil fuels

Chapter 3

3.1-5. 2.25 km

3.1-6. 34.7y

3.1-8. (f) If a and b are contained in a set A, then the sum of a and b is contained in A.

3.1-12. Yes

3.1-14. (a) 1 (b) 0.5 (c) $(1 - \pi)^{-1} = -0.467$ (d) undefined

3.1-16. (b) $-5x^{-6}$

3.1-17. (d) $\ln x$

3.1-19. (a) scalar (b) vector (c) vector (d) scalar (e) scalar (f) vector

3.1-27. 1 in 28,000

3.1-29. $F = ma$

3.1-34. $x(t) = \int v(t') \, dt'$

3.1-40. Big trouble in 1990!
3.2-6. (a) 6.9 (b) 8.45 (e) 67.24
3.2-9. (a) 4.14×10^1; 41.4 (e) 2.38×10^1; 23.8
3.2-13. 3×10^{14} watts or 300 terawatts (TW)
3.2-19. (a) 381 mm (f) 251 J (j) 105 kPa (n) 310 K
3.2-23. 2 N
3.3-2. 158 m
3.3-10. 27% of actual speed
3.3-13. 541 N
3.3-17. 888 N
3.3-19. 18 750 N
3.3-22. 1 373 400 J; 38.15 W
3.3-26. 37 C; 310 K
3.3-28. 504 kPa
3.3-34. 589 J
3.3-37. 7.6%
3.3-41. 8.3%
3.3-46. 100 A
3.3-53. H_2O (based on oxygen-16) $= 18.01602$
3.3-59. 11.88 g O_2
3.3-61. 1.15×10^9 Pa
3.3-62. 0.0006

Chapter 4
4.2-1. 5512 (octal); B4A (hexidecimal)
4.2-5. (a) 10010; 24 (c) 11001.01001100; 19.4C (f) 10.101101111110; 2.B7E
4.2-6. $2^{32} - 1 = 4294967295$
4.2-9. About 1.8 megabytes
4.2-15. 5 bits for 28, 7 bits for 28.25
4.3-6. RPN: 55.78, $\boxed{\text{ENTER}}$, 45.36, $\boxed{\text{X}}$
ALG: 55.78, $\boxed{\text{x}}$ 45.36, $\boxed{=}$
4.3-10. ln 2 $= 0.6931347573$ compared to 0.6931471806 (exact)

Chapter 5
5.0-2. (b) The effect of an ''ether'' on light propagation.
(d) The quantization of electric charge.
5.1-9. (a) Dependent: thermal conductivity
Independent: temperature difference across material
(d) Dependent: strength
Independent: composition
5.2-5. 10.6 ± 0.2 μm
5.2-10. 53 ± 0.1 cm or $\pm 0.19\%$
5.3-1. $\overline{R} = 5.016$ Ω , $\sigma = 0.195$ Ω
5.3-9. (b) PPM $= 211$
(c) 45

5.3-13.   Plot F vs. $1/r^2$

5.3-16. $R = 11.7 (1 + 4.47 \times 10^{-3} \, T)$

5.3-21. $FC = -29.8 + 0.67v$ (m/s)

5.4-8.   9.5% are discarded.

Chapter 8

8.1-2.   \$1240/month

8.3-2.   \$45,454

8.3-6.   10.5%

8.3-11. $T = (\ln 2)/r$

8.4-1.   (a) \$14,693, (b) \$7866, (c) \$4496

8.5-2.   \$9811

Credits

Part I   Illustration by William Blake, from *Europe*, *A Prophecy*, 1794.

Chapter 1   Ken Karp.

Part II   © 1976 by B. Kilban.

Chapter 2   Richard Wood/Taurus Photos.

Part III   Courtesy Bell Laboratories.

Chapter 3   Ellis Herwig/Stock, Boston.

Chapter 4   Ken Karp

Chapter 5   Courtesy Bell Laboratories.

Chapter 6   Courtesy Bell Laboratories.

Chapter 7   Ellis Herwig/Stock, Boston.

Part IV   Michal Heron/Woodfin Camp.

Chapter 8   Edward C. Topple, New York Stock Exchange.

Chapter 9   Ken Karp.

Chapter 10   Stella Kupferberg.

Index